T0300994

Radioactivity

This book provides an accessible introduction to radioactivity. The first in a two-volume set, this volume is presented in two parts, covering radiation physics and natural radiation exposure.

It first explores the discovery and physics of the phenomenon of radioactivity, covering the discovery of radioactive decay and the historical development of the physics and applications of radioactivity through to 1940. Chapters then present descriptive summaries of the physics of the atom and the atomic nucleus, mass and energy conditions, the nature of isotopes, and the different decay patterns. Chapter three discusses decay laws and introduces natural origins of radioactivity as well as methods for producing radioactive isotopes through nuclear reaction processes in reactor and accelerator. The book then provides an introduction on dosimetry, radiation chemistry, and impact of radiation on biological systems.

The second half of the book details natural radioactivity and the role of radioactivity in the formation of the planetary system and our Earth. The author describes how the inner radioactivity of our planet determines its dynamics and how it could have contributed to the origins of life. The volume concludes with an exploration of the external and internal radioactivity to which humans are exposed and their possible side effects.

This book will be of interest to non-science undergraduate and physics graduate students alike, as well as to interested lay-people looking for an introduction to radioactivity.

Key Features:

- Written in an accessible style, to be understood by readers without a formal scientific education
- Highly illustrated throughout
- Authored by an expert in the field, drawing from decades of experience in experimental nuclear physics

Michael Wiescher is Freimann Professor for Physics and Director of the Institute for Structure and Nuclear Astrophysics at the University of Notre Dame, Adjunct Professor at Michigan State University, and Visiting Professor at the University of Edinburgh, UK.

Radioactivity

Volume I

A Natural Phenomenon

Michael Wiescher

CRC Press
Taylor & Francis Group
Boca Raton London New York

CRC Press is an imprint of the
Taylor & Francis Group, an **informa** business

Designed cover image: © Shutterstock

First edition published 2025
by CRC Press
2385 NW Executive Center Drive, Suite 320, Boca Raton FL 33431

and by CRC Press
4 Park Square, Milton Park, Abingdon, Oxon, OX14 4RN

CRC Press is an imprint of Taylor & Francis Group, LLC

ISBN: 9781032565156 (hbk)
ISBN: 9781032564005 (pbk)
ISBN: 9781003435907 (ebk)

DOI: 10.1201/9781003435907

Typeset in Times
by Deanta Global Publishing Services, Chennai, India

Contents

Preface

The discovery and interpretation of radioactivity are the most exciting achievements of modern science. This discovery is not only a major scientific step forward but also a symbol for the success of modern science with its enormous impact on the progress and history of mankind. The phenomenon of radioactivity is more than a physical process that provides the foundations of modern science by experimentally confirming the theoretical predictions of modern quantum physics. Radioactivity has also had an enormous impact on societal, economic, and military developments that have fundamentally altered the way science and scientific developments and applications are being seen today.

This book is not intended to be a textbook on nuclear physics or a historical account of science but rather seeks to present the existence of radioactivity and the role of both radioactivity and nuclear physics in people's daily lives, in a way that is as free of bias as possible. The natural phenomenon of radioactivity, like everything that occurs in nature, offers advantages in small doses but is dangerous in large quantities. Although often disputed, the sentence attributed to Paracelsus is also valid for radioactivity: "Omnia sunt venena, nihil est sine veneno. Sola dosis facit venenum" (All things are poison, and nothing is without poison. The dose alone makes it poisonous).[1]

Radioactivity is part of the chemistry of our bodies and might even be responsible for the origin of and our development as biological beings. We are also subjected to radioactive influence every day. While we can use radioactivity, we have to deal with its dangers when exposed to huge quantities. For these reasons, it is important that we understand the phenomenon itself as well as the close and increasing overlap that it has with us and our daily lives. This book aims to present the origin and role of radioactivity in a way that can hopefully be understood by everyone who is interested.

This book was first published in Germany where it seemed appropriate to present an analysis of the phenomenon of radioactivity to a country with a strong "green" movement, which fiercely opposes everything that carries the label "nuclear." While there may be reasons to oppose nuclear weapons, because of their enormous destructive power, the reasons for opposing nuclear energy seem less obvious, particularly if one considers the benefits of this energy source with only a limited CO_2 footprint. There are certainly concerns that need to be addressed, particularly the long-term storage of highly radioactive waste materials. However, this problem needs our focus to solve it and not fear-based opposition.

The first part of this book presents a short history of radioactivity and its discovery, which includes fast – probably too fast – development as a tool in medicine and also for use in a broad range of other applications. The science of radioactivity is a quantitative science – there are, therefore, quite possibly too many numbers in this book – but quantity is directly related to the dangers that radioactivity poses. Numbers can be confusing particularly in a topic like this and for this reason, the basic physics laws of radioactivity, its interaction with matter and materials as well as its impact on chemical and biological systems, will be presented at some length in the following three chapters before coming to the topic of radioactivity itself.

Most people view radioactivity as something evil that is mainly used to produce nuclear bombs and nuclear reactors. Radioactivity is, however, a much older phenomenon that had already emerged in the form of a neutron, the first and smallest radioactive particle in our universe, in the first second after the Big Bang. Radioactivity has not left us since and has developed in parallel with all of the stable elements as "natural" radioactivity. It was, therefore, part of the universe long before the solar system and planet earth were formed from interstellar dust. But as we will show, radioactivity also played a key role in their formation.

Radioactivity was not only an important and long-neglected ingredient for the formation of planet earth but continues to have existential importance as the planet's main internal heat source. This keeps earth and its inhabitants alive, as the chapter about our radioactive earth will demonstrate. Radioactivity's interaction with biological systems is less clear and more research needs to be done, but it seems clear that the fear of radioactivity as a deadly life-threatening force is exaggerated. Multiple studies of different sections of the population point in this direction; more studies will, however, be performed because evidence remains elusive. Parallel to natural radiation exposure – specifically, radiogenic exposure from the ground and cosmogenic radiation from the skies – the average human dose from anthropogenic radiation has continued to grow over the decades. Was it exposure to fall-out from nuclear weapons testing and leakage from nuclear reactor incidents that put mankind on alert? Is it the average individual dose from medical treatment and other applications that carries the load for exposure? These concerns and applications will be discussed and although many hazards have been eliminated or replaced, others remain and will be part of an in-depth analysis in the second volume of this book.

NOTE

1. Theophrastus Bombast von Hohenheim, called Paracelsus (1493–1541), was a German doctor and medical reformer.

Acknowledgments

It has been an enormous effort collecting this data, some of which will be outdated by the time it appears in print. Yet, as a whole, this book aims at painting a more balanced picture of present radiation exposure as it forms in nature and is created by man.

A book is never written alone and arises from learning, experience, and the social environment. Especially the latter has always been important to me and helped to guide my considerations around the question of radioactivity, its origin, its dangers, and its applications. In addition to my family, who often view my immersion in radioactive topics with astonishment, many colleagues and friends have contributed to the completion of this book on many levels! They helped with both information and discussion, and I name them here in alphabetical order: Prof. Ani Aprahamian (Freimann Professor of Physics at Notre Dame and presently Director of the Alikanyan Armenian National Laboratory (AANL) in Erevan, Armenia) who I thank for her friendship as well as numerous discussions and arguments on questions of nuclear energy use, especially in questions of isotope production for medical applications at cyclotrons. Prof. Dr. Edward Calabrese of the University of Massachusetts at Amhurst and Dr. Jerry Cuttler, CEO of Cuttler@Assiociates, INC, were exceptionally helpful in discussions on the issue of the LNT and hormesis models for predicting radiation damage at low radiation dose, an issue that is of critical importance but still remains unresolved due to a lack of reliable data. Both scientists, with their insights and views, have given me much help in discussing ionizing radiation effects on the human body. I also want to thank Jack Devanney, engineer and co-founder of the ThorCon liquid salt reactor company, who introduced me to interesting and most likely controversial ideas and concepts regarding a meaningful nuclear energy concept and policy.

Likewise, my thanks go to my friends and colleagues Dr. Roland Diehl of the Max Planck Institute for Extraterrestrial Physics in Garching, who helped me a great deal in questions of gamma astronomy and by critically reading the paper; Prof. Dr. Dieter Frekers of the Institute for Nuclear Physics at the University of Münster was helpful with various references to new information on the effects of radioactivity and also on questions of neutrino physics. I would like to thank Dr. Franz Käppeler of the Research Center Karlsruhe, who over the years made clear to me the secrets of neutron capture, from the reactor to the stars, and who over many years gave me the opportunity to participate in experiments at the Research Center Karlsruhe – he was a true friend and I miss him dearly; I am also grateful to Professor Dr. Walter Kutschera of the VERA Institute of the University of Vienna, one of the fathers of radiocarbon age determination at accelerators, for his (very) critical comments and constant help. They have all been good friends to me. Prof. Dr. Karlheinz Langanke, Research Director at GSI, long-time friend from student days, I thank for his numerous and helpful comments and improvements and his great, not to say enthusiastic interest in the subject, which was discussed over many hours by phone, e-mail, Zoom, and WhatsApp.

In addition, I would also like to thank Professor Suzie Lapi of the University of Alabama for her guidance and information on the production of radioisotopes for radiomedicine. I thank Dr. Jay LaVerne of the Radiation Laboratory at the University of Notre Dame for years of critical collaboration, especially for his advice in radiation chemistry and biology. I also thank him for several of the illustrations shown here. I am also especially grateful to Dr. Khachatur Manukyan, materials scientist at the Nuclear Science Laboratory, for developing the program in cultural heritage research and for co-authorship of our joint book, *Scientific Analysis of Cultural Heritage Objects*, much of which has been incorporated into this work. Dr. Alberto Mengoni, who for decades has been a collaborator and friend in much work at the CERN Research Center and at the Karlsruhe Research Center, I thank for the information he provided as Science Attaché of the Italian Embassy in Tokyo, and also Prof. Dr. René Reifarth of the University of Frankfurt, who brought me to Frankfurt as Heraeus Visiting Professor to do joint research and to give the lecture on Applications of Nuclear

Physics in All Circumstances, much of which has also gone into this book. I had many discussions with Dr. Dieter Schneider of Livermore National Laboratory on the question of plasma and fusion physics as well as numerous conversations on the topics covered here.

Special thanks to my old student friend and critic Martin Stadtler, who helped me in editing the original German version of the book. Also, thanks to my editor Elizabeth Jo who worked her way through the English version of the book, improving my sentence structure and attempting to understand the concepts and principles of nuclear physics. Finally, last but not least – as we Americans say – I thank Dr. Armenuhi Yeghishyan for her tireless and prompt help in creating the graphic representations shown here.

About the Author

Michael Wiescher is the Freimann Professor of Physics at the University of Notre Dame. He received his Ph.D. at the University of Münster, Germany in 1980. After several years as postdoc and lecturer at Ohio State University, the University of Mainz, Germany, and Caltech, he accepted in 1986 a faculty position at the University of Notre Dame, where he developed a program in nuclear astrophysics using stable and radioactive beams. Dr. Wiescher's research interests are in low-energy nuclear physics, with focus on nuclear astrophysics and nuclear applications. His research is being pursued mainly at the Notre Dame Nuclear Science Laboratory and at several other national and international research institutions. Between 2003 and 2015, Wiescher served as director of the Joint Institute for Nuclear Astrophysics of the University of Chicago, Michigan State University, and Notre Dame. For the following decade he was director of the Institute for Structure and Nuclear Astrophysics at Notre Dame. He has published approximately 500 scientific and review papers and has served on the organizing and advisory committees of more than 100 national and international workshops and conferences. Dr. Wiescher is a Fellow of the American Physical Society and the American Association of the Advancement of Science. He is also an elected member of the Academia Europaea. In 2003, he was awarded the Hans Bethe Prize in Nuclear Physics and Astrophysics of the American Physical Society, and in 2018 he received the Laboratory Astrophysics Award of the American Astronomical Society. In 2021, Wiescher received the Wolfson fellowship of the British Royal Society and in 2023, he was selected as EMMI visiting professor at the Helmholtzzentrum für Schwerionenforschung in Germany.

1 Discovery and Interpretation

1.1 DISCOVERY AND INTERPRETATION

Every science has a history, a history of discovery and interpretation that is characterized by and reflected in the enthusiasm that drives scientists forward, followed by the calmer history of further development toward applications. The science of nuclear physics as the source of radioactivity has a rather checkered history, which seems to malign its accomplishments in retrospect. Nuclear physics is the science of the atomic nucleus, and radioactivity is the energy released in nuclear decay and reaction processes. These are closely intertwined physical phenomena. The first paragraph of this history will be discussed and its heroes will be remembered once more. There are many more heroes than these pages can name, since science after all is a community effort based on the exchange of ideas and information. In this vein, alongside the heroes, there are many more who did work in their shadow and contributed to the overall success and also to the identification of new avenues and directions for the field. Although this is a general statement, nuclear physics and the study of the new phenomenon of radioactivity during the first decades of the 20th century are a prime example of such collaborative development.

1.1.1 SCIENTIFIC DEVELOPMENT IN THE EARLY 20TH CENTURY

The history of the discovery of radioactivity is traditionally focused on the history of the researchers involved. This often gives the impression that its history is characterized by a series of chance observations and the incoherent discoveries and findings of outstanding personalities. The study of radioactivity is, however, the history of effective information exchange within a closely networked small group of researchers, which has led to new experiments and discoveries. This approach was and still is the way that work is done within the scientific community. The difference today, in comparison with the past, is the e-mail and internet exchanges that have considerably accelerated internal scientific exchange. Nonetheless, even in the absence of the internet, it is astonishing how quickly information was communicated in the late 19th and early 20th centuries when exchange was completely dependent on handwritten letters and scientific publications in journals. Specialized journals as a means of scientific communication emerged mainly at the beginning of the 19th century but quickly developed into the ideal medium for announcement, dissemination, and discussion of new scientific results and theses. Toward the end of the 19th century, this system had been firmly established; information transfer was only limited by the speed of ship and railway, which were soon to be replaced by telegraph and telephone. Today the speed of knowledge transfer is only dictated by the time it takes researchers to publish their results on the internet.

An analysis of scientific correspondence was published in the renowned scientific periodical *Nature*, in which letters of probably the two most famous research personalities of the 19th and 20th centuries were considered, Charles Darwin (1809–1882) and Albert Einstein (1879–1955).[1] The paper shows that communication behavior in the 19th century can be described mathematically with a similar formalism to that used when evaluating modern communication behavior via the internet. Darwin wrote at least 7,591 letters and received 6,530; Einstein sent more than 14,500 letters and received more than 16,200 – in each case, during the last thirty years of his scientific career. This exchange could only be maintained through strict prioritization, with letters considered important being answered first and others not at all. In both cases, the rate of response was 25–30%,

DOI: 10.1201/9781003435907-1

with the time scale for response scaled to the level of importance the researchers attached to the problem at hand and to the writer.

The study demonstrates that as early as 1900, researchers were in closely networked groups and followed suggestions and ideas within their specific research groups. This is no different today when similar behavior characterizes our e-mail communication. The size of research groups also remains roughly constant but with increasingly specialized focus, which is directly correlated to the growing complexity of the research field. At the beginning of the 20th century, only a few scientists were working on radioactivity, while today there are thousands of researchers working in hundreds of small, highly specialized groups focused on various aspects of the subject. Within such small groups, communication time can be optimized to a great extent.

In the nuclear physics community, the density of such communication is exemplified in the discovery made by the couple Irène Joliot-Curie (1897–1956), daughter of Marie and Pierre Curie, and Frédéric Joliot-Curie (1900–1958) in January 1934. These researchers found that radioactive nitrogen-13 can be produced by bombarding boron with alpha particles.[2] This result was published in the weekly reports of the *Académie des Sciences* on January 15, 1934, and in *Nature* on the 10th of February in the same year. Within a few weeks, at the end of February 1934, three independent institutes had already repeated the experiment and published results confirming the observation: Cambridge in Great Britain, California Institute for Technology (Caltech) in Pasadena, California, and also in Berkeley, California, United States.

The discovery of nuclear fission by Otto Hahn (1879–1968) and Fritz Straßmann (1902–1980) in December 1938 was also quickly communicated to the United States. At the end of the same month, Lise Meitner (1878–1969) in exile in Sweden and her nephew Otto Frisch (1904–1979) in the UK interpreted the chemical observations of Hahn and Straßmann to be the result of a nuclear fission reaction. Niels Bohr (1885–1962), informed by Frisch in Copenhagen, traveled to the United States in January 1939 and communicated the results to nuclear physicists at Columbia University, New York, as well as to Italian nuclear physicist Enrico Fermi (1901–1954), who had recently emigrated to the United States. Hahn and Straßmann published their discovery in the German journal *Naturwissenschaften* in early January 1939. In Paris, Joliot recognized the importance of the discovery after receiving the journal. On January 26, 1939, the experiment was repeated in Copenhagen, Paris, and at Columbia University, where in each case the results were confirmed.

Research on radioactivity also exemplifies how scientific results can be translated into practical or economically advantageous applications. Wilhelm Conrad Röntgen (1845–1923) observed that certain, previously unknown rays are differently absorbed in materials of different densities: the absorption level can be visualized by the different exposure levels on a photographic plate. This phenomenon was immediately recognized as a possibility for medicine, as a way to examine body parts without surgery. Within a few years, X-ray diagnostics developed into a globally flourishing medical application and business. However, the dangers of intensive X-ray radiation were not immediately recognized and were widely underestimated. Over time and in many cases, this meant serious damage to the diagnosing physicians and technicians (Figure 1.1), an early sign of the dangers of extensive radiation exposure.

X-rays originate as atomic excitation processes in the outer electron shell and, upon discovery, radiation from the decay of the atomic nucleus also quickly found applications. Pierre Curie (1859–1906) observed burn marks on his skin as a consequence of intensive radiation exposure. For long periods of time, he wore radium probes on his body to determine the duration and effect of the radiation as part of a self-experiment. The most famous example of his research is the burn mark on his inner arm (Figure 1.1). This observation gave Pierre Curie the idea to pursue possible applications for medical purposes, for example, the destruction of undesirable tissue growth ranging from warts to tumors.

With this work, the history of radioactivity discovery and its applications began. As we can see from its development, a small but international, closely networked community of researchers based their investigations on the latest research results in order to understand the newly discovered

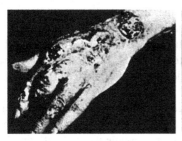 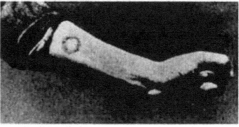

FIGURE 1.1 Radiation damage: Left: Skin cancer on the left hand of Mihran Kassabian, an early radiologist after more than fifteen years of work with X-ray sources for medical examination. Right: Burn on Pierre Curie's inner arm after self-experiment with a radium sample that he wore on his arm for several days. Source: Mould, R. F. (1993). *A Century of X-Rays and Radioactivity in Medicine: With Emphasis on Photographic Records of the Early Years.* CRC Press.

phenomenon of radioactivity as well as the physical principles under which it operated and to develop a broad range of societal and industrial applications for what they hoped was the benefit of mankind.

1.1.2 The First Generation of Researchers and Their Results

The discovery of radioactivity is often presented as a happy coincidence: French physicist Henri Becquerel (1852–1908) procured faintly luminous phosphorescent minerals, including uranium salts. He kept this carefully wrapped material in his drawer, coincidentally placed next to photographic plates since he wanted to carry out further experiments on the nature and origin of X-rays, which Wilhelm Röntgen had discovered in 1895. It had been observed that X-rays produce fluorescent effects in certain materials, i.e., they make them glow. Becquerel wanted to find out whether sunlight also contains X-rays and would thus increase the luminosity of materials. In fact, experiments with a sensitive photographic plate did show a faint glow after exposure to sunlight. Today, this effect is specifically used as thermoluminescence in archaeometry to determine the age of crystalline materials.

The events were, however, not coincidental. In spring of 1896, Becquerel locked his uranium samples together with light-sensitive photographic plates in his drawer. When he removed the plates a few days later, he found them exposed to a much higher degree than expected. They clearly showed the silhouette of the uranium samples. He concluded that this exposure must have been directly caused by invisible radiation from the minerals. For this reason, the discovery should not be described as a lucky coincidence since Henri Becquerel followed a plan and procedure inspired by Röntgen's discovery of X-rays and their properties. Becquerel quickly interpreted his observation as a sign of a new type of radiation. The physicists' community was highly motivated by these results and so continued researching the novel phenomenon and asking questions about the properties and origins of this new type of radiation. Becquerel's correct interpretation of the exposed photographic plates as the signature of a new type of radiation was the decisive step of discovery.

Maria Salomea Skłodowska, better known as Marie Curie (1867–1934), had arrived in Paris from Warsaw in 1891 to study physics. While searching for a job and a topic for her doctorate, she met and married the young French physicist Pierre Curie. Marie Curie was fascinated by Henri Becquerel's results and, together with her husband, tried to find other substances besides uranium that emit the new, mysterious radiation. This work required time-consuming chemical preparation of mineralogical samples, including handling large quantities of raw materials containing uranium. Marie Curie discovered that, in addition to uranium, the element thorium also emitted the mysterious radiation. Thorium had been found in 1815, in rock from a copper mine in Norway, and named after the Norse god Thor. Curie's discovery suggested that radioactive substances could be embedded in ancient rocks. The Curie couple extended their search to other minerals. They discovered that pitchblende often radiates more intensively than pure uranium or thorium. This uranium-containing material was being used by the Czech glass industry to color crystal glass. Marie Curie and her husband had large quantities sent to them from the Bohemian mining town of St. Joachimsthal (today: Jáchymov, Czech Republic) and began extensive experiments to determine the nature and behavior of the radiation.

The Curies did not use photographic plates, but a method based on the ionization[3] of gases by radiation that they had invented. It was found that high levels of radioactivity produce an electric current in a gas, which can be measured directly. The research couple worked with purely chemical methods and discovered that, in addition to uranium, two other components of the pitchblende radiated. The first was a metallic element that the Curies called *radium*, the radiating element. They named the second element polonium, in honor of Marie Curie's country of origin. When it came to determining special properties, such as atomic weight and chemical bonding behavior, large quantities of pitchblende were required. These were prepared in batches of 20 kg each from the Radium Institute in Vienna[4] and shipped to Paris. Marie Curie used various chemical processes to extract the metallic components from the pitchblende. Pierre Curie examined the radioactivity of the respective metals. In addition to the ionization meter, he used fluorescent materials such as zinc sulfide (ZnS) which when excited by radiation begins to glow.

Marie Curie presented the results of this work in June 1903 as a doctoral thesis with the simple title *Recherches sur les substances radioactives* (Research on radioactive substances). She not only described the chemical separation of the new radioactive substances but also discussed the detection techniques developed for this purpose as well as observations of the temporal variation of the radioactive components and the characteristics of the radiation itself. In addition, she described the luminescent properties of the radiation as well as its penetration of materials and deflection in electric fields. Following the work published in 1898 by the New Zealand physicist Ernest Rutherford (1871–1937), the results of the Curies' work provided the basis for the classification of radiation into α- and β-radiation. The expert committee wrote that the work was the greatest scientific achievement ever presented in a doctoral thesis. In the same year in Stockholm, Henri Becquerel and the Curie couple received the Nobel Prize in Physics (Figure 1.2). A year later, Marie Curie's dissertation, translated into many languages, was presented to the international scientific community.

Pierre Curie was particularly interested in possible radiation applications, primarily motivated by the great success of X-rays in medical diagnostics. Curie thought about direct radiation application for cancer therapy and carried out various experiments, exposing himself to considerable amounts of radiation. For example, he tied a container filled with radium salts to his arm for ten

FIGURE 1.2   Henri Becquerel (left) and the Curie couple shared the Nobel Prize for Physics in 1903. Source: https://www.filosofiaagressiva.blogspot.com.

hours and then over the next several weeks observed the development of a burn caused by the intensive radiation. He noted how the tissue was destroyed and after fifty days how only a gray scar remained. From this he concluded that radiation could also destroy cancerous tissue, which in the following years led to the first applications of so-called brachytherapy[5] in cancer treatment.

To quantify the radiation, Marie Curie introduced the first unit for measuring the intensity of radiation from a radioactive source: 1 curie (Ci) is the amount of radiation emitting from 1 g of radium. This definition gives only the amount of radiation and says nothing about the nature of the radiation or its effect on chemical and biological materials such as the human body. Today, in addition to Ci, a different unit of measurement applies to radiation that is based less on chemical terms, such as the quantity of material emitting radiation, and more on the physical process of an element's radioactive decay. This physical unit is the becquerel (Bq), named after the discoverer of radioactivity.

One becquerel corresponds to one decay process per second and thus one radiation event per second. One gram of radium has 37 billion decays per second, usually expressed as $3.7 \cdot 10^{10}$ decays/s. This corresponds to $3.7 \cdot 10^{10}$ Bq or 37 giga-becquerel (GBq).[6] Assuming that the Curies had chemically processed about ten tons of pitchblende with an activity of 158 kilo-becquerel per gram material (kBq/g), they unconsciously exposed themselves to radiation of about $1.58 \cdot 10^{12}$ Bq or 1.58 tera-becquerel (TBq). The 1.58 TBq is probably an underestimation because the Curies carried out countless experiments with the extracted substance, which probably increased their radiation exposure considerably. This is proven by their legacy: The Curies' log books and notes, recording their experiments and results, are still highly contaminated and are not made accessible without protective clothing even today, due to legal radiation-protection regulations. Even Marie Curie's cookbook is radioactively contaminated, which suggests that the food was not left unaffected either.

The consequences of this exposure did not go unnoticed. Burns on fingers and hands, fatigue, and other symptoms were apparent that are now attributed to so-called radiation sickness. Colleagues like Ernest Rutherford observed that Pierre Curie's hands increasingly trembled and could hardly hold any materials. Pierre Curie died in 1906 as the result of a traffic accident. Absent-minded, he fell under the wheels of a heavily loaded horse-drawn cart while crossing a road. It is suspected that this carelessness may have been the result of an excessive radiation dose to which he was exposed through self-experiments, with large amounts of radium attached to his body, in order to directly investigate the burning effect of the radiation.[7] Marie Curie's hands also were increasingly scarred by permanent radiation exposure during the chemical work. Toward the end of her life, they were considered crippled and could hardly hold any instruments. Marie Curie died of leukemia twenty-eight years after her husband's death, on July 4, 1934, possibly as a long-term consequence of the enormous radiation dose she was being exposed to during her many tireless hours working with highly radioactive substances.[8]

1.1.3 Physical Interpretation of Radium Chemistry

Just as the work of Marie and Pierre Curie dominated the first chemical studies on radioactivity, New Zealand–born physicist Ernest Rutherford played a comparable role in the physical interpretation of the phenomenon. Even at the time, many researchers were already trying to unravel the mysteries of radioactivity. Their contributions should not be underestimated, but only Rutherford drew the right conclusions from the interpretation of his data. It is thanks to him that radioactivity was quickly characterized as the physical decay of an atomic nucleus. He also described the atomic nucleus itself as a small central particle in the micro-planetary void of the electron shell of the atom. He is therefore rightly regarded as the father of modern nuclear physics.

As a doctoral student of Joseph John Thomson (1856–1940) at the Cavendish Laboratory in Cambridge, Rutherford had already been critically involved with the ionization of gases by radiation particles.[9] As part of this research, he also examined the ability of radiation resulting from radioactive decay to penetrate materials. On the basis of his results, he postulated the existence of two radiation components in 1898, after accepting a position at McGill University in Montreal. He called these components α (alpha) and β (beta) radiation for the sake of simplicity – with α-radiation, in contrast to β-radiation, having very low penetration capacity. One year later, it was possible to prove that the two particles had different charges since magnetic fields deflected them in different directions.

In 1900, French physicist and chemist Paul Villard (1860–1934) discovered a third component that could not be deflected by magnetic fields and had a very high penetration capacity. Rutherford called this third type of radiation γ (gamma) radiation. In 1907 he returned from Canada to accept a professorship at the University of Manchester (UK). There, together with his students, Rutherford concentrated his research on determining the identity of the different types of radiation. In the following years, the positively charged α-radiation was identified as the nucleus of the noble gas element helium. It was also shown that the negatively charged β-radiation consists of electrons. Shortly before the beginning of the First World War, Rutherford confirmed the assumption that γ-radiation, similar to X-rays, is a very high-energy electromagnetic wave.

In addition to this classification of radiation, there was, however, another observation that puzzled Henri Becquerel, Marie and Pierre Curie, Ernest Rutherford, and other researchers. When studying the radiation behavior of uranium, radium, and thorium, they noticed the temporal variation of radiation. In some cases, radiation intensity seemed to decrease, while in others it seemed to increase. Radiation also seemed to escape from the samples as a chemically inactive gas. These changes were precisely measured; results showed that both the decrease and the increase in one type of radiation had a characteristic exponential time dependence.

Already at McGill University, Rutherford and his student Frederick Soddy (1877–1956) were looking for a solution, in order to understand not only the mechanism that caused the radiation but also the reason for these time-dependent changes in the type of radiation and intensity. Early observations of uranium by Becquerel and other researchers, as well as Rutherford's own experiments with thorium, seemed to show that the quantity of the different radioactive elements changed. It was also noted that this change affected the nature of the radiation. Based on these observations, Rutherford developed the hypothesis that radioactivity is linked to the transformation of elements.

According to this hypothesis, a radioactive element emits radiation and, in the case of α-radiation, the decay changes its mass and charge number so that a new element is gradually formed. The same applies to β-radiation, although here only the charge number is changed while the mass remains more or less constant. In each case, this also leads to the formation of a new element. The decaying element is called the radioactive *mother nucleus*, and the element that forms is called the *daughter nucleus*.

In 1904, Ernest Rutherford published a summary of the state of research on radioactivity in a book that is still worth reading today.[10] In it, he describes the physical principles, characteristics, and technical design of the instruments used to measure radiation as well as countless experiments and also the information that contributed to the rapid development of the state of radiation research, particularly in Europe. Rutherford summarizes the results of the observations and, in addition to the law of decay, also provides a first interpretation of the decay series, although many mother and daughter elements remained unknown and unnamed.

What Rutherford and other researchers observed as variation in the radiation and the elements associated with it, we know today as the natural decay series. Starting from the slow decay of long-lived uranium and thorium atoms, new radioactive elements are built, which in turn decay. This forms a chain of decays that only reaches an end with formation of a stable atomic nucleus of the element lead. The decay chain itself consists of sequences of α- and β-decays. This explains the difference in mass and charge number between the heavy uranium and thorium atomic nuclei and the much lighter lead nuclei. Rutherford's student, chemist Frederick Soddy later formulated the so-called radioactive displacement law, which describes the step-by-step sequence of decay processes in the decay series, resulting in changed mass and atomic numbers. This method is still used today in the search for super-heavy elements. Soddy is therefore regarded as one of the fathers of nuclear chemistry.

This discovery seemed to fulfill the dream of medieval alchemists who sought to create new elements such as gold by the transmutation of elements. What was observed was not, however, artificially stimulated transmutation but rather a natural, slow process in which heavy elements slowly transform into less heavy elements by emitting high-energy radiation. This process takes place in all uranium-bearing minerals as they are found in rock.

Rutherford and Soddy were aware of the importance of their discovery when they proved in 1901 that radioactive thorium was transformed into radium by decay.[11] When Soddy realized this, he spontaneously exclaimed: "Rutherford, this is transmutation!" Rutherford yelled at him: "For Christ's sake, Soddy, don't call it transmutation. They'll have our heads off as alchemists."

For the first time in 1933, Irène and Frédéric Joliot-Curie artificially created radioactive elements – a scientific revolution. By bombarding various materials such as boron, aluminum, and magnesium with α-particles from radioactive sources, they were able to create new radioactive products which, on closer analysis, they identified as the new isotopes nitrogen, phosphorus, and silicon. Due to their short half-lives, these elements do not occur in nature. This was the first step toward the generation of artificial or anthropogenic radioactivity, which is not based on natural processes such as the decay of long-lived actinides[12] in rock material (radiogenic radioactivity) or production by high-energy cosmic radiation (cosmogenic radioactivity).

The artificial production of radioactive elements through nuclear reactions heralded a new era. Artificial (or anthropogenic) radioactivity has become more and more a part of humankind's development and experience. Today, at the beginning of the 21st century, about 50% of the radiation that an average person is exposed to comes from radioactive elements, or radioisotopes for short, that are artificially produced in nuclear reactors or particle accelerators.

In 1928 in earlier experiments, the Joliot-Curie couple had produced a new type of β-decay in which positrons – electron-like particles with a positive charge – are emitted instead of negatively

FIGURE 1.3   The 1935 Nobel Prize award ceremony – from left: James Chadwick (physics) for the discovery of the neutron, the couple Irène and Frédéric Joliot-Curie (chemistry) for the discovery of artificial radioactivity, Hans Spemann (medicine). Source: Photo Fulgur Agency; Musée Curie (Coll. ACJC).

charged electrons. They did not, however, realize the importance of this discovery. Then in 1932, young physicist Carl David Anderson (1905–1991) identified positrons as the antiparticles of electrons during the study of cosmic rays at the California Institute for Technology (Caltech). This was the first discovery of the phenomenon of antimatter.[13] Since then, radioactive decay has been distinguished between β^+ and β^- radiation.

During their experiments, the Joliot-Curie couple also overlooked another particle that, although extremely difficult to measure experimentally, had long been expected and predicted by Ernest Rutherford. This particle is the neutron with a mass similar to that of the proton but with no electrical charge, making its detection extremely difficult. Without an electric charge, the neutron had little direct effect on the instruments commonly used at the time. Yet, this was the key to the first understanding of the atomic nucleus' structure.[14]

1.1.4 Early Atomic Models and the Structure of the Atomic Nucleus

At the beginning of the 20th century there was great uncertainty about the microscopic structure of matter. The scientific community was not even absolutely sure whether the ancient idea was correct – that the basic structure of matter was atomic in nature, i.e., composed of the smallest indivisible particles. This idea had been postulated by ancient Greek philosophers, but its nature or existence remained nebulous. The humanistic character of the late 19th century had, however, given great importance to the ancient ideas, and this very much influenced questions and interpretations, as reflected in the debate between two great Austrian physicists, Ernst Mach (1838–1916) and Ludwig Boltzmann (1844–1906). While Mach took an anti-atomistic view, postulating energy as the decisive medium of the universe, Boltzmann maintained the atom as the smallest particle on which matter is built.

If we assume the atom as the basic building block, questions immediately arise: How is matter made up of these atoms? What constitutes the structure of the atom as the smallest elementary particle? Since it was understood that the atom was neutral, it therefore had to be equally comprised of oppositely charged particles. Joseph Thomson interpreted the observation of the cathode rays – by Julius Plücker (1801–1868) and his student Johann Wilhelm Hittorf (1824–1914) in the 19th century – as a stream of light negatively charged particles with small mass. These negative particles, electrons, had to be opposed by positively charged particles of large mass, in order to achieve the necessary atomic weight.

Joseph Thomson, one of the leading thinkers of his time, is regarded as the teacher of the entire subsequent generation of researchers, particularly his doctoral student Ernest Rutherford. In 1904, Thomson introduced the so-called plum pudding model of the atom, in which negative electrons were embedded in a mass of positively charged particles – like raisins in dough. In this way, opposite charges were neutralized. Thanks to Thomson's great reputation and even greater intellectual influence on the physicist community, this model remained the generally accepted model of the atom for almost ten years, despite various disagreements, until it was replaced by the Rutherford model (Figure 1.4).

Nevertheless, voices of doubt were increasingly heard. In 1908, Austrian physicist Arthur Erich Haas (1884–1941) postulated a modified form of the Thomson model: In it, electrons were arranged on the surface of a positively charged sphere. This was the first time that he was able to show a connection between the spectral lines of hydrogen and an atomic model. He also gave the radius of the hydrogen atom, which corresponded well with the Bohr radius that was later formulated.[15]

At the same time, French physicist Jean Perrin (1870–1942) speculated about a special arrangement of the electrons. In his opinion, similar to planets around the sun, electrons should orbit around an atomic nucleus. This was a beautiful theory, but according to the understanding of classical electrodynamics that was held at the time, such circling electrons would continuously emit energy in the form of an electromagnetic radiation, thus losing the ability to maintain their orbit and crashing into the nucleus.

FIGURE 1.4   From left: Joseph J. Thomson and the two physicists he brought to Cambridge, Ernest Rutherford and Francis Aston. In 1921, Aston (1877–1945) invented the mass spectrometer, which in the following decades made it possible to measure the masses of numerous atoms with hitherto unattainable precision. Source: AIP Emilio Segrè Visual Archives, Gift of C.J. Peterson.

Ernest Rutherford was well acquainted with all of these theories. He was considered a gifted experimenter and set out to gain direct information about the structure of the atom through experimental studies. In the experimental setup he developed, he irradiated a thin gold foil with particles from a uranium sample. An aperture bundled the particles into a precisely defined beam. As part of this experiment, which was actually carried out as a series of several experiments between 1908 and 1913, Rutherford, his German assistant Hans Geiger (1882–1945),[16] and his student Ernest Marsden (1889–1970) observed the scattering behavior of the particles as a function of the scattering angle. As proof, they used zinc sulfide, which had already been used by the Curies to measure radiation. The α-particles passed through the gold foil mostly unhindered, but the beam, concentrated by apertures, expanded. In addition, the researchers also observed a few α-particles that were backscattered.

Rutherford compared this observation to a bullet bouncing off a thin piece of paper. From this he concluded that the mass of the gold foil cannot be evenly and densely distributed but must be confined to a few high-density point centers that are sometimes hit by the α-radiation, thus causing the backscattering. From data obtained about the angular distribution of the scattered particles, Rutherford concluded that the mass of the gold atom is concentrated in its interior, the so-called atomic nucleus. Further, the atom's radius could be estimated from the scattering data. According to this, the atomic nucleus of gold had to be about 10,000 times smaller than the gold atom with its full electron shell. These were revolutionary results that still shape our understanding of the atomic nucleus. Today, the phenomenon of Rutherford scattering is used for analysis of materials and thin films, mainly in the electronics industry.

The internal structure of the atom can be described as a small positively charged nucleus with a negatively charged electron shell. The structure of this shell was derived from the Bohr model. Developed in 1913, Danish physicist Niels Bohr conceptualized the model after his one-year visit to the Cavendish laboratory established by Rutherford. There, Bohr built on Rutherford's concept of the hydrogen atom and assigned the electrons to fixed orbits in the shell around the atomic nucleus. For his model of a nucleus, he took the positively charged hydrogen nucleus, with orbits fixed by the quantum rules formulated by Max Planck. Accordingly, energy could only be changed by the transition of an electron from one orbit to the other (see Chapter 2.2).

This model elegantly avoided the difficulties of Jean Perrin's previously postulated model, in which his orbits could not be stabilized within a classical electrodynamic framework. Bohr did cut the Gordian knot by simply deciding that the orbits had to be stable and must therefore be fixed by quantum rules and numbers. The Bohr model also clarified fundamental questions about the atomic shell, which had the same radius as predicted by Arthur Haas. Countless spectroscopic observations of emitted and absorbed electromagnetic radiation confirmed this insight in the following years and decades.

The question still remained open regarding the positively charged nuclei of other atoms. In 1919, Rutherford postulated that all atomic nuclei must consist of a conglomerate of positively charged hydrogen nuclei and called these positive particles *protons*. However, there was a considerable discrepancy when comparing the associated masses of the particles involved; the mass of all protons in the atomic nucleus was only half the mass of the atom. Since the electrons had only a small mass, about 2,000 times smaller than the mass of the proton, they could be largely neglected. Therefore, something was missing. In 1920, Rutherford postulated a particle that should have approximately the same mass as a proton but with no electrical charge, thus fulfilling the conservation of mass and the neutral charge.[17] He called this particle a *neutron*.[18]

For more than ten years the neutron remained an imaginary particle. Since it should have no electrical charge, it could not be detected by deflection in electric or magnetic fields. It was not until 1932 that James Chadwick (1891–1974), a student of Rutherford, succeeded in detecting it by means of nuclear reactions, the same reactions that had already been performed by the Joliot-Curies.

An informal group of old friends on the occasion of a meeting of the Bunsengesellschaft in Münster, 1932—a snapshot taken by the late F. Paneth.
From left to right: Chadwick, v. Hevesy, Frau Geiger, Geiger, Lise Meitner, Rutherford, Hahn, Stefan Meyer, Przibram.

FIGURE 1.5   A meeting of old friends and colleagues at the 1932 conference of the German Bunsen Society (*Bunsentagung*) on the topic of radioactivity in Münster, Germany. From left: Rutherford students James Chadwick, Georg de Hevesy, Hans Geiger with his wife Lili Geiger, also Lise Meitner, Ernest Rutherford, and Otto Hahn. On the right, two Austrian physicists and radium researchers Stefan Meyer and Karl Przibram, who supplied the nuclear physics community with radioactive materials from the rich holdings of the Institute of Radium Research at the University of Vienna. Source: wikipedia.org. Note: Wolfgang Reiter, *Aufbruch und Zerstörung – Zur Geschichte der Naturwissenschaften in Österreich 1850 bis 1950* (Berlin: LIT Verlag, 2017).

Chadwick had worked with Hans Geiger in Germany before the First World War to learn about and use the Geiger counter. Surprised by the outbreak of war, Chadwick was interned in 1914 and could not return to Cambridge until after the war. He did not, however, hold any grudges and maintained close contact with Geiger, who provided him with one of his counters in 1930. In the experiment, Chadwick bombarded a thin foil of light beryllium material with α-particles from a uranium source and registered high-energy radiation. This had been observed in similar experiments by the Joliot-Curies and other nuclear researchers and interpreted as γ-radiation. Chadwick let this radiation hit a paraffin layer behind which he had placed the Geiger counter. Radiation knocked protons out of the paraffin, which the Geiger counter detected as charged particles. However, the strong transfer of momentum and energy associated with these particles could not have been caused by γ-radiation and this high-energy radiation must have been neutral particles, with a mass comparable to that of the protons. The long-predicted neutron had been found by the international collaboration pictured in Figure 1.5.

This established the basic concept of the atomic nucleus: It is composed of positively charged protons and neutral neutrons. With this model, the properties of the atomic nucleus could largely be described (see Chapter 2).

1.1.5 EARLY APPLICATION OF RADIOACTIVITY

In addition to basic research on understanding the atom and atomic nucleus, a second line of research also developed. It had already been taken up with the work of Wilhelm Conrad Röntgen and Pierre Curie, who identified possible radiation applications for medicine. While X-rays, because of their penetrating properties, contributed primarily to improved medical diagnostics, higher-energy radiation seemed to offer countless therapeutic applications.

This notion of radioactivity as a new miracle cure led to the uncontrolled growth of a radiation industry. From radon spas that appeared in Europe and the United States, for the treatment of rheumatic diseases, to direct tumor irradiation with radioactive sources, radiation was promoted to the general public. In addition, a health-driven cosmetics industry was also established that supplied, for example, radioactive toothpaste whose rays were supposed to gently massage the gums. Instruments were developed, such as for the removal of unwanted body hair through extensive radiation exposure. This industry offered a rich and diverse field of possible business enterprises and a multitude of technical development opportunities with sometimes devastating consequences for unsuspecting customers (cf. Chapter 15, Volume 2).

Within this uncontrolled environment, damage often only occurred after many years. The consequences of exposure depended on the dose of the respective radiation – and so the possible dangers of intensive radioactive irradiation only slowly entered the consciousness of the public. When these dangers were realized, euphoria turned into the general fear with which radioactivity has been met in the following decades. Yet initially, the biochemical effects were not generally realized and so what caused biological radiation damage remained largely unknown. The new science of genetics, which developed parallel to and independent of nuclear physics, provided the explanation. In the 1920s, the effect of ionizing radiation on genetic mutation processes was discovered, greatly influencing the general perception of radioactivity in the following decades (cf. Chapter 4).

Alongside and perhaps also because of the countless commercially motivated applications, serious research into the interaction between radiation and biological systems advanced. The focus was on questions concerning the biological effects of radiation as well as the uptake and distribution of radioactive elements in the biochemical processes of plants, animals, and humans. Pierre Curie had only seen the destructive power of this radiation, which he believed to be beneficial for the targeted treatment of cancerous tissue and tumors. The secondary effects, which carried much further-reaching consequences, were largely unknown in the early 20th century.

Before secondary effects could be fully realized, more detailed studies were required on the various possible interactions between biological systems and radiation as well as radioactive materials. The former directly relates to the biochemical processes that are triggered by a sudden energy transfer into the molecular structure of a biological system, the latter to the slower physiochemical processes involved in the uptake of radioactive materials by biological systems. Questions about effects on the human body were of course the most important, but for obvious reasons such studies were not available.[19] Experiments were therefore mostly performed on bacteria, insects, and plants. It was hoped that the results would lead directly to more complex biological organisms.

Two researchers stood out and had great influence on the subsequent development of the field: Hungarian nuclear physicist Georg de Hevesy (1885–1966) and American geneticist Hermann Muller (1890–1967). At times, Hevesy worked closely with Rutherford and received important suggestions from him that informed research on plant uptake of radioactivity. Muller made a name for himself with his investigation of radiation-induced mutations in bacterial cultures and fruit flies.

Georg de Hevesy[20] was a young Hungarian chemist who graduated in 1911 from Freiburg in Germany and moved to Manchester where he worked with Ernest Rutherford. Rutherford challenged him to find a chemical method for separating an unknown radium-decay product, known as radium D, from the lead substrate: "If you are worth your salt, you will separate radium D from all that nuisance of lead." Hevesy tried this for two years without success, which is not surprising in retrospect. Today, radium D is identified as lead isotope ^{210}Pb and therefore has the same chemical properties as stable lead. However, during these investigations, Hevesy came up with the idea of using radioactive isotopes as an indicator – since they could not be separated chemically – for tracking materials as they move through dynamic body processes of various kinds, such as tracking food being processed in the gastrointestinal system as it moves into the bloodstream.

A frequently told example[21] is Hevesy's food experiment. His landlady made promises and assurances about serving fresh meat every day. She was not, however, maintaining that aspect of the contract and instead presented the uneaten meat from the previous day in a new form, as meatloaf or goulash. Hevesy placed a small but detectable amount of radioactive material into the uneaten meat on the edge of his plate and the following day, he triumphantly proved that the meatloaf served was not made from fresh meat but rather the previous day's irradiated meat. Hevesy used these methods with increasing success to trace plant and animal processes.

In the 1920s, when working for Niels Bohr in Copenhagen, Hevesy used radioactive-lead salt solutions to track water uptake in plants, measuring how long it takes to distribute radioactive ^{212}Pb in stems and leaves. He injected rabbits with radioactive bismuth ^{210}Bi and lead ^{210}Pb to determine the distribution and deposition of these substances in various organs. Hevesy also did not shy away from self-experiments when he used the newly discovered deuterium, an isotope of hydrogen, to study the water cycle in the human body. He drank water enriched with deuterium HDO (instead of H_2O) and detected it in his urine. From this he determined that a typical water molecule remains in the body for an average of 12 to 14 days before being excreted – thus documenting the body's continuous water cycle. In addition, he used radioactive phosphorus ^{32}P to monitor various metabolic processes in the human body.

For all of these experiments Hevesy used Geiger counters, developed and continuously improved by Hans Geiger, to locate the radioactive isotopes in the body. Through his efforts, Hevesy showed that the body is a dynamic biological system that is in constant change and exchange with the external environment. The methods he developed laid the foundation for radiation medicine as part of today's medical diagnostics. Georg de Hevesy received the Nobel Prize in 1943 for his use of radioisotopes as indicators of chemical processes.

Hermann Muller was not a chemist but had studied biology at Columbia University in New York, receiving his doctorate in 1916. His education was, therefore, outside the closely knit community of nuclear chemists and nuclear physicists around Rutherford.[22] Even as a student, Muller was very interested in the new science of genetics, the work of Gregor Mendel (1822–1884) and the laws of inheritance. In the early 20th century, genetics was associated with questions about mutations and their causes.

Throughout his life Muller pursued the problem with lengthy studies on fruit flies (Drosophilae). In the late 1920s, he began experimenting with X-rays and radioactive sources to study the effects of radiation on mutation rates. In 1927, he showed that a high dose of X-rays caused an increased rate of lethal mutations, compared to un-irradiated fruit flies. These experiments were probably motivated by the idea that the natural mutation rate Muller had found in his Drosophila studies could have been caused by the radiogenic and cosmogenic radioactivity of our environment.

In later experiments, Muller showed that the number of mutations increased linearly with the radiation dose. Based on this observation he postulated that the linearity was also valid in areas he could not investigate experimentally.[23] This he considered less true for somatic, internally reparable damage, than for hereditary cells, where damage by radiation inevitably leads to mutations in the genetic material. If the measured linearity curve of external radiation exposure is projected to a zero value, mutation still occurs. It was thought that this mutation rate might be caused by natural radiation exposure. However, this value is below the mutation rate observed by Muller in the fruit fly. He concluded from this that natural radiation cannot explain the observed mutation rates.

Muller determined that during cell division other statistical processes must be the cause of mutation (cf. Chapter 8.1). He rejected natural radiation as the cause because his calculations suggested that background radiation could only account for 1/1,300 of the mutation rate in the control group.[24] It had been believed that Muller exposed flies to radiation doses some 200,000 times higher than the background level; however, a recent publication suggests that Muller treated the flies at a level that was 95,000,000 times higher than background levels.[25] This puts Muller's conclusions into question; natural exposure may contribute to at least some of the observed natural-mutation rate. Also, Muller's simple-minded ansatz of linearity for extrapolation toward very low exposure remains in question and needs to be questioned and confirmed. While the impact of "natural" low-level radiation remains a hotly debated topic, it is beyond doubt that increased radiation exposure will lead to mutation.

In 1946, Hermann Muller was awarded the Nobel Prize for Medicine for his discovery that X-rays can lead to mutations in genetic make-up. His scientific findings convinced him of the dangers of radiation from radioactive decay. In the post-war years, especially under the impression of atomic bombs dropped on Hiroshima and Nagasaki, Muller became a strong critic of radioactivity use and application. He used his reputation in the scientific community to advocate legal regulations and limits that are still valid today – in the United States and worldwide.

1.1.6 COLLABORATION AND COMMUNICATION

As we have noted, the history of nuclear physics and its applications, in the first decades of the 20th century, was a history built on close scientific communication and the free exchange of ideas and results. However, this flow of information is not limited to scientific results alone and also includes scientific equipment and technologies. A classic example is the already mentioned case of German physicist Hans Geiger. After obtaining his doctorate in Erlangen in 1906, he became a scientific assistant to Ernest Rutherford in Manchester.[26] There he developed the Geiger counter, probably the most famous and influential instrument in the history of radioactivity – and still used today. Geiger also took part in the famous Rutherford scattering experiment, which revealed the structure of the atom, before returning to Germany.

Rutherford's student James Chadwick followed Geiger to Germany in 1913 to work with the Geiger counter. This work was interrupted by the First World War, which Chadwick had to spend in an internment camp in Germany. However, this did not interrupt the friendship nor the scientific exchange. In 1928, Hans Geiger gave Chadwick an improved version of the counter, the Geiger-Müller counter, which he used in his experiments at Cavendish Laboratory in Cambridge. In Berlin, Chadwick also met Walther Bothe (1891–1957) who was Hans Geiger's assistant. In 1930, like the Joliot-Curie couple, Bothe experimented with α-radiation on beryllium and discovered a new, intense, neutral radiation, which he – perhaps influenced by misguided information exchange – also interpreted as γ-radiation.

As already mentioned, James Chadwick experimented using paraffin, as a secondary target, and the new Geiger-Müller counter. With this method he identified the unknown radiation as neutrons, which his teacher Rutherford had been speculating about since 1920. Bothe's student Wolfgang Gentner (1906–1980) had received an assistant position in Paris with the Joliot-Curies, after completing his doctorate. There he worked at the cyclotron and convinced Walther Bothe – when he returned to Germany – to also develop a cyclotron. Thus, exchange and close international scientific cooperation shaped the development of nuclear physics in the first half of the 20th century, a level of interchange that was not even interrupted by war and war propaganda in the years after 1939. In 1940 for example, Wolfgang Gentner, as a German, was able to help in the release of several leading French physicists including the Joliot-Curies who had been arrested by German occupation forces. He even arranged for them to be reinstated in their former positions as directors of the cyclotron institute. Gentner also provided technical assistance to make the cyclotron fully operational so that Joliot and his staff could use it for medical research.

These examples underline the small but closely networked international community of researchers who moved the field forward, guided in their investigations by the latest results. They not only wanted to understand the physical principles of radiation but were also interested in developing possible applications. These communication and collaboration structures still exist, although research institutions have become larger and more complex, with questions focused on much more specific problems.

While the fundamental principles of radioactive decay are well understood, new applications have developed over the last decades. Some of these applications stem from fundamental research questions, such as in geophysics and astrophysics, while others have emerged from industrial applications, ranging from materials science to medical applications and biological research. In particular, medical diagnostics and analysis as well as radiation therapy drive further developments.

Reservations against this development have come from radiation biology and radiation medicine, which have developed into a more and more independent discipline. Although for a long time Georg de Hevesy continued to cultivate the relationship with his teacher Rutherford and was in direct contact with nuclear physics and radiochemical institutes such as the Radium Institute in Vienna, the need for radioactive material for experiments meant that intellectual questions diverged.

In particular, these developments and dynamics also serve to exemplify Hermann Muller's role. Although he was also a heavy consumer of radioactive materials, he remained an outsider in terms of his training and connections. This outsider role was intensified when Muller – who worked for a long time in Berlin in the 1930s, with a group of German and Russian geneticists – moved to the Soviet Union because of the increasing National Socialist influence on his field of research.[27] Disillusioned by his experiences in the Soviet Union, Muller returned to the United States and, despite political concerns and reservations about him and his life during the Second World War, he became a consultant on the Manhattan Project, although not informed of its true purpose. Thus, despite his involvement, he remained an outsider.[28]

Yet, even in the midst of these questions and experiences, scientists continued to exchange information and ideas. This network has brought new techniques and applications, and has particularly influenced the role of nuclear physics in the second half of the 20th century, developing from a small research field probing the nature of radioactive decay to large-scale industrial use.

1.1.7 BASIC RESEARCH AND APPLICATION

Many aspects of radioactivity and its effects are still a field of research that is gaining new results and insights every day. Nuclear physics research focuses on the complex interaction of forces within the association of protons and neutrons that represent the atomic nucleus. More applied research examines the countless possible reactions that control the origin of the elements. Information about these reactions, occurring inside stars, influences how nuclear reactors function and determines the radiation exposure of reactor materials. Study of nuclear and elementary particle reactions at extremely high energies leads researchers to the rare conditions of matter, including theoretical models predicting the first seconds of the universe and even the center of extremely dense neutron stars.

This type of basic research is primarily carried out at instruments that are widely known as particle accelerators. These accelerators give selected elementary particles or entire atomic nuclei the energies they need to initiate collisions with other particles, causing nuclear reactions that emit well-known or new kinds of radiation. These reaction products are documented using specific detectors for particle or γ detection. With the help of theoretical models, conclusions can be drawn about the nature of nuclear or elementary particle reactions as well as the particles and reaction products involved or produced.

Basic research has little to do with nuclear physics applications in everyday life, from energy production to medical radiation therapy. Basic research deals with fundamental issues concerning the origin, function, and change of our world. However, from the results of basic research, new techniques, developments, and applications may also emerge. This is particularly the case in the field of nuclear physics, whose applications permeate many aspects of human existence. Apart from the release of nuclear energy, whether through a reactor or a bomb, this application goes almost unnoticed by the public, although it is reflected in all areas of medical and technical worlds. The resulting academic-industrial complex, which modern natural scientists are increasingly having to deal with, is reminiscent of this quote from Goethe's famous Faust monologue:

No dog would put up with such existence!
So I am seeking magic's assistance, [...]
To grant me a vision of Nature's forces
that bind the world, all its seeds and sources.

Development and construction of the technical facilities required for nuclear physics, both in terms of accelerator and detector technologies, can be extremely expensive and can only be financed through international cooperation. The best-known example is the large hadron collider (LHC), the largest and most powerful accelerator in the world. The LHC is part of the European nuclear research center CERN near Geneva. Smaller accelerators and detectors are often also built in close cooperation between universities, state research institutions, and industrial companies. The latter are often so-called spin-off companies, founded by enterprising scientists who want to put their new knowledge to applied, commercial use.

Examples of these applications and their implications are considered with greater depth in volume two of this book. Here we will consider a broader view of these developments, for example spin-off companies such as the High Voltage Engineering Corporation (HVEC), which emerged from accelerator research at the Massachusetts Institute of Technology (MIT), and the National Electrostatics Corporation (NEC) founded by researchers at the University of Wisconsin. In Europe, the best-known example is ion beam applications (IBA), a spin-off of the cyclotron research center of the Université de Louvain-la-Neuve (ULLN) founded in Belgium in 1986. These companies now operate on an international basis, producing accelerators primarily for materials research and increasingly for pharmaceutical and medical applications. In addition to these newly founded companies, long-established companies such as Siemens in Germany are also involved, with a focus on medical technology. These comprise just one example of the academic-industrial complex. Many others are to be found in the detector and measurement industry, in companies for electrical and magnetic engineering, electronics, data processing, and digitalization, in the computer and software sector, and in vacuum technology.

Close cooperation between academic, governmental, and industrial institutions has led to an economic boom over the last forty years, the sociological, technological, and economic effects of which have not yet been thoroughly investigated. Many of the components developed in this process – in electronics, data processing, software technology, and material development – are subject to rapid commercialization since such components can also be used in the application-oriented research areas of the energy, computer, automotive, aerospace, and, undoubtedly, defense industries. As a result, a dense network of socio-economic relationships has been established, ensuring the rapid and effective transfer of new experimental data and developments.

Targeted application of nuclear physics methods and technologies is usually based on well-known nuclear reactions or decay processes, which can provide extremely precise and reliable results on the composition or age of natural or artificially produced materials. Today, the sensitivity of these methods allows for a wide range of applications outside of traditional nuclear physics. These applications range from geochemical analysis in oil or mineral exploration to archeometric analysis of historical materials as well as isotope analysis for forensic and climatological research. Nuclear physics techniques are mainly used by the radio-medical community in diagnostics and radiation therapy. Application possibilities as well as the precision of the methodology have increased considerably over the last twenty years. In particular, techniques for data production and visualization, originally developed at universities and government research institutes, are now widely commercialized with appropriate hardware and software.

The carbon-14 method (^{14}C) – originally intended to determine the age of historical-anthropological materials such as the mummy of the Egyptian pharaoh Ramses II – is now used in forensic research to determine forgeries in the art world, but the range of applications does not stop there. The pharmaceutical industry needs ^{14}C in order to determine if bodies are breaking down or excreting the by-products of new drugs, since it works quickly and without adverse side effects. With the accuracy achieved today, the ^{14}C method can even determine the lifespan and structure of cells in the human body[29] (Figure 1.6). The production of other radio-isotopes and radiopharmaceuticals, for visualization in diagnostics and for organ-specific application in therapy, is being carried out at nuclear reactors and increasingly at accelerator facilities.

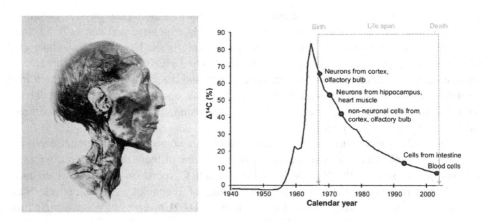

FIGURE 1.6   The carbon-14 method (^{14}C) can be used to date artificial and organic materials. This method is mainly used in archaeology for dating civilization products and human remains. The age of the unwrapped mummy of Ramses II, photographed in 1889 by the German Egyptologist Emil Brugsch (1842–1930) (left, Wikimedia Commons) corresponds almost exactly to a half-life of carbon-14. Today, applications have expanded to include medical examination. This is mainly due to the so-called bomb peak but also to the rapid increase of carbon-14 in our atmosphere and thus in the human body, as caused by the atomic bomb tests from 1945 to 1968. The right-hand figure shows a visualization of how the birth date of different cells in an individual of a known lifespan can be determined retrospectively by comparing the measured ^{14}C signal from DNA extracts with the ^{14}C bomb peak calibration curve. Source: Kirsty L Spalding, Ratan D Bhardwaj, Bruce A Buchholz, Henrik Druid, Jonas Frisén, "Retrospective birth dating of cells in humans," *Cell* 122 (1), (2005): 133–43. .

NOTES

1. João Gama Oliveira and Albert-László Barabási, "Human dynamics: Darwin and Einstein correspondence patterns," *Nature* 43, 12517 (2005).
2. This was the first time that radioactivity had been artificially generated by human activity, which is commonly referred to as anthropogenic radioactivity.
3. Ionization means that an electron is knocked out of the shell of a neutral atom and thus generates an electric current as an independent, negatively charged particle. This current can be directly measured.
4. Wolfgang Reiter, *Aufbruch und Zerstörung – Zur Geschichte der Naturwissenschaften in Österreich 1850–1950* (Wien: LIT Verlag, 2017), 155–218.
5. In this approach, radioactive material is placed as close as possible to the tumor in order to achieve maximum radiation and destructive effects.
6. Typical metric prefixes for physical units: $10^{-15} \equiv$ femto (f); $10^{-12} \equiv$ pico (p); $10^{-9} \equiv$ nano (n); $10^{-6} \equiv$ micro (µ); $10^{-3} \equiv$ milli (m); $10^{3} \equiv$ kilo (k); $10^{6} \equiv$ mega (M); $10^{9} \equiv$ giga (G); $10^{12} \equiv$ tera (T); $10^{15} \equiv$ peta (P).
7. Richard F. Mould, "Pierre Curie, 1859–1905," *Current Oncology* 14 (2), (2007): 74–82.
8. *The Book of Blanche and Marie*, a novel by Per Olov Enquist published in 2004, describes the story of Blanche Wittman (1859–1913), a medium for hypnosis and séance sessions in turn-of-the-century Paris. In her second career, she is said to have been the laboratory assistant and later the lover of Marie Curie. The book describes that the radiation exposure resulting from her work was so extreme that first Blanche Wittman's arms and later also one of her legs had to be amputated and she only continued to exist as a torso until her agonizing death. This book caused great excitement. However, subsequent investigations revealed that Blanche Wittman was never employed as an assistant at the Marie Curie Institute. The *Frankfurter Allgemeine Zeitung* of March 16, 2005, wrote that Blanche's diaries, which form the gravitational center of the novel, never existed and that the story of an agonizing death by radiation was probably due more to poetic freedom than to historical truth.
9. For this work, J. J. Thomson was awarded the Nobel Prize in Physics in 1906.
10. Ernest Rutherford, *Radio-Activity* (Cambridge: Cambridge University Press, 1904).
11. Muriel Howorth, *Pioneer Research on the Atom: The Life Story of Frederick Soddy* (London: New World, 1958), 83–84; Lawrence Badash, "Radium, radioactivity and the popularity of scientific discovery," *Proceedings of the American Philosophical Society* 122 (1978): 145–154; Thaddeus J. Trenn, *The Self-Splitting Atom: The History of the Rutherford-Soddy Collaboration* (London: Taylor & Francis, 1977), 42, 58–60, 111–117.
12. The term *actinides* includes a group of similar heavy elements. The list contains actinium and the next fourteen elements in the periodic table: thorium, protactinium, uranium, and the subsequent transuranium elements neptunium, plutonium, americium, curium, berkelium, californium, einsteinium, fermium, mendelevium, nobelium, and lawrencium, all of which can usually only be produced artificially by irradiation.
13. The atoms of antimatter are composed of antiparticles such that while positrons with a positive charge correspond to electrons with negative charge, antiprotons, which are of the same mass as protons, have a negative charge; neither neutrons nor antineutrons have a charge. In addition to their respective charge, particles of normal matter differ from their antiparticles by the reversal of other important quantum characteristics. Particles and antiparticles annihilate each other and are completely converted into γ-radiation when they meet. According to $E = m \cdot c^2$, the energy of this radiation corresponds to the mass of the two particles. This is the basis of positron emission tomography (PET), currently an important method in medical diagnostics. In addition to positrons, heavier antiparticles can also be produced today, although not in the quantities predicted by Dan Brown in his novel *Illuminati*.
14. Carlo Cercignani, *Ludwig Boltzmann: The Man Who Trusted Atoms* (Oxford: Oxford University Press, 1998).
15. Michael Wiescher, "Arthur E. Haas, his life and cosmologies," *Physics in Perspective* 19 (2017): 3–59, and *Arthur E. Haas – The Hidden Pioneer of Quantum Mechanics* (Berlin: Springer Verlag, 2021).
16. Hans Geiger developed the Geiger counter, which is still today considered the standard instrument for measuring the level of radioactivity.
17. With this suggestion Rutherford also introduced an approach that has been successfully used for many decades in nuclear and particle physics, the postulation of a new particle to explain some of the discrepancies in the existing model description of quantum physical phenomenon. This led to the prediction of the neutrino, by Wolfgang Pauli (1900–1958) in 1930, to explain β-decay as well as to the prediction of the Higgs particle by Peter Higgs (*1929) in 1964, explaining lepton masses. The neutrino was

confirmed in 1956 in an experiment at the Savannah nuclear reactor, and the Higgs particle was confirmed by collider experiments at CERN in 2012. Both theorists received the Nobel Prize, Pauli in 1945 and Higgs in 2013. These examples demonstrate not only the predictive power of theoretical physics but also the ever-increasing necessity of experimental confirmation for the ever-more complicated features of the quantum world.

18. E. Rutherford, "Nuclear constitution of atoms," *Proceedings of the Royal Society A* 97 (686), (1920): 374.

19. The only direct sources, regarding the effects of radiation of varying intensity on the human body, are the long-term studies on victims of the atomic bomb explosions in Hiroshima and Nagasaki. Since the exact radiation dose to which victims were exposed was unknown or difficult to determine, the results were supplemented with animal experiments. To this end, various animal species were used to test the radiation effect from explosions in the atomic bomb test program (cf. Chapter 8 and Chapter 12, Volume 2).

20. Hilde Levi, *George de Hevesy, Life and Work* (Bristol: Adam Hilger Ltd, 1985).

21. William G. Myers, "Georg Charles de Hevesy: The father of nuclear medicine," *The Journal of Nuclear Medicine* 20 (6), (1979): 590–594.

22. Elof Axel Carlson, *Genes, Radiation, and Society: The Life and Work of H. J. Muller* (Ithaca: Cornell University Press, 1981).

23. However, new research points to threshold effects at low doses, e.g., T. Koana, Y. Takashima, M. O. Okada, M. Ikehata, J. Miyakoshi, and K. Sakai, "A threshold exists in the dose-response relationship for somatic mutation frequency induced by X-irradiation of Drosophila," *Radiation Research* (2004): 161, 391.

24. Hermann J. Muller and L. M. Mott-Smith, "Evidence that natural radioactivity is inadequate to explain the frequency of 'Natural' mutations," *Proceedings of the National Academy of Sciences* 16 (4), (1930): 277–285.

25. Edward Calabrese, "Muller's Nobel Prize data: Getting the dose wrong and its significance," *Environmental Research* 176, 108528 (2019).

26. In modern scientific jargon, Geiger was a postdoc.

27. Although genetics was strongly promoted in National Socialist Germany, research and research funds were primarily reserved for so-called racial research.

28. Elof Axel Carlson, *Genes, Radiation, and Society: The Life and Work of H. J. Muller* (Ithaca: Cornell University Press, 1981).

29. Michael A. Malfatti, Bruce A. Buchholz, Heather A. Enright, Benjamin J. Stewar, Ted J. Ognibene, A. Daniel McCartt, Gabriela G. Loot, Maike Zimmermann, Tiffany M. Scharadin, George D. Cimino, Brian A. Jonas, Chong-Xian Pan, Graham Bench, Paul T. Henderson, and Kenneth W. Turteltaub, "Radiocarbon tracers in toxicology and medicine: Recent advances in technology and science," *Toxics* (2019): 7, 27.

2 The Natural Phenomenon of Radioactivity

2.1 NATURAL PHENOMENON OF RADIOACTIVITY

Radioactivity is the result of a physical process and is described mathematically, like all experimentally observed physical phenomena, within the framework of models. This procedure helps and enables us to make predictions about possible consequences and effects. The mathematical description is, however, only a tool for quantifying qualitative observations and statements. An absolutely unambiguous quantification is for the most part not possible due to the complexity of physical/chemical/biological processes and event sequences. Therefore, statistical methods are often applied to determine the probability of a physical process or event. For example, microscopic processes that determine the behavior of atoms and atomic nuclei cannot be described within the framework of classical physics. In particular, the processes of motion, modes, and the configurations of subatomic particles can no longer be described within the framework of classical mechanics. They are therefore considered within the framework of quantum mechanics, a model in which mode or configuration changes no longer occur continuously, as in the case of macroscopic motion processes, but rather take place only within a framework that measures small discontinuous jumps, so-called quantum leaps.[1]

Quantum mechanical processes are always described in terms of probability. The Heisenberg uncertainty principle formulated by Werner Heisenberg (1901–1976) states that no quantum configuration can be determined with 100% certainty. The smallest particles can only be observed through their interaction with other particles or waves, but observation is itself an interaction that inevitably changes the state of the observed particle. This not only applies to observations for determining the location and velocity of any elementary particle but also its energy and the time during which a quantum system remains in a given energy state, which is especially important when considering physical decay processes.

Today, radioactivity is generally understood as decay of the unstable components of chemical elements that transition to a more stable final state by emission of radiation. Depending on the radioactive element, this transition can be rapid or occur very slowly over time. This variance in the decay process is essentially due to the energy difference between the radioactive state and the final state of decay as well as the nature of the decay radiation. In order to understand this, one must first introduce and explain in more detail some basic concepts of nuclear physics and the interactions that determine the inner workings and structure of a nucleus as a multi-particle system, as well as the interaction of different nuclei with each other through nuclear reaction mechanisms.

2.1.1 FOUR FUNDAMENTAL FORCES

Natural processes are subject to four fundamental forces, which are in their function often referred to and described as interactions. These are the gravitational force, the strong force, the electromagnetic force, and the weak force. Modern cosmology associates the origin of these forces with different phases in the early expanding universe, the so-called Big Bang. The goal of modern cosmology and particle physics is to uniformly describe the development of these four forces within the framework of Grand Unified Theory or GUT.

DOI: 10.1201/9781003435907-2

Modern GUT predictions have determined that during the earliest moments of the universe – at less than 10^{-43} seconds, the so-called Planck era – all four forces were unified. With the end of the Plank era, it is expected that gravitational force was frozen out of the equilibrium of forces due to the early universe's expansion and the associated cooling. During the following period, the GUT period, the universe was controlled by two forces: the now independent gravitational force and the GUT force. Theoretical predictions claim that the subsequent GUT era also lasted for only a few fractions of a second, from 10^{-43} to 10^{-38} seconds. Near the end of this era, the universe had cooled to the point that the nuclear strong force was frozen out of the GUT force, leaving three fundamental and independent forces ruling the physics of the early universe: gravity, the strong force, and the still combined electroweak force. This "phase transition" was a critical moment in the development of the early universe because the associated release of energy caused space to rapidly expand, a process called *inflation*. With rapid inflationary expansion, the temperature of the universe was still more than 10^{15} K and only consisted of photons. By 10^{-10} s, the temperature had cooled below 10^{15} K. At that point, the last of the fundamental forces, electromagnetic and nuclear weak forces, froze out, leaving four no longer unified forces to govern the physics of the entire universe.

During the subsequent development of the universe, the four forces behaved very differently. Gravitational force describes attraction between masses and is considered a weak but long-range force. It has little influence on elementary particles but can determine interaction between galaxies that are hundreds of thousands of light years apart. Electromagnetic force describes the interaction between electrically charged particles, which carry either a positive or negative charge, where like-charged particles repel each other and oppositely charged particles attract each other. This force is much stronger than the gravitational force but has a much smaller range, within micrometers, and thus determines the size of an atom.

Strong interaction holds the atomic nucleus together. It is, as the name implies, considerably stronger than the electromagnetic force but has significantly shorter range, corresponding to the size of an atomic nucleus. The last of the forces is the weak interaction: It plays a role in certain decay processes, specifically when a charge exchange occurs within an atomic nucleus. Weak interaction is many orders of magnitude weaker than the strong interaction and has a very short range. Detailed description of these forces, in the context of modern physical theories, is beyond the scope of this work. Simplified, it can be said that the four fundamental forces are classified as so-called exchange forces, where respective components are bound to each other through the exchange of particles. Modern quantum and particle physics aim to understand the fundamentals of these interactions in order to explain fundamental quantities in nature, such as mass, charge, and their origins. However, to discuss radioactive decay processes, only the strong, weak, and electromagnetic forces play a role, since gravity has no influence on the microphysics of the atomic nucleus.

2.1.2 BOHR ATOMIC MODEL

In the discussion of radioactive decay phenomena, their natural occurrences, and their effects on daily life, we can use a simple model to describe the atom and the atomic nucleus: the Bohr model,[2] which is based on analogy to our planetary system. With this model, negatively charged electrons travel on specific paths as they orbit a positively charged atomic nucleus. The number of electrons determines the chemical properties of the atom and is a basis for assigning the chemical elements. The number of electrons is therefore called the *atomic number* (Z). The positive charge of the atomic nucleus must correspond to the negative charge of the electron shell, since the atom is electrically neutral. A charged atom from which an electron has been removed or added by external forces is called a *positive* or *negative ion*.

While in the solar system the planets are kept in their orbit around the sun by the gravitational force, the electromagnetic force (often called the Coulomb force) plays this role in the atomic system. Such a model was first proposed by French physicist Jean Perrin in 1901 and, as discussed in Chapter 1, Ernest Rutherford experimentally confirmed it with his students Hans Geiger and

Ernest Marsden in the famous scattering experiment of α-particles on a gold foil. The α-particles, emitted from a radioactive source, passed through the gold foil mostly unhindered and only the particle beam expanded somewhat. This means that the particles moved mainly through empty space and were therefore hardly hindered in their trajectory. However, researchers also observed a few α-particles that were completely backscattered. Rutherford compared this to a rifle bullet bouncing off a thin piece of paper. From this he concluded that the mass of the gold foil's density cannot be uniformly distributed but rather is confined to a few punctual centers of high density that are hit by only a few particles, and this then leads to the rebound.

From the data obtained on the angular distribution of the scattered α-particles, Rutherford concluded that the mass of the gold atom is concentrated inside the atom, in the so-called atomic nucleus whose radius could be estimated from the scattering data. This showed that gold's atomic nucleus is about 10,000 times smaller than the gold atom with its full electron shell. However, such a model was unstable according to classical ideas because, in the framework of electrodynamics, the orbiting electrons would have to continuously radiate electromagnetic energy in order to proceed on their paths and would not therefore be able to keep fixed orbits due to this energy loss. The dilemma was solved when Danish physicist Niels Bohr suggested fixed orbits for the electrons, which were determined by certain quantum numbers and thus constant. This idea had no physical basis, and violated all laws of classical physics, but was able to provide theoretical predictions for numerous atomic effects, which were confirmed again and again by experiments. Today, this empirical model has been considerably improved but is still perfectly adequate for discussing radiation generation without major quantum mechanical calculations and derivations, as seen in Figure 2.1.

In the framework of Bohr's atomic model, each electron orbit corresponds to a particular quantum mechanical energy state defined by integer quantum numbers.[3] As seen in Figure 2.2, change of energy state can only take place in a quantized mode, by change of a quantum number, and cannot therefore occur in a continuous state. This means that electrons can only jump from one orbit with fixed quantum numbers to another that is also fixed by quantum numbers. Intermediate jumps are not allowed.

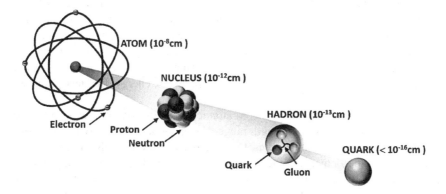

FIGURE 2.1 Schematic representation of an atomic nucleus whose outer dimensions are determined by the electron shell, i.e., a negative charge distribution around the atom's much smaller positively charged nucleus. Today the Bohr atomic model describes the negative charge distribution as negative electrons moving on circular or elliptical orbits that are firmly defined by quantum numbers. The positively charged atomic nucleus consists of positively charged protons and neutral neutrons of comparable mass. They are held together by the so-called strong residual interaction. Protons as well as neutrons belong to the class of baryons, which in turn consist of three even-smaller quark particles that are held together by gluons as exchange particles. The gluons are considered carriers of the strong interaction, photons as carriers of the electromagnetic interaction. The figure gives the respective order of magnitude for the different building blocks of the atomic bond.

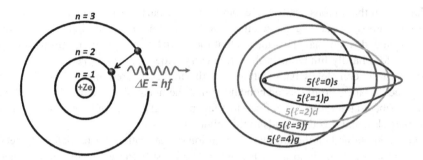

FIGURE 2.2 Illustration of Bohr's atomic model and identification of the quantum numbers that define the energy state of an electron in its orbit around the atomic nucleus. The principal quantum number n determines the radius and the angular momentum quantum number ℓ determines the deviation from the circular radius (left representation), while the corresponding eccentricity of the elliptical orbit is shown on the right. At transitions between the different orbits, energy is absorbed or emitted in quantized form as wave packets. The energy of these wave packets corresponds to the energy difference of the two quantum states.

During such transitions, corresponding quantum numbers change according to certain rules. When electrons transition from an orbit with a higher quantum number configuration – an orbit with a larger radius or greater eccentricity, which corresponds to a higher energy state – to an orbit with a lower energy configuration, electromagnetic energy is emitted in quantized form as an energy packet or photon. Here, the photon energy corresponds to the energy difference between the initial and final quantum states. Conversely, an electron can jump from a lower energy state to a higher state if it absorbs the appropriate transition energy from an external energy carrier such as light or particle radiation, as shown in Figure 2–2.

According to the duality principle of quantum mechanics, photons can be described both as particles and as electromagnetic waves with a certain frequency ν, or as wave packets with a certain wavelength λ .[4] Thus, the correlation between the transition energy E and frequency of the emitted or absorbed radiation can be described according to the famous formula of Max Planck (1858–1947) $E = h\nu$, where $h = 6.62607015 \cdot 10^{-34}\ m^2kg/s$ stands for the so-called Planck's quantum constant. The correlation between the frequency ν and the wavelength of the photons λ is determined by another fundamental constant, which corresponds to the speed of light $c = 299{,}792{,}458\ m/s$. This is the speed at which electromagnetic radiation travels through space, and it corresponds to the product of frequency and wavelength: $c = \lambda \cdot \nu$. By these relations, energy, frequency, and wavelength of a photon are fixed.

The electromagnetic radiation emitted from atomic transitions covers a wide energy spectrum, respectively a wide frequency or wavelength range. Visible light of relatively low frequency is produced from low-energy transitions in atoms. Other transitions produce high-energy photons called *X-rays*. Wilhelm Conrad Röntgen discovered these X-rays in 1888 by bombarding metals with high-energy electrons, which were at that time called *cathode rays*. This corresponds to an energy transfer from the cathode rays to the metal, which in turn excites electrons in the metal's atomic structure to higher energy states, which, under emission of X-rays, jump back to their original quantum state. Thus, this is an energy conversion process from the initial kinetic energy of the electron beam, which is accelerated in electric fields, to atomic excitation energy and further, upon subsequent decay, to photon energy as X-rays. This is essentially the process that takes place during any X-ray irradiation, even at the dentist's or orthopedist's office. X-rays thus come from atomic excitation processes, whereas radiation associated with radioactive decay is based on excitation processes in the nucleus of the atom.

2.1.3 STRUCTURE OF THE ATOMIC NUCLEUS

Atomic nuclei consist of protons and neutrons, which are called *nucleons* – components of the atomic nucleus or nucleus – and belong to the elementary particle class of baryons. Baryons are classified as elementary particles consisting of three quarks coupled together by the strong interaction.[5] This strong attraction gives the atomic nucleus its stability. As individual particles, protons and neutrons are somewhat different, which is related to the configuration of the quarks. Protons have a positive elementary electric charge, which corresponds to the negative electric charge of an electron.

The electron belongs to the elementary particle class of leptons, whose behavior toward each other determines the weak interaction. However, the mass of protons is about two thousand times greater than the mass of electrons. Neutrons are electrically neutral particles, but as baryons they have a mass that is only slightly higher than the mass of protons. This difference is approximately equal to the electron mass. Since, according to Einstein's well-known relation $E = mc^2$, the energy E is directly proportional to the mass m of a particle, with the speed of light c as the constant of proportionality, this means that the neutron corresponds to an energetically higher baryon state than the proton and thus should decay to a proton under energy emission. This is the simplest example of radioactive decay.

Based on early studies of radioactivity, Ernest Rutherford introduced the classification of α-, β-, and γ-radiation, which is still in general use today. By measuring the penetration depth of radiation and its deflection in magnetic fields, he and other physicists of the early 20th century were able to identify the nature of these three radiation components. They found that α-radiation is the positively charged nucleus of a helium atom, consisting of two protons and two neutrons; because of its large mass, this type of radiation has little penetration depth or range and is rapidly absorbed by even thin materials. β-radiation was identified as free negatively charged electrons – often called β^--*particles* for convention reasons. Later, positively charged electrons were also discovered, the positrons, which are the antiparticles of electrons and therefore often called β^+-*particles*. An interaction between positrons and electrons leads to the annihilation of both particles with the release of radiation energy, which according to Einstein's relation corresponds to the mass of both particles. Electrons like positrons have a much longer range in materials than do the massive α-particles. The third type, γ-radiation, is again high-energy electromagnetic radiation resulting from transitions between the various quantum configurations of the atomic nucleus. This is similar to the atomic origin of X-rays, except that the energy of the radiation is typically hundreds to thousands of times higher than that of X-rays.

Similar to the atom, excited states of a nucleus are determined by certain quantum numbers that correspond to certain particle correlations, deformations, and dynamic collective motions (vibration, or rotation) of the nucleus, which can only be described in the context of microscopic-macroscopic nuclear models. The problem with an accurate quantitative description of the excitation structures of the atomic nucleus is that, unlike our Sun-planet system or Bohr's atomic model, the nucleus is not a system in which the interacting force acts as a central force between a massive nucleus and the orbiting particles but rather, is a multitude of quasi-identical baryons bound together to form a nuclear bond by multiple interactions with neighboring baryon components. This is called the *many-body problem*, and even with today's supercomputers, such systems cannot be precisely calculated because of the high dimensionality of the interaction. Instead, one relies on the empirically observed emerging regularities of the nucleus, which can be described as phenomenological models such as the liquid drop model of the nucleus. This affects the reliability of our theoretical predictions for decay processes because the probability for decay, and hence the lifetime of a radioactive nuclear state, can only be determined by the nuclear structural configurations of the initial and final states.

Certain conservation laws do apply to decay processes. In the law pertaining to charge conservation, the electric charge of the initial state of a decay process must equal that of the final state. The same is true for total energy or mass (conservation of energy) and for conservation of momentum.

Here, the product of mass and velocity – the momentum – of the initial state must be equal to the momentum state of the final state (conservation of momentum). In addition to these three conservation laws, there are a number of other conservation laws such as for angular momentum, which keeps the rotational conditions of particles constant, and other more quantum-mechanical parameters.

These conservation laws govern the decay conditions and therefore allow theoretical predictions about radioactive decay. For example, in the case of a neutron decaying to a proton, this means that the decay product must be a negatively charged electron so that mass and charge are conserved. However, when decay of the neutron was observed, it was realized that the electron was not observed with a sharp momentum but rather with a momentum distribution. This suggested the emission of another, nearly massless, neutral particle necessary for momentum conservation. This particle was called a *small neutron* or *neutrino* because of these properties. The decay process in which a neutron transforms into a proton, an electron, and an anti-neutrino particle is called β^--*decay* and is subject to the weak interaction.

However, within the atomic nucleus itself, a proton can also decay into a neutron. This takes place especially when there are too many protons compared to the number of neutrons and the repulsive Coulomb force restricts the strong interaction. This decay of the proton takes place with emission of a positron β^+ and a neutrino to ensure charge and momentum conservation. Physicists refer to this process as β^+-decay because it removes a positively charged particle from the nuclear assembly. Since the positron is considered the antiparticle to the electron, the neutrino released during β^--decay is called the *anti-neutrino* for symmetry reasons.[6]

The number of protons in an atomic nucleus is equal to the atomic number of the electron shell, which determines the chemical behavior of the element. The number of neutrons N and the number of protons Z result in the mass number $A = Z+N$; A corresponds to the number of nucleons and thus to the mass of the atomic nucleus.[7] The nucleus of the carbon atom has 6 protons and 6 neutrons, mass number 12; in the nomenclature of nuclear physics this carbon nucleus is therefore called ^{12}C, where C is the short form for the chemical element. The designation $^{12}_{6}C_6$ is also often found, where the two subscripts indicate the number of protons and neutrons in the nucleus, respectively. The general form for an element X with mass number A, atomic number Z, and neutron number N would then be $^{A}_{Z}X_{N}$.

The exact mass and charge number of different atoms or ions can be determined by so-called mass spectrometry in which the atomic nuclei are accelerated by electric fields and deflected by magnetic fields onto certain trajectories, then detected by position-sensitive particle counters. Deflection paths in a magnetic field correspond to the respective mass and charge, which can thus be precisely determined. This technique has been largely perfected so that mass measurements can be made with great precision even for the smallest trace elements. Today, this classical technique is, however, supplemented by ion traps in which electrically charged particles are trapped and stored by various combinations of electromagnetic fields. By storing the charged particles, it is possible to investigate their physical properties with high precision. Another development is so-called laser traps. Here, a few single, neutral atomic nuclei are slowed down, in the crossfire of laser beams, by cooling (energy extraction) and kept at a well-defined location in order to measure mass and decay properties in detail.

When atoms have nuclei with the same number of protons but a different number of neutrons, they cannot be chemically distinguished from one another. This is because the number of protons equals the number of electrons in the atomic shell, and these are responsible for chemical interaction processes. Such atoms are called *isotopes*. Each chemical element consists of a large number of isotopes, atoms that only differ by the number of neutrons in the nucleus. For the element carbon, for example, there are two stable isotopes $^{12}_{6}C_6$ and $^{13}_{6}C_7$: The first has an identical number of protons and neutrons, and the second has one additional neutron in the nuclear bond. Another example is the element gold (Au), which is characterized by 79 protons. There is only one stable gold isotope, $^{197}_{79}Au_{118}$, with 118 neutrons and mass number 197. For the element lead (Pb) with atomic number

82, there are three stable isotopes: $^{206}_{82}Pb_{124}$, $^{207}_{82}Pb_{125}$, and $^{208}_{82}Pb_{126}$, obviously all with the same atomic number (proton number) but different neutron and mass numbers.

The chemically observed elements on our planet are a mixture of different stable isotopes. The average mixing ratio of protons, neutrons, and mass can be found in tables for the physical properties of the elements but, even so, there is often considerable variation due to local geological-physical or chemical conditions. This variation is called *fractionation*, which depends on the fine differences in the physical properties of the various isotopes, such as mass or size. Fractionation is an observation that has found enormous application as isometry in the fields of archaeometry, forensics, geology, and climate research through isotopic analysis.

2.1.4 ENERGY AND BONDING OF THE ATOMIC NUCLEUS

An important question when considering radioactive decay events is the energy state of both the radioactive nucleus and the decay products. Physical systems always want to adopt the most energetically favorable state, even in the physics of atomic nuclei. Stable nuclei correspond to the energetically most favorable configuration of the proton-neutron system of an atomic nucleus, the ground state, while radioactive nuclei are energetically more unstable. The purpose of decay is therefore to achieve a more stable configuration. Decay modes, such as the emission of α-, β- and γ-radiation, provide the means to accomplish this.

The binding energy holds nucleons together in the atomic nucleus and mainly arises from the strong interaction that couples the nucleons together over a short distance. This energy is reduced by the electromagnetic or Coulomb force between protons, where positive charges repel them from each other. For nuclei with a large number of protons, strong internal repulsive forces act to weaken the bond through the strong interaction, while the weak interaction balances out the effects of excess protons or even neutrons in the nucleus.

These individual terms are shown schematically in Figure 2.3 using the example of a nitrogen isotope, $^{18}_{7}N_{11}$, with seven protons and eleven neutrons. The so-called volume term corresponds to the strong interaction that each nucleon has with all neighboring nucleons. Nucleons near the surface have reduced binding energy due to the absence of neighbors. Therefore, the corresponding binding-energy fraction must be subtracted when measuring the strong interaction. In addition, positively charged protons repel each other and this electromagnetic interaction reduces the

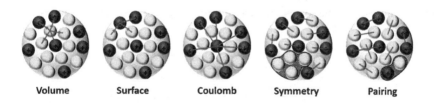

| Volume | Surface | Coulomb | Symmetry | Pairing |

FIGURE 2.3 Various contributions that determine the binding energy of a nitrogen atomic nucleus, $^{18}_{7}N_{11}$. The *volume* term describes the interaction of each baryon with neighboring baryons. The *surface* term describes reduction in the volume term because baryons near the surface of the nucleus have a reduced number of neighbors. The *Coulomb* term describes the electromagnetic repulsion of positive neutrons from each other, so it acts only between charged particles. The *symmetry* term accounts for the fact that neutron-proton pairs are more strongly bound to each other, and unpaired neutrons (or even protons) contribute less to the overall binding. Finally, *pairing* is the last component and, unlike the symmetry term, takes into account the pairing of like baryons as they contribute to the total binding. From the last two contributions, it follows that nuclei with an even number of protons and neutrons are more strongly bound than nuclei of similar mass but with odd proton and neutron numbers, which in turn are more strongly bound than nuclei with either odd proton and even neutron numbers or even proton and odd neutron numbers.

binding energy and must therefore be subtracted. Further, the strong interaction force acts especially between protons and neutrons so that, if there is an excess of one type of nucleon, the binding
energy is reduced by the lack of strong interaction between the excess nucleons. Finally, pairing
energy describes a particularly tight coupling between like pairs of nucleons. This term increases
the binding energy when Z and N, and hence A, correspond to an even number. If A is even but both
Z and N are odd, then the pairing energy is absent, reducing the binding energy. If A is odd, the
pairing energy is zero and the term can be neglected.

Binding energy $B(Z,N)$ of a nucleus with Z protons and N neutrons is equal to the difference
between the mass of the nucleus $m(Z,N)$ and the masses of its individual components m_p for protons
and m_n for neutrons. To convert energy into mass, Einstein's mass-energy equivalence $E = mc^2$ or $m
= E/c^2$ is used. The binding energy is then

$$B(Z,A) = \frac{m(Z,N) - (Z \cdot m_p + N \cdot m_n)}{c^2} \tag{2.1}$$

In practical terms, the binding energy corresponds to the amount of energy one must pump into
a nucleus in order to decompose it into its individual components. On the other hand, if one man-
ages to fuse individual protons and neutrons together to form a nucleus, the corresponding binding
energy of the nucleus is released. This would accomplish the dream of an inexhaustible source of
fusion energy, which is obtainable if one succeeds in fusing hydrogen nuclei – i.e., protons – to
heavy nuclei. To do this, however, one must first overcome the strong repulsive Coulomb force, and
this has not yet been achieved despite the greatest efforts and enormous costs.

Due to the complexity of the interaction conditions between individual nucleons within an atomic
nucleus, it has not yet been possible to calculate the binding energy directly, except for smaller
atomic nuclei up to carbon. For heavier nuclei, the computational capacity of current computer systems is insufficient. Therefore, binding energies of currently known nuclei are approximately calculated by means of empirical formulas or are determined by detailed measurement of the nuclear
masses. In the case of unknown nuclei, one must rely on the empirical mass formulas because,
despite considerable effort and enormously improved measurement methods, many of these nuclei
cannot be produced in sufficient quantities for reliable mass measurement.

Binding an atomic nucleus is therefore often schematically represented as a potential well, shown
in Figure 2.4, where the depth of the potential corresponds to the binding energy. Unlike the strong
and weak interaction forces, the electromagnetic force acting inside the nucleus is long-range and
the Coulomb fraction of the binding energy goes far beyond the radius of the atomic nucleus given
by the mass number A and the strong interaction force. However, the Coulomb force only applies to

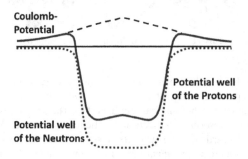

FIGURE 2.4 Representation of the nucleus' potentials. The depth of the potential corresponds to the amount
of energy that is required to separate the particles in the nucleus. The figure shows that the binding potential of
the uncharged neutrons reduces the binding potential of the positively charged protons by an amount equivalent to the Coulomb potential.

positively charged protons, which are subject to electromagnetic interaction. Therefore, the binding potential is different for protons and neutrons.

The radius of an atomic nucleus depends directly on the number of tightly bound nucleons, i.e., the mass number A of the nucleus, as well as on the size of a single nucleon. For the approximate radius r of an atomic nucleus, the following relation holds (in units of Fermi with $1 fm = 10^{-13} cm$),

$$r = r_0 \cdot A^{(1/3)} \quad r_0 \cong 1.26 fm \tag{2.2}$$

where r_0 is approximately the radius of a single nucleon.

The best-known – and simplest – empirical mass model for the atomic nucleus is the so-called liquid drop model, which Carl Friedrich von Weizsäcker (1912–2007) established, as early as 1936, as the Weizsäcker mass formula. Here, the atomic nucleus is treated like a drop of water. The volume energy of the atomic nucleus corresponds to the condensation energy of the water drop which, reduced by a surface term, holds the drop together. Included are the repulsive Coulomb forces and the so-called symmetry energies, which are supposed to correct strong imbalances between the number of protons and neutrons. Despite its simplicity, Weizsäcker's mass formula has been extremely successful in predicting binding energies globally; even today's mass formulas are still often based on this basic formalism. Thus, binding energy increases with mass number: The more nucleons to be bound, the larger the expenditure of binding energy.

For mathematically interested readers, Weizsäcker's mass formula – more precisely, the formula for the binding energy of an atomic nucleus with mass number A and charge number Z – is presented as follows:

$$B(Z,A) = a_v \cdot A - a_0 \cdot A^{2/3} - a_c \cdot \frac{z^2}{A^{1/3}} - a_s \cdot \frac{(A-2Z)^2}{A} \tag{2.3}$$

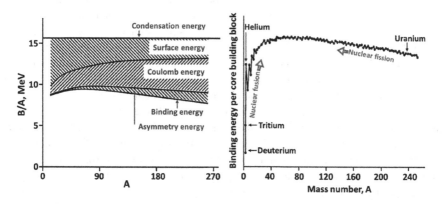

FIGURE 2.5   On the left, calculated according to the Weizsäcker model, different fractions of the binding energy $B(Z, A)$ of an atomic nucleus are shown schematically as a function of the mass number A. The different components are subtracted from the constant volume term of the binding energy as expressed in the Weizsäcker formula. Subtracting the surface energy, the Coulomb energy, and the asymmetry energy from the volume energy gives the binding energy. Pairing effects are neglected. The measured total binding energy per nucleon $B (Z,A)/A$ is shown on the right. The curve demonstrates that the maximum binding energy per unit mass is in the middle mass range of iron $A = 56$ and nickel $A = 58$. In addition to the particularly stable nuclei – i.e., ^{4}He and ^{12}C – the curve also shows that binding energy can be obtained in the fusion of light nuclei, such as deuterium and tritium, to heavier nuclei (up to ^{56}Fe). Similarly, binding energy can be released in the fission (fission) of heavy nuclei such as uranium to lighter nuclei.

Here the terms $a_V = 15.8\ MeV$, $a_O = 18.3\ MeV$, $a_C = 0.714\ MeV$, $a_S = 23.2\ MeV$[8] are empirical parameters that fit the strength of the binding energy's different components (volume term, surface term, Coulomb term, and symmetry term) to the experimental data. These parameters depend on the quality of the experimental data set and the quality of the fit with the empirical mass formula, and they therefore appear in the literature with slightly different numerical values. However, in order to better represent subtleties, which are mostly due to nuclear structure factors, binding energy per nucleon $B(Z, N)/A$ is usually used as a practical quantity and is shown graphically in Figure 2.5.

As given by Weizsäcker's formula, Figure 2.5 shows the binding energy per nucleon $B(Z, A)/A$ as a function of nuclear mass A. While the volume contribution to the binding is constant, the baryons located at the surface have fewer neighbors and therefore weaker bonding. To correct for this, a so-called surface term is subtracted. Binding energy is also reduced by Coulomb energy, which accounts for equally charged particles repelling each other. Finally, since the interaction between protons and neutrons is slightly different, the asymmetry term accounts for the inequality between the number of protons and neutrons.

2.1.5 RADIOACTIVE DECAY

In radioactive decay, an unstable energetically excited configuration of protons and neutrons within a nuclear assembly changes to a more stable, energetically more favorable configuration by emitting radiation of certain energy. In terms of particle α- or β-emission, the energy corresponds to the mass m of the particle ($E = m \cdot c^2$) plus its kinetic energy, which depends on the particle velocity: ($E_{kin} = \dfrac{1}{2} mv^2$). For decay by γ-emission, the energy corresponds to the frequency v of the emitted electromagnetic wave ($E = h \cdot v$).

The decay mode depends on the balance between the number of protons and neutrons. Unstable isotopes are characterized by a neutron number that is either too low or too high relative to the proton number of the nucleus. This affects internal forces such as the Coulomb force and the weak interaction, which, along with the strong interaction, determine the stability of the nucleus. Simplified, if there are too few neutrons compared to protons, the nucleus is unstable due to an excess of repulsive Coulomb force between the equally charged protons. A proton then mutates to a neutron by β^+-decay to regain stability.[9] In this process, the initial state is energetically higher and therefore unstable, but then decays by weak interaction into an energetically more favorable final state. Since the atomic number Z is reduced by one charge state, a new element with the atomic number Z-1 is created. In this process, a positron and an electron from the atomic shell also compensate and in effect neutralize each other, so the number of electrons is also reduced.

The case is similar for nuclei that are too rich in neutrons. These nuclei are also unstable because the balance between the baryon pairs, for the strong interaction, is no longer given. In this case, the nucleus seeks to reach a more stable state by transforming a neutron into a proton through β^--decay. Again, this changes the atomic number Z and thus causes transformation to a new element with atomic number Z+1. In β^+-decay as in β^--decay, the corresponding neutrinos are emitted in addition to the positrons or electrons, respectively.

A well-known example is the relatively long-lived isotope $^{14}_{6}C_8$, which transforms into a $^{14}_{7}N_7$ nitrogen isotope by β^--decay. Thus, the original element with atomic number $Z = 6$ changes by decay into a new element with atomic number Z+1 = 7. However, it can be seen that the number of nucleons or baryons remains constant with $A = 14$. The energy released during decay is equal to the difference in energy between the initial system of the ^{14}C nucleus and the final state of the ^{14}N nucleus plus decay products. This energy is converted into the kinetic energy of the electron and anti-neutrino.

In a few, often very critical cases, there is a third decay possibility based on the weak interaction. In a sense, this decay is similar to β^+-decay, in which a proton in the nuclear bond is converted to a neutron, since the same result is achieved when the nucleus captures an electron from its electron shell. Here, the positive charge of a proton is compensated by the negative charge of a captured electron, thus forming a neutral neutron. This so-called electron capture can proceed in parallel with β^+-decay. Which possibility will occur again depends on the probability for quantum mechanical transition from the initial radioactive state to the final stable nucleus. Particularly important radioactive elements that decay by electron capture are beryllium isotope $^7_4Be_3$, potassium isotope $^{40}_{19}K_{21}$, and titanium isotope $^{44}_{22}Ti_{22}$.

The radioactive potassium isotope $^{40}_{19}K_{21}$ is a particularly interesting case since it can enter three different decay channels. The β^--decay sequence can transform a neutron in the nuclear bond into a proton, producing calcium isotope $^{40}_{20}Ca_{20}$: $^{40}_{19}K_{21} \Rightarrow {}^{40}_{20}Ca_{20} + \beta^- + \nu$. However, β^+-positron decay to the argon-noble gas isotope $^{40}_{18}Ar_{22}$ is also energetically possible, with a proton in the nuclear bond transformed to a neutron: $^{40}_{19}K_{21} \Rightarrow {}^{40}_{18}Ar_{22} + \beta^+ + \nu$. The third possibility is electron capture into the ^{40}K nucleus, converting a proton to a neutron. The $^{40}_{18}Ar_{22}$ isotope is created in an excited state that transitions to the stable ground state by emitting 1.46 MeV radiation. Figure 2.6 shows a so-called decay scheme. The horizontal lines represent energetic states for the various isotopes represented by rectangular potential wells; arrows show the decay potentials. The vertical axis is the energy axis, and the horizontal axis represents the number of protons in the nucleus.

In α-decay, a helium nucleus is emitted consisting of two protons and two neutrons. Helium's nucleus has a particularly strong bond and based on this strong binding, special structures are formed within an atomic nucleus, the so-called α-cluster structures. In Figure 2.7, highly unstable beryllium $^8_4Be_4$ has a structure that looks like two α-particles coupled together, a configuration that immediately decays into two free α-particles. On the other hand, the $^{12}_6C_6$ carbon nucleus exhibits a distinct α-cluster structure of three α-particles in its higher excited states.[10] Such alpha-cluster configurations can be found in many light nuclei.

In general, α-decay occurs in very massive nuclei that simply cannot, because of size, be kept together by short-range strong interaction and therefore transition to a less massive state with emittance of an α-particle. Since the α-particle has a mass number $A = 4$ and an atomic number of $Z = 2$, α-decay of isotope $^A_ZX_N$ forms a new isotope that is reduced by two protons and two neutrons, characterized as $^{A-4}_{Z-2}X_{N-2}$ where X stands for the respective element symbol.

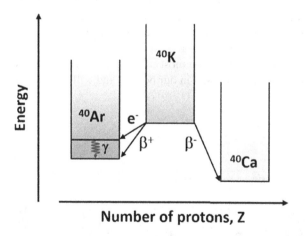

Number of protons, Z

FIGURE 2.6 Decay scheme of $^{40}_{19}K_{21}$. The scheme demonstrates three decay possibilities: β^--decay that transforms a neutron into a proton and forms the calcium isotope $^{40}_{20}Ca_{20}$. Alternatively, β^+-decay is allowed, which transforms a proton into a neutron and forms the argon isotope $^{40}_{18}Ar_{22}$ as the final nucleus. The third possibility is formed by electron capture from the shell into the ^{40}K nucleus; this leaves an excited $^{40}_{18}Ar_{22}$ nucleus that, by subsequent γ-decay with an energy of 1.46 MeV, passes into the ground state of ^{40}Ar.

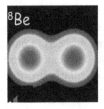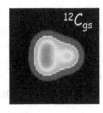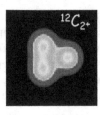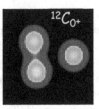

FIGURE 2.7 Nuclear structure calculations for ^{8}Be (left) and three ^{12}C configurations: ground state; first excited state with spin 2$^+$; and the excited 0$^+$ state, the so-called Hoyle state. All ^{12}C configurations exhibit three α-particles coupled to each other.

The most famous example of the α-decay of heavy isotopes is provided by the work of Marie Skłodowska-Curie and Pierre Curie, who conducted intensive studies on radioactivity through their work with pitchblende. Pitchblende is a shiny blackish mineral composed of more than 80% uranium and the Curie couple studied the radiation it emitted, a phenomenon that had first been observed by Henri Becquerel in 1896.[11] Through elaborate chemical analysis, the Curies studied more than ten tons of pitchblende and succeeded in extracting several new elements, which were analyzed for their characteristic properties.

One of these elements was the already known thorium with atomic weight $A = 232$ and atomic number $Z = 90$. In addition, they found two other unknown radiating elements, which they named radium with atomic weight $A = 226$ and atomic number $Z = 88$ and polonium with atomic weight $A = 210$ and atomic number $Z = 84$. All of these observed isotopes decay by α-emission. The mother-daughter combinations are $^{232}_{90}Th_{142} \Rightarrow {}^{228}_{88}Rd_{140} + \alpha$; $^{226}_{88}Rd_{138} \Rightarrow {}^{222}_{86}Rn_{136} + \alpha$; and $^{210}_{84}Po_{126} \Rightarrow {}^{206}_{82}Pb_{124} + \alpha$. Thus, the heavy thorium isotope decays to a radium isotope with atomic weight $A = 228$. The radium isotope with atomic weight $A = 226$ discovered by the Curies decays to radon, which in fact was later identified as a noble gas escaping from the pitchblende. The relatively light polonium isotope decays to a stable lead isotope. Using this decay arithmetic, Ernest Rutherford and his collaborator Frederic Soddy were later able to establish the so-called natural decay series, by which heavy unstable isotopes such as uranium and thorium slowly decay to stable lead isotopes.

Figure 2.8 shows the decay series of radioactive thorium isotope $^{228}_{90}$Th$_{138}$ as depicted by Ernest Rutherford.[12] There are several such decay series, but the best known are triggered by decay of heavy uranium isotopes $^{235}_{92}$U$_{143}$ and $^{238}_{92}$U$_{146}$, thorium isotope $^{232}_{90}$Th$_{142}$, and neptunium isotope $^{237}_{93}$Np$_{144}$. These are the so-called actinides[13] whose slow decay gradually feeds other isotopes along the chain. The phenomenon of these natural decay series, evolving from the associated long-lived radioactive isotopes, represents one of the most important components of so-called natural radioactivity in our universe, on our earth, and in our bodies, and will be discussed in detail throughout this book.

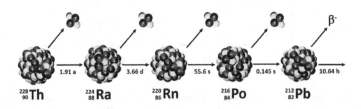

FIGURE 2.8 Thorium's natural decay chain. Through a series of successive α-decays, the original $^{228}_{90}$Th isotope transforms first to radium, then radon, and polonium, and finally to the β$^-$ unstable lead isotope $^{212}_{82}$Pb.

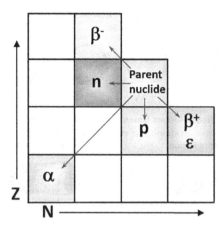

FIGURE 2.9 The vertical axis indicates the number of protons, i.e., the atomic number Z, and the horizontal axis indicates the number of neutrons N. The diagram shows changes in the number of protons and neutrons during α-decay as well as β⁻- and β⁺-decay and electron capture, respectively. Also drawn are the changes in neutron and proton emission, which can, however, only occur in cases of extreme neutron or proton excess in the nucleus.

Figure 2.9 schematically shows the various possibilities for radioactive decay. The α-decay removes two protons and two neutrons; also shown are the β⁻, β⁺, and electron capture decays, converting a neutron into a proton and vice-versa. Particle decays such as neutron and proton emission do exist but can only occur at the limits of particle stability, in nuclei with an extreme excess of neutrons or protons, respectively. The schematic demonstrates changes in the initial isotope's atomic number and mass number as associated with each type of decay.

The third type of decay, γ-decay occurs without any change in atomic number or mass number. This means that the decaying atomic nucleus only passes from an energetically excited, i.e., more massive, deformed, or otherwise unstable state to another energetically lower state while emitting high-energy photons. This is similar to the emission of X-rays during electron transitions in the atomic shell except that the emitted energy is about a thousand times higher and therefore more destructive than that of X-rays. Excited states are not, however, to be understood by an analogy to planetary orbits as in the Bohr atomic model, with larger radii or greater eccentricity. Rather, these excited states are mostly interpreted as certain quantum configurations of the atomic nucleus, which classically correspond to a deformation or deviation from the spherical shape, or as oscillations that resemble the trembling of a water drop. Other configurations with a high angular momentum quantum number are interpreted as rapid rotations of the nucleus about its own axis. These are analogies from classical mechanics and, in the case of an atomic nucleus, are described by the mathematical apparatus of quantum mechanics. For each configuration, only one energy state is allowed. Transitions from one state to the other thus follow certain selection rules that determine the transition probability. The energy difference between different states is emitted as γ-photon with its characteristic energy, wavelength, or frequency.

Besides the three best-known types, α-, β-, and γ-decay, there is, among others, proton decay which occurs in nuclei that have such a large excess of protons or neutrons that even β⁺-decay cannot preserve stability. In addition, there is also neutron decay, which occurs in the case of extremely neutron-rich nuclei that can no longer be stabilized by β⁻decay (Figure 2.7). However, these cases rarely occur in nature because the production of such nuclei requires enormous energies that occur at most in exploding stars.

What can occur on Earth is natural nuclear fission or simply, fission. Nuclear fission as a decay process occurs only in strongly deformed nuclei. Among light nuclei this includes already mentioned $^8_4Be_4$, which practically speaking consists of two helium nuclei $^4_2He_2$, thus forming a

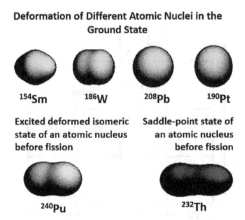

Deformation of Different Atomic Nuclei in the Ground State

^{154}Sm ^{186}W ^{208}Pb ^{190}Pt

Excited deformed isomeric   Saddle-point state of
state of an atomic nucleus   an atomic nucleus
before fission   before fission

^{240}Pu ^{232}Th

FIGURE 2.10 Heavy atomic nuclei characterized by different deformations in their ground state - the energetically most stable state of the special nucleus. The particularly stable double-magic lead isotope $^{208}_{82}$Pb is spherical, while the unstable isotopes have large deformations that can lead to fission or emission of an α-particle.

so-called α-cluster that then also spontaneously decays into two independent α-particles or $^4_2$He^2 nuclei. However, it is usually extremely heavy nuclei that are so deformed, by forces acting internally between the nucleons, that normal α-decay is not sufficient to preserve their stability. These deformed nuclei can by fission pass into an energetically more favorable state of two approximately equal-sized nuclear fragments. Figure 2.10 illustrates the shape of various heavy atomic nuclei in the ground state as well as excited long-lived states. Such excited states are called *isomers*. The shapes were determined by experimental studies of the individual quantum numbers for the respective states.

Nuclear fission decay products are pairs of various nuclei in the medium mass range. Specific decay probabilities determine which end products are formed. In addition to two large nuclear fragments, neutrons are released because heavy nuclei often have a relative excess of neutrons to compensate for the repulsive effect of the Coulomb force within the nuclear assembly. In most cases, the nuclear fragments produced during fission are also not stable. They represent highly excited configurations and transition to a stable state through further β$^-$-decay coupled with γ-decay processes. An example of this process is the fission of radioactive isotope $^{252}_{98}$Cf$_{154}$ of the heavy element californium, discovered in Berkeley, California in 1950. Two of the various decay possibilities are $^{252}_{98}$Cf$_{154}$ ⇒ $^{145}_{56}$Ba$_{89}$ + $^{104}_{42}$Mo$_{62}$ + 3n, releasing a barium and a molybdenum isotope in addition to three neutrons, or $^{252}_{98}$Cf$_{154}$ ⇒ $^{128}_{50}$Sn$_{78}$ + $^{122}_{48}$Cd$_{74}$ + 2n. In the latter case, only two neutrons are produced in addition to a tin and a cadmium isotope. It is easy to see that in both cases the atomic number $A = 252$, the atomic number $Z = 98$, and the neutron number $N = 154$ remain constant when the respective sizes of the decay products are added up $A = 145 + 104 + 3 = 128 + 122 + 2 = 252$, $Z = 56 + 42 = 50 + 48 = 98$, and $N = 89 + 62 + 3 = 78 + 74 + 2 = 154$. In addition, there are several other possibilities but all of these fission products must follow the same selection rules for mass and charge conservation. In some cases, neutron capture causes the fission process. The best-known example is neutron-induced fission of $^{235}_{92}$U$_{143}$ in which the uranium isotope decays to various nuclei in the intermediate mass range, emitting excess neutrons.

Figure 2.11 shows the so-called fission product curve, which represents distribution of the two heavy fission products of ^{235}U fission as a function of mass. The fission products are distributed over two mass ranges $A ≈ 95$ and $A ≈ 137$ separated by a minimum. The neutron-rich fission products

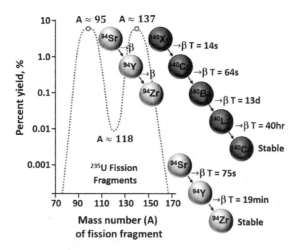

FIGURE 2.11　Distribution of the massive products in ^{235}U nuclear fission. The distribution is determined by the decay probability. Fission produces mass distribution around a lighter component, in the mass range between $A \approx 80$ and $A \approx 110$, and a heavier component between $A \approx 120$ and $A \approx 160$. Nuclei outside this mass range are only produced with negligible probability of less than 1%.

are mostly highly radioactive and decay via a sequence of β^--processes and also neutron emission to form stable nuclei, also schematically shown.

These decay processes represent a transition between two energy states, from the deformed configuration of the original nucleus to a particular distribution of fission products. Bonding energy is released in the process and corresponds to the difference between the total binding energy of the initial state and that of the final state, thus to the difference between the mass of the radioactive nucleus and the mass of its decay products (cf. Figure 2.5). If the masses of the involved nuclei are known exactly, the released energy can be easily calculated. The mathematical details of these calculations will not, however, be discussed here. Interested readers are referred to more detailed textbooks on nuclear physics.

2.1.6　ISOTOPES, ISOTONES, AND ISOBARS

Nuclei of the known elements are represented by an isotope table in which individual isotopes are arranged according to their number of protons and neutrons. Figure 2.12 shows the isotope map, sometimes called the *nuclide map*, for all elements and isotopes known to date. The vertical axis indicates the atomic number Z by which the elements are identified, and thus the number of protons. The horizontal axis shows the number of neutrons N in the respective nucleus. Each box has a certain Z, N combination corresponding to a certain atomic nucleus.

In Figure 2–12, isotopes of an element, characterized by atomic number Z, are arranged along the horizontal line while isotones, as nuclei with the same neutron number N, are arranged along the vertical line. Of further importance are isobars, nuclei with a constant mass number A which are located along the diagonal line increasing from left to right. Also characterized are the so-called magic numbers Z, $N = 2, 6, 8, 14, 28, 50, 82, 126$, each corresponding to a certain number of protons or neutrons, where the nucleus has a special stability due to internal couplings of nucleons. The most stable nuclei are the double magic nuclei such as $^{4}_{2}He_{2}$, $^{12}_{6}C_{6}$, $^{16}_{8}O_{8}$, $^{28}_{14}Si_{14}$, $^{56}_{28}Ni_{28}$, and $^{208}_{82}Pb_{126}$. Double magic nuclei such as $^{100}_{50}Sn_{50}$ and $^{132}_{50}Sn_{82}$ decay by β^+- and β^--decay but possess greater internal stability than their respective neighboring isotopes and isotones because of their symmetry and pairing energies.

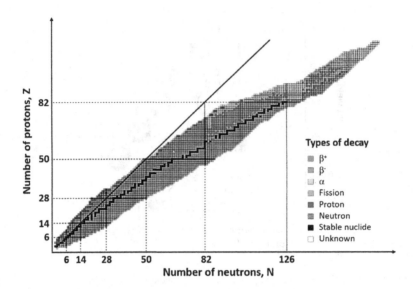

FIGURE 2.12 Table of isotopes known today, ordered by the number of protons (Z) against the number of neutrons (N) in the respective atomic nuclei. Isotopes marked in black are stable and will not undergo radioactive decay. Isotopes marked in color are radioactive. Coloration indicates the dominant decay mode in each case. Some nuclei can decay in different ways, for example ^{40}K and especially the heavy nuclei with masses heavier than double-magic lead $^{208}_{82}$Pb. From: https://www.isotope_table.svg: Table_isotopes_en.svg.

Figure 2.12 illustrates some of the decay characteristics described above. The diagonal line rising from left to right corresponds to the $N = Z$ axis, signifying nuclei that have the same number of protons as neutrons. Boxes marked in black are stable nuclei that do not change configuration as a result of radioactive decay. For light nuclei, stable isotopes lie along the $N = Z$ line but turn into the neutron-rich region as the proton number increases; this is due to the repulsive Coulomb force between positively charged protons in the nucleus, which must be compensated by additional neutrally charged neutrons to maintain stability. The orange boxes in the area to the left, above the stable nuclei, have too many protons or too few neutrons to maintain their stability. They decay to a less proton-rich nucleus by β^+ or positron emission along the isobaric axis until they reach the stability axis; in each case a neutrino is released in addition to the positron. Thus, an extremely proton-rich isotope does not decay directly to a stable nucleus but through a sequence of β^+-decays. This results in the formation of several radioactive nuclei along the isobar line; these gradually build up, by decay of the parent nucleus, and are degraded by their own decay until the decay sequence reaches the stability line. This is like the natural α-decay series, where there is also a sequence of radioactive parent and daughter nuclei in parallel.

The extreme case of proton-rich nuclei is marked in red along the upper edge of the nucleus. They decay by emission of a proton to achieve a more favorable proton-neutron balance. These nuclei mark the boundary of β^+ stability or what is known as the proton drip line. Beyond this line there are no stable isotopes since such nuclei, if formed, decay immediately by proton emission to nuclei located on the right side of the drip line.

The situation is similar for nuclei depicted in blue, lying to the right of the stability line, in the lower region of the isotope map. These neutron-rich or proton-poor nuclei decay by emitting a β^- particle or electron as well as an anti-neutrino. In this process, a neutron in the nucleus is converted into a proton. The decay again proceeds along the diagonal isobaric axis, since the atomic weight is preserved and only the atomic and neutron numbers change by one unit each. Again, there are β^--decay sequences when a very neutron-rich parent nucleus slowly decays, over a series of successive β^--instable daughter nuclei, along the isobaric line toward the stability line. The decay chain is

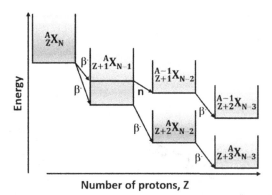

FIGURE 2.13 Decay chain of neutron-rich nuclei. Shown is a sequence of pure β⁻-decays along an isobaric series with constant mass number A. Also shown is a decay sequence with β⁻-delayed neutron decay; this shifts the decay chain into the adjacent isobaric sequence with mass number A-1.

schematically shown in Figure 2.13 where the energy state of a neutron-rich isotope (Z,N) sequentially decays by β⁻-decay into less energetic isobaric configurations.

However, in these β⁻-decay chains on the neutron-rich side, there is the possibility of branching by emission of a neutron since the neutron has no electric charge and can therefore leave the nuclear bond more easily than the positively charged proton. This so-called β-delayed neutron decay occurs when the β⁻-decay populates an excited state in a daughter nucleus that is energetically high enough to emit a free neutron, rather than decaying via a γ-decay to the ground state or via a subsequent β⁻-transition. This shifts the decay chain from the isobaric line, characterized by atomic weight A, to the adjacent isobaric series A-1 (see also Figure 2–13). However, this case only occurs for extremely neutron-rich radioactive nuclei.

The limit of stability is characterized by the neutron drip line, which represents nuclei that contain too many neutrons to maintain stability and therefore decay by neutron emission to a more favorable configuration. However, the neutron drip line is only known for the light boxes as shown in purple in Figure 2.12 and is largely unexplored for larger masses.

Radioactive atomic nuclei in the upper mass range transition primarily by α-decay to a more energetically favorable state, as marked by yellow boxes in Figure 2.12. In addition to members of the natural decay series already discussed in the massive range between uranium $(Z = 92)$ and lead $(Z = 82)$, there exist several other α-unstable nuclei. These include massive nuclei along the proton drip line as well as nuclei that lie close to the stability line, where other decay possibilities would be expected. The possibility of β-decay exists but, for reasons of nuclear structure, α-decay is favored, with only a very weak fraction decaying into the β-decay channel.

Similarly, spontaneous fission of heavy atomic nuclei is marked by green boxes in Figure 2.12. For example, the uranium isotope $^{235}_{92}U$ can decay by α-decay into the thorium isotope $^{231}_{90}Th$ but can also decay by spontaneous fission into two isotopes in the intermediate mass range, emitting a number of neutrons. Figures 2.11 and 2.14 show, in different plots, the so-called fission yield produced by spontaneous nuclear fission of ^{235}U. While Figure 2.11 only shows the distribution as a function of mass number, Figure 2.14 shows two-dimensional distribution as a function of neutron and proton number within the framework of a nuclide map.

One can see a distribution of nuclei mostly on the neutron-rich side, since ^{235}U already has a high neutron excess. This is reflected in the neutron-rich daughter nuclei as well as in the additional emission of free neutrons during fission. These released neutrons play an important role because they

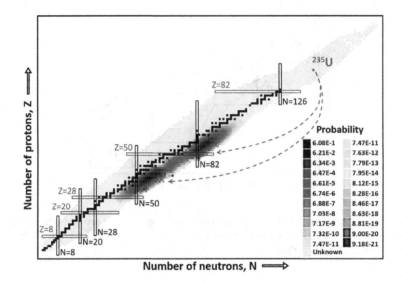

FIGURE 2.14 Fission yield for fission of uranium isotope $^{235}_{92}$U. The fission products are distributed over two mass ranges around $A = 95$ and 135 in the neutron-rich region to the right of the stability line. Color coding indicates the relative yield of each decay product. Source: https://www.nndc.bnl.gov/.

initiate a special process, that of neutron-induced fission, which will be discussed in more detail in the next chapter.

In summary, radioactive decays balance deviations in the configuration of an atomic nucleus. The β-decays balance large deviations in the proton-neutron ratio and restore an equilibrium that is favorable for stability. For massive nuclei, in which the internal repulsive forces against the strong force become too great, a new lower-mass state of stability is reached through α-decay or through fission. In the range of natural decay chains, the heavy unstable uranium and thorium nuclei decay through a series of processes to stable lead isotopes. The γ-decay corresponds to internal excitation of an atomic nucleus from high-energy states that correspond to, for example, deformation, rapid rotations, or vibrations, to low-energy rest states of the nucleus with little or no deformation, rotation, or vibration.

NOTES

1. Although classical movement can also be considered in terms of quantum mechanics, quantum leaps are so infinitesimally small that they are essentially not observable without spectroscopic tools. In conversational jargon, politicians and the media incorrectly apply the notion of "quantum leap" by reversing the meaning. Instead of referencing to an incredibly small shift, as in the quantum mechanics of an atom, the term is used to refer to a massive qualitative shift in a system.

2. The idea of separate, positively and negatively charged particles arose after the discovery of cathode rays by German physicist Julius Plücker and his student Johann Wilhelm Hittorf. Later, these rays were found to be free electrons by British physicist Joseph John Thomson. Wilhelm Wien (1864–1928) later hypothesized a positive component, which later came to be known as protons. With the Thomson model of the atom, discrete positively and negatively charged particles were first discussed as protons and electrons, respectively. Furthermore, this model proposed that these particles were so tightly packed together that their charges canceled each other out.

3. In the Bohr model, these are the principal quantum numbers, $n = 1,2,3...$ that define the energy state and orbital radius. Orbital momentum quantum numbers, $\ell = 0,1,2,3...(n\text{-}1)$ determine the rotational state, i.e., the eccentricity of the orbit, and spin quantum numbers $s = \pm 1/2$ determine the intrinsic rotation of the electrons. Only integer changes of the quantum numbers are possible, since fractional changes would allow continuous states.

4. Particle and wave are merely two different ways of mathematically describing certain microscopic phenomena. This duality allows for the full and self-consistent application of the quantum mechanical model in the description of atomic and nuclear phenomena and processes.

5. Through the binding energy of the strong interaction, atomic nuclei are held together. This is similar to the way that electromagnetic force determines the binding of atoms and molecules.

6. It is not yet clear whether neutrinos and anti-neutrinos are two separate types of particles or identical particles in the grouping Majorana. This question is currently being targeted by research in the direction of the weak interaction, particularly with regard to the development of the early universe.

7. The mass number is often used to represent the mass of the entire atom. A goal in nuclear physics is to create increasingly more massive elements and also to study the properties of material with high density. This has already been expressed, in the early periods of science fiction, in the book *Atomgewicht 500 (Atomic Mass 500)* by Hans Dominic. Today the record for most massive element is oganesson with $Z = 118$ and $A = 294$. Heavier elements up to $Z = 120$ and $A = 320$ have been proven to exist, although they haven't received official names.

8. The physical unit MeV stands for the mega-electron volt, where 1 eV is the kinetic energy generated by an electron that moves across an electric potential difference of 1 volt. The eV unit is used extensively in nuclear and particle physics, and all microscopic variables are usually given in this unit. Conversion to the classical energy unit joule is given by the following: $1 \text{ eV} = 1.602 \cdot 10^{-19}$ J. This corresponds to the charge of an electron, $1.602 \cdot 10^{-19}$ Coulomb.

9. Should the imbalance between neutrons in relation to the number of protons be too great, more drastic decays take place whereby the unstable nucleus simply emits a whole proton rather than the positron discussed in Chapter 1, section 1.1.3.

10. The configuration on the right-hand side corresponds to the so-called Hoyle state. Without this state, no carbon could have formed in the universe and so, it was critical for the origin of all life forms. Many scientists, such as Britain's Astronomer Royal Martin Rees (b. 1942), consider the existence of this configuration as an example of the anthropic principle, according to which the structure of the universe, in all its components, has been designed to allow for the observer's conscious life. Martin Rees, *Just Six Numbers* (New York: HarperCollins Publishers, 1999).

11. Marie Skłodowska-Curie, *Recherché sur les Substances Radioactives* (Sorbonne, PhD Thesis, 1903).

12. Ernest Rutherford, *Radiations from Radioactive Substances* (Cambridge: Cambridge University Press, 1930).

13. The collective term *actinide*, plural *actinoids*, comes from the element actinium and encompasses the fourteen elements following it in the periodic table. These include: thorium, protactinium, uranium, also the transuranic neptunium, plutonium, americium, curium, berkelium, californium, einsteinium, fermium, mendelevium, nobelium, and lawrencium. The nuclei of these actinides have between 89 and 103 protons and between 117 and 159 neutrons. Actinides are metallic elements with comparable chemical properties that set the parameters for their grouped classification. Only the lighter actinides, up through plutonium are naturally occurring. Actinides from plutonium through fermium were first identified as products of nuclear weapons testing. The heavier elements can only be created artificially here on earth. While a row of heavier elements – for example, meitnerium, roentgenium, darmstadtium – have been identified, their chemical properties are not yet known. They are therefore denoted as transactinides.

3 The Decay Laws

3.1 THE DECAY LAWS

Radioactive decay follows certain laws that were observed by Marie and Pierre Curie and mathematically formulated in 1904 by Ernest Rutherford in his first book on radioactivity.[1] These laws generally apply to any type of decay, which means that radioactive decay laws do not depend on the type of emitted radiation and are only determined by *the probability of the decay event*. The decay laws also describe *the relationship between decay probability and the lifetime of a particular radioactive isotope*. This means that the lifetime is not a fixed span of time but rather, a statistical number that shows some variation depending on the physical mechanism of the decay.

The decay laws form the basis for calculating the amount of radioactivity and serve in predicting the radioactive burden on the environment. Therefore, all concepts of long-term exposure to radiation are essentially based on these laws. For this reason, we will discuss the decay laws in more detail. We will also consider how the decay laws impact the abundance and production of radioactive isotopes. This information essentially provides the foundation and explanation for the radioactive isotopes observed in nature, which determine all our lives.

3.1.1 ACTIVITY OF A RADIOACTIVE SUBSTANCE

If you have a certain number N of radioactive isotopes, the number of radioactive decay processes – or the activity A of the radioactive sample per unit of time[2] – can be easily calculated. This is necessarily the case since the activity of a radioactive sample correlates with the number of radioactive isotopes. However, each isotope decays with a certain probability per time unit; this probability is called the decay constant λ. The decay constant is a characteristic parameter for each radioactive isotope, reflecting the probability of its decay. In turn, the probability of decay depends on the mass difference between the radioactive nucleus and the decay products as well as the internal nuclear structure of the radioactive nucleus, which have each been discussed in Chapter 2.[3] This underlines the fact that each of the radioactive isotopes shown in the previous chapter has its very own characteristic decay constant λ.

A large decay constant corresponds to a high probability that radioactive decay will occur, while a small decay constant corresponds to a small likelihood of decay. Regarding the activity of a particular radioactive particle, a high probability of radioactive decay corresponds to high decay activity. Thus, the number of decays or activity can be described as the product resulting from the decay probability and the number of decaying isotopes: $A = \lambda \cdot N$. Given that the number of decays per second corresponds to the decay constant λ, and calculating the number of isotopes N in a cubic centimeter cm^3 of material, we find that activity A corresponds to the number of decays per second and cubic centimeter (these standard units appear in rectangular brackets):

$$A\left[s^{-1}cm^{-3}\right] = \lambda\left[s^{-1}\right] \cdot N\left[cm^{-3}\right] \tag{3.1}$$

Using this relation, one can easily calculate the respective activity of a radioactive substance with known decay constant, as for uranium 238 in Figure 3.1.

Physical quantities usually have a unit that allows quantitative conclusions to be drawn. This is particularly interesting in the case of activity, where the classical unit was defined by Marie Curie.

DOI: 10.1201/9781003435907-3

FIGURE 3.1 Activity of uranium material shown on a container labeled *radioactive* as advertised on Amazon (Source: https://boingboing.net/2007/11/30/uranium-ore-for-sale.html, as of March 11, 2024) and as a greenish fluorescent substance by Etsy.

She worked with large quantities of the mineral pitchblende, in order to extract a few grams of radioactive radium by chemical methods as described earlier. While Marie Curie was aware of the high level of decay radiation she was working with, she underestimated the long-term and dangerous effects of radiation.

As a chemist, she defined activity in relation to the amount of radium. She named the physical unit for the decay activity after herself, thus the curie (Ci). According to her definition, a curie is equal to the amount of radiation emitted by one gram of radium in one second. Converting that to the number of decays per second: $1 \text{ Ci} = 3.7 \cdot 10^{10} \text{ s}^{-1}$. This is 37 billion decays per second in one gram of pure radium material. This is an enormous number of decays. Today, it is extremely difficult to obtain the necessary permission from risk and safety authorities to handle such large amounts of radioactive material.

For simplicity, the curie unit has been exchanged for a new unit of activity, the becquerel (Bq), named after the discoverer of radioactive decay, Henri Becquerel. The definition is very simple, $1 \text{ Bq} = 1 \text{ s}^{-1}$. Therefore, one becquerel corresponds to one decay per second and $1 \text{ Ci} = 3.7 \cdot 10^{10}$ Bq or 37 GBq; conversely, $1 \text{ Bq} = 2.7 \cdot 10^{-11} \text{ Ci}$.[4]

The use of these different units for the same amount of radioactivity has interesting socio-psychological implications. Critics and opponents of nuclear power plants prefer to use the becquerel. With this, the amount of radioactivity appears to be enormous because the associated numbers appear to be huge. Using the curie, on the other hand, gives the impression of much lower activity and so suggests a lesser amount of radioactivity.

For example on August 20, 2013, Reuters news agency reported that – according to a spokesman for TEPCO, the Japanese company operating the Fukushima-Daiichi reactor complex – 80 million Bq per liter of contaminated water had leaked into the groundwater. Converting that to the unit Ci, groundwater contamination was 2.16 milli-Ci per liter; these are equal amounts but the curie appears to be considerably less, giving a significantly different impression!

This semantic difference always needs to be kept in mind when considering quantities of radioactivity, in order to be clear about the real extent of radioactivity present. It is also important to keep in mind that different types of radiation have different effects. Radiation levels can therefore be interpreted differently depending on the source of the radiation, which dictates the level of danger.

We can consider, for example, natural radioactivity in the human body (see Chapter 7). Each human being's internal γ-activity is primarily generated by the decay of radioactive isotope $^{40}_{19}\text{K}_{21}$, as presented in the previous chapter. This particular activity corresponds to a value of about 3,000 to 5,000 Bq, depending on personal weight.[5] By contrast, our body's internal α-activity is generated by the radioactive polonium isotope $^{210}_{84}\text{Po}_{126}$, discovered by Marie Curie, and corresponds to only about 60 Bq.

 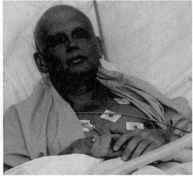

FIGURE 3.2   On the left, Palestinian leader Yasser Arafat; on the right, Russian dissident Alexander Litvinenko. Both are considered victims of targeted poisoning, using minute amounts of the highly radioactive element ^{210}Po. In nature, this material occurs only in very small quantities. Therefore, for such a poisoning, a large enough quantity of ^{210}Po could only be produced in a nuclear reactor. In the case of Yasser Arafat, slow poisoning would have required very small, hard-to-detect quantities over a long period of time. Therefore, the accusation of deliberate poisoning by ^{210}Po is not without controversy. Source: BBC reports.

Within the human body, if the α-activity of $^{210}_{84}$Po$_{126}$ were comparable to the γ-activity of ^{40}K, the outcome would be life-threatening. Allegedly, just such a small activity of $^{210}_{84}$Po was involved in slowly poisoning the former Palestinian leader Yasser Arafat (Figure 3.2). While extremely small amounts of 10 picograms are sufficient to end a life, detection of such miniscule amounts is extremely difficult. Therefore, various medical examinations did not yield a conclusive cause of death.[6]

In the case of Russian dissident Alexander Litvinenko, a considerably larger amount of a few nanograms led to radiation death within two weeks (see Figure 3.2). The dose he received corresponded to an activity of 17 Mega-Bq ($1.7 \cdot 10^7$ Bq). In this case, the dissident's death by ^{210}Po radiation poisoning is proven beyond a doubt, although the origin of the material, which can only be produced in sufficient quantity in nuclear reactors, has not been clearly established.[7]

3.1.2 DECAY LAWS

Radioactive material decays. Thus, the amount N is not constant but rather, decreases with time. With the decay of material, the characteristic activity A decreases. This means that both the amount of material and the activity are time-dependent quantities or functions. Hence, they are referred to as amount with respect to time $N(t)$ and activity with respect to time $A(t)$, where t stands for time. The time dependence of both quantities follows the decay law.

The decay law was observed experimentally and empirically determined from chemical and physical analyses of radioactive materials. Today the decay law is considered a fundamental law for all natural decay processes and is therefore concerned with not only radioactive systems but also the decay of unstable chemical and biological materials.[8] The decay of naturally unstable materials occurs with an exponential time dependence starting with an initial amount of N_0 at an arbitrary starting point for counting $t = 0$:

$$N(t) = N(0) \cdot e^{-\lambda \cdot t} \qquad (3.2)$$

Since the amount of activity is proportional to the number of radioactive isotopes, this can be reformulated as:

$$A(t) = A(0) \cdot e^{-\lambda \cdot t} \qquad (3.3)$$

The quantities $N(0)$ and $A(0)$ correspond to the initial quantity or initial activity, thus $t = 0$ for the radioactive material at an arbitrarily chosen time.

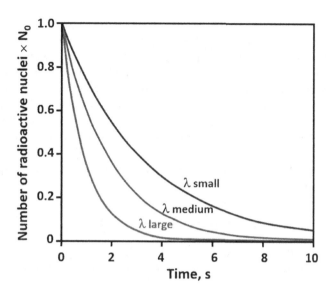

FIGURE 3.3 The calculated decay of the activity of three radioisotopes with different decay constants as a function of time.

The speed of decay of the radioactive material depends on the decay constant λ. Figure 3.3 shows the decay of different materials, each having the same initial quantity but different decay constants. The larger the decay constant, the more probable the decay and the faster the decay occurs, as reflected in the slope of the curve.

Therefore, decay probability can be correlated with the so-called lifetime of a radioactive product τ, which is calculated directly from the decay constant: $\tau = 1/\lambda$. This is not, however, an absolute lifetime, since some radioactive nuclei live longer. This is similar to some people living longer while others have shorter lives than the suggested statistical life expectancy of almost eighty years. So, lifetime is a probability value, measuring the time at which about 69% of the material in a radioactive sample has decayed.

More prevalent than life expectancy is, however, the concept of half-life. This corresponds to the time in which 50% of the radioactive nuclei in a particular material has decayed or the time at which 50% of the original activity is still present. Using the decay equation, we see that the half-life is directly proportional to the lifetime and inversely proportional to the decay constant. The proportionality factor is the natural logarithm of 2, expressed numerically: $\ln 2 = 0.69315$:

$$t_{1/2} = \tau \cdot \ln 2 = \frac{\ln 2}{\lambda} \tag{3.4}$$

According to this, the half-life $t_{1/2} = 0.69315 \cdot \tau$ corresponds to only 69% of the lifetime and so, the lifetime is always shorter. This is understandable since at the half-life only 50% of the radioactive material has decayed, while at the lifetime it has already decayed by 69%.

The lifetimes of the nuclear state range from fractions of seconds to billions of years, since each radioactive nucleus has its own characteristic lifetime that only depends on the quantum mechanical characteristic of the nuclear wave-functions involved. Because of the complexity of the quantum systems, there is no clear way to reliably predict the decay times.

3.1.3 DECAY FORMALISM

A single decay only exists if the end product is a stable nucleus, as in the β-decay of carbon isotope ^{14}C which decays to nitrogen isotope ^{14}N. As described in the previous chapter, a decay process is frequently not limited to a single decay of a radioisotope. Rather, the decay of a parent nucleus often triggers an entire decay series. The initial decay builds to a daughter nucleus, which in turn decays to a daughter. Thus, entire decay chains can form, extending over several daughter generations, toward the stable final product. In this process, an equilibrium is often established between the buildup and the decay of individual isotopes in the chain. This equilibrium can also be determined from the decay equations.

The formation of a stable isotope N_2, from the radioactive decay of isotope N_1, follows directly from the decay equation. Accordingly,

$$N_2(t) = N_1(0) \cdot \left(1 - e^{-\lambda_1 \cdot t}\right) \tag{3.5}$$

The symmetry of the decay and buildup curves, as shown in Figure 3.4, demonstrates the correlation between the exponential decay of the parent isotope and the exponential buildup of a new daughter isotope.

However, if the daughter isotope is also radioactive, then the equation for calculating the abundance $N_2(t)$ must also consider its decay. This results in the following, much more complicated expression, which takes into account both decay constants λ_1 and λ_2:

$$N_2(t) = N_1(0) \cdot \frac{\lambda_1}{\lambda_2 - \lambda_1} \cdot \left(e^{-\lambda_1 \cdot t} - e^{-\lambda_2 \cdot t}\right) \tag{3.6}$$

This can be expressed in terms of the corresponding activity $A_2(t)$ of the daughter nucleus by the following expression,

$$A_2(t) = A_1(0) \cdot \frac{\lambda_1}{\lambda_2 - \lambda_1} \cdot \left(e^{-\lambda_1 \cdot t} - e^{-\lambda_2 \cdot t}\right) \tag{3.7}$$

which shows that the time-dependent evolution, of the activity of the radioactive daughter isotope, depends directly on the ratio of the two decay constants. If λ_2 is greater than λ_1, i.e., the decay

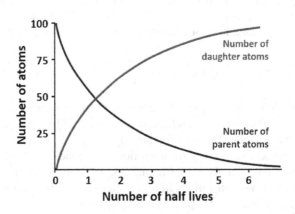

FIGURE 3.4 Decay of a parent isotope and the formation of a stable daughter isotope, as a function of the half-life. After a half-life, 50% of the radioactive isotope has decayed and now forms the stable daughter isotope.

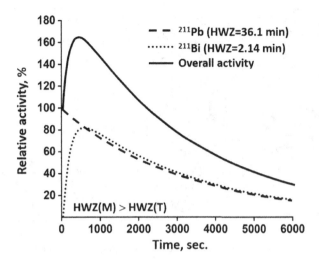

FIGURE 3.5 Activity of radioactive isotopes within a decay chain in which the lifetime or half-life of the parent isotope ^{211}Pb is longer than that of the daughter isotope ^{211}Bi. This is generally the case with natural-decay chains, which are an important component in the radioactivity of our planet and thus of humans. This activity is called *transient equilibrium*. Source: wikipedia.org.

probability of the radioactive daughter is greater than that of the mother – meaning that the half-life of the mother is longer than that of the daughter – then the daughter activity builds up relatively quickly to a value that is in equilibrium with the activity of the mother and so decays in parallel with the decay of the mother. Conversely, if λ_2 is less than λ_1, i.e., the decay probability of the radioactive daughter is less than that of the mother, then the mother's activity gradually transitions to that of the daughter and then decays according to the daughter's half-life (Figure 3.5).

Within the framework of this formalism, one can follow the decay, the time-dependent buildup, and then the decay of all members of a decay chain. The abundance of the mother isotope and that of the different generations of daughters are thus correlated over time. Therefore, once the decay constants are known, the age of a given material can be determined from the relative abundance of the different isotopes of a decay chain – as they exist at the time of measurement – in the material.

3.1.4 Natural Decay Chains

From the existence of decay chains, it follows that there can be a continuous decay and build-up of radioactive isotopes within a particular chain. This was already noticed by Marie and Pierre Curie when they identified several new radioactive elements, besides uranium, during their chemical analysis of pitchblende. These were not only thorium, radium, and polonium but also radon, a noble gas that escaped from the pitchblende without any chemical reaction.

This was the first sign that elements are not unchangeable fundamental chemical materials but rather can be transformed into each other. The Curies even observed how, after separation from the pitchblende, the various radioactive components degraded and daughter nuclei built up. This showed that the transmutation of elements is a natural occurrence, taking place in natural minerals such as pitchblende (Figure 3.6).

The best known and also, for our discussion, most significant decay chains are the decay sequences of the actinides ^{232}Th, ^{235}U, ^{238}U, and to a lesser extent ^{237}Np, which were already briefly introduced in Chapter 2. Applying the formalism described above, the structure of the daughter

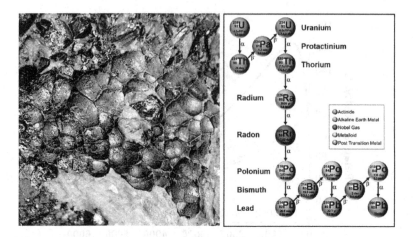

FIGURE 3.6 Pitchblende (uraninite) is a crystalline form of uranium oxide containing 88% uranium. The slow decay of radioactive uranium isotopes gradually builds up various amounts of other elements, thus activating daughter elements that are also radioactive. The diagram on the right shows the natural decay chain from ^{238}U to stable ^{206}Pb. Within this chain, other long-lived isotopes are built up, such as ^{234}U, ^{230}Th, ^{226}Ra, as well as the noble-gas isotope ^{222}Rn and the previously mentioned ^{210}Po. These isotopes decay by α-decay, like the parent isotope ^{238}U, but have a much shorter half-life than ^{238}U, whose activity therefore determines the total activity in the chain.

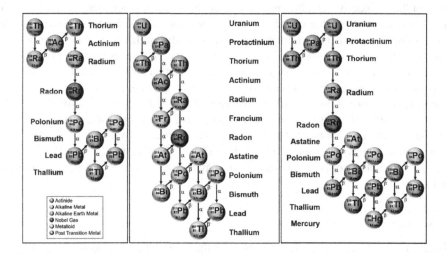

FIGURE 3.7 Schematic representation of the three most important, natural actinide-decay series: the ^{232}Th thorium series, the ^{235}U actinium series, and the ^{238}U uranium series. Prepared after Wikipedia.

nuclides of these isotopes can be easily described. Since, in most cases, the respective daughter isotopes have a much shorter half-life than the actinide parent isotopes, equilibrium abundances between the different components are established after a certain time. These equilibrium abundances determine the total activity of the chain, which results from the respective α- and β-decays.

Figure 3.7 illustrates the three most important natural-decay series, each containing between 10 and 15 decay processes including also some decay branches that split the decay sequences. Two decay processes, here mostly α- or β-decay, offer equally strong decay possibilities. This complicates the mathematical formalism but does little to change the respective abundance ratios.

The ^{232}Th series has ten consecutive decays with a bifurcation (two-branched division) to the stable endpoint ^{208}Pb; six α-particles are emitted in this decay series. The ^{235}U series has eleven consecutive decays with six bifurcations over which seven α-particles are emitted; the endpoint of the decay series is the stable lead isotope ^{207}Pb. Finally, the ^{238}U uranium series has fourteen successive decays, emitting eight α-particles with four branches, all of which lead to the stable ^{206}Pb lead isotope as the endpoint of the series.

The equilibrium abundances of the different radioactive isotopes in a decay chain are given by the decay constants, according to the formalism developed above. Here, the temporal evolution of the isotope abundances – i.e., the gradual depletion of the mother by decay and the concomitant buildup of the daughters – is described by a network of decay reactions, which also consider the various branching that occurs through different decay channels.

If linear decay chains are assumed, the relative abundances of the daughter isotopes to the mother isotope can be described in a simplified way using the so-called Bateman equations.[9] With these equations, ratios are mainly determined by the ratio of the decay constant of the mother relative to that of the daughter isotope. The half-lives of the different actinides are in the range of several millions to billions of years, while the respective daughters are very short-lived in comparison with half-lives of seconds to years. Therefore, the abundances of the daughter isotopes are many orders of magnitude lower than those of the respective parent elements, since – practically speaking – with build-up, the daughters decay again immediately. This means that the activity of the successor daughter isotopes occurs within a short period of time after the decay of the mother.

3.1.5 Nuclear Reactions and Artificial Radioactivity

At the time, the idea of the transmutation of elements was dismissed as alchemy; however, the discovery of nuclear reactions generating new elements re-introduced the old alchemist dream back into the sciences as a revolutionary way of seeing the world. It emerged as a natural process, following the new rules of nuclear physics – not as a result of secretive chemistry for turning mercury into gold. As early as 1933, the hitherto unthinkable transformation of the elements by human influence succeeded and opened a new chapter in the sciences.

Irène Curie, Marie Curie's daughter, and her husband Frédéric Joliot-Curie – after irradiating boron, magnesium, and aluminum with α-particles from a radioactive polonium source – demonstrated the formation of new, short-lived, radioactive isotopes. These were, according to their chemical analysis, isotopes of nitrogen, silicon, and phosphorus. Thus, for the first time, elements were produced as demonstrated in a laboratory experiment. These elements were, however, also radioactive.

The new isotopes were produced by reactions between the incident helium nucleus $^4_2\text{He}_2$ (α) and the isotopes $^{10}_5\text{B}_5$, $^{24}_{12}\text{Mg}_{12}$, and $^{27}_{13}\text{Al}_{14}$. This new type of reaction made the transmutation of elements possible, a transformation that alchemists had hoped to achieve by chemical reaction.[10] These transmutation processes, today called *nuclear reactions*, took place by capture of the α-particle followed by emission of a neutron:

$$^{10}_5\text{B}_5 + {^4_2}\text{He}_2 \rightarrow {^{13}_7}\text{N}_6 + n; \; {^{24}_{12}}\text{Mg}_{12} + {^4_2}\text{He}_2 \rightarrow {^{27}_{14}}\text{Si}_{13} + n; \text{ and } {^{27}_{13}}\text{Al}_{14} + {^4_2}\text{He}_2 \rightarrow {^{30}_{15}}\text{P}_{15} + n \qquad (3.8)$$

For this work, the Joliot-Curie couple received the Nobel Prize in 1935. The significance of this discovery opened unimagined possibilities as scientists envisioned a wide range of applications for radioisotopes, particularly in medicine. Nuclear reactions also opened the possibility of producing large quantities of radioactive isotopes, specifically for experimentation and application.

The Joliot-Curies were still using radiochemically extracted polonium, as a source of α-rays, whose energy could be reduced by thin metal foils. Yet already in the 1930s, particle accelerators of various kinds were developed and particularly applied to the study of nuclear reactions and the production of various radioactive isotopes.

There are two types of nuclear reactions: In the first kind, energy is released as a consequence of the reaction process; this process is called *exothermic*. During the second type of reaction, energy must be supplied to the reaction system in order for the process to occur. This reaction, requiring additional energy, is called *endothermic* and the energy needed can be provided by the kinetic energy of the particles involved. This kinetic energy, of the participating particles, either results from decay processes, as it did in the Joliot-Curie experiments, or must be produced by accelerating the particles in so-called particle accelerators.

As with decay processes themselves, the energy that is either released during or required for a nuclear reaction is equal to the difference in binding energy or mass of both the input system (the target particle and projectile + the kinetic energy of the projectile) and the output system. The output system is again the total binding energy or mass of the final configuration – the residual nucleus and reaction product + their kinetic energy. This difference in energy, between the input system and output system, is generally referred to as the reaction Q-value. If the Q-value is positive, energy is released and the reaction is exothermic. If the Q-value is negative, the reaction is endothermic and only proceeds when the kinetic energy of the projectile provides the necessary energy.

In exothermic reactions of positive Q-value, energy is either released as the kinetic energy of the reaction products or as the electromagnetic energy of γ-radiation, as described earlier. This means that exothermic nuclear reactions produce radiation with a specific, quantifiable (measurable and predicted) energy release. The amount of radiation depends on the mass difference between the initial and final systems as well as the additional kinetic energies of the interacting particles.

The reaction probability, or the cross section of the nuclear process, depends (in first approximation) on the probability that an incident particle will hit the nucleus. Therefore, the probability is given in size units of an area (cm^2). This probability is small, since the nucleus is 10,000 times smaller than the atom itself. For this reason, the standard unit for the probability of a nuclear reaction is taken to be approximately equal to the sectional area of an atomic nucleus.

Assuming the radius of an average nucleus to be $r = 4$ - 6 fm, a typical nucleus' area is about 100 $fm^2 = 10^{-24}$ cm^2; this is defined as 1 barn, *barn* being derived from the English term for barn or barn door. The barn is, however, only the hit probability; most incident particles bounce off the nucleus or scatter in another direction due to interaction with the electric field. This scattering probability has the largest effective cross section and is therefore most likely to occur.

The cross sections of reactions leading to the formation of another nucleus depend on direct quantum-mechanical interaction between the incident projectile and the atomic nucleus. Their cross section for such a process is much smaller since the hit probability needs to be multiplied by the probability of particle exchange between the projectile and the nucleus. Here, as described earlier, three of the four fundamental forces play a role (Figure 2.3).

Reactions that take place via the strong interaction have the greatest probability of occurring. These reactions are essentially based on an exchange of nucleons, between the projectile and the nucleus; the effective cross section is typically a thousand times smaller than the barn and so in the millibarn range.

Reactions that proceed via the electromagnetic interaction are much weaker, by another three orders of magnitude. These are reactions in which the nucleus captures the projectile and emits γ-radiation in the process. Here, the effective cross section is typically in the microbarn range.

Again, reactions which occur via the weak interaction are even smaller by several orders of magnitude. These are associated with a charge-exchange process between protons and neutrons. Such reactions are exceedingly slow and therefore are not used for producing isotopes. Finally, the gravitational force has no influence on these microscopic systems with their extremely low masses.

It is worth mentioning that the cross section is frequently subject to strong variations that depend on the energy being introduced. This is mainly due to the energy dependence of quantum-mechanical transition probabilities associated with particle unbound states of the compound system that has formed in the fusion process. This causes a sudden increase in cross section, called *resonance*.

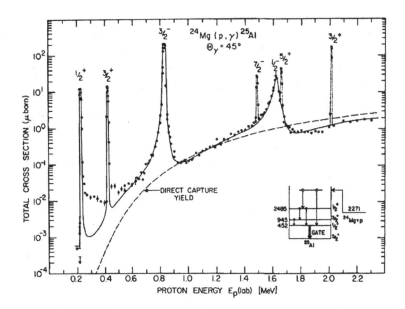

FIGURE 3.8 The resonance structure of the low-energy reaction cross section of the ^{24}Mg(p,γ)^{25}Al proton-capture reaction at low energies. Courtesy and with permission of Hanns Peter Trautvetter, PhD Thesis, University of Toronto, 1973.

These resonances can sometimes be extraordinarily complex as visualized in Figure 3.8. In general, however, the effective cross section drops sharply toward low energies in charged-particle reactions because the reaction probability drops exponentially due to the repulsive Coulomb barrier (discussed in Chapter 2) between the positively charged nucleus and the positively charged projectile, as indicated by the dashed line which depicts the non-resonant component in Figure 3.8.

Nuclear reactions with neutrons (which have no electrical charge) are much different, since in these cases the effective cross section mostly increases with low energy. This behavior is described by the so-called $1/v$ law which states that with decreasing velocity v, the neutron is more likely to be absorbed. According to this law, the effective cross section is inversely proportional to the particle velocity v. In a few cases, however, this law is not applicable. For example, in neutron capture at ^{14}C or ^{238}U, the effective cross section behaves differently due to angular momentum conditions which actually cause a decrease with low energy due to the so-called angular momentum barrier.

To describe nuclear reactions, a particular notation is used. For example, the first artificial reaction to produce the radioactive isotope $^{13}_{7}$N$_{6,}$ by bombarding a $^{10}_{5}$B$_{5}$ foil with α-particles, is described in simplified form: ^{10}B(α,n)^{13}N. In the case of isotopes, only the mass number A (10) is given since element identification, B (for boron), has already determined the charge number, which is known to be Z (5), so that the neutron number N (5) can be easily determined from: $N = A - Z$. The projectile and light-reaction product are inside the parenthesis (α,n); the heavy residual nucleus makes the end of the formula: ^{13}N. The reaction as a whole is based on the strong interaction.

An alternative way to produce ^{13}N, by proton irradiation of carbon isotope ^{12}C, was also discovered by Joliot-Curie in 1933. This reaction, based on electromagnetic interaction, is formulated in standard notation as ^{12}C(p,γ)^{13}N with (p,γ) signifying proton and γ-radiation product. It therefore has a much smaller cross section (from $A = 12$ to $A = 13$) than the ^{10}B(α,n)^{13}N ($A = 10$ to $A = 13$) reaction.

3.1.6 PARTICLE ACCELERATORS

After these principles became clear in the 1930s, the predominant goal of experimental physicists has been to develop accelerator systems that will bring particles to the high energies necessary for reactions to occur. During this process, a wide variety of accelerator systems have been developed and are still in use today. The three most important acceleration principles are summarized here.[11]

The first particle accelerators were based on two different principles. DC accelerators generate megavolt high voltage, which accelerates ionized particles, with either positive or negative charge, to mega-electronvolt energies. This high voltage can be achieved by internal amplification of AC voltage, using an electrical circuit (Cockroft-Walton generator). High-volt energies can also be achieved by rapid charge transport, via a non-conductive charge carrier, to an electrically isolated ion source (Van de Graaff generator).

This generator, still in common use today, was named for the inventor of the accelerator principle: Robert Van de Graaff (1901–1967). In their first year of development, DC accelerators were installed at a growing number of research institutes, in Europe as well as in the United States, in order to study the new phenomenon of nuclear reactions. Today, these accelerators still serve in many cases as pre-accelerator systems for modern large-scale accelerator facilities around the world.

Almost simultaneous with these first DC accelerators, American physicist and later Nobel Prize winner Ernest Lawrence (1901–1958) developed the cyclotron in 1930 at Berkeley. With the cyclotron, charged particles travel along a spiral path held by an external, perpendicular magnetic field. The particles are accelerated – after each half orbit, by a voltage pulse – as they pass through a gap between two electrodes. With this the particles achieve higher velocity, which translates into a gradually increasing radius of their trajectory. Cyclotrons can achieve much higher particle energies because increased energy only depends on the number of orbits and thus the size of the cyclotron.

The first successful examples of these three types of accelerators are still widely used (Figures 3.9 and 3.10). The acceleration principle has remained the same, although it has been considerably improved and refined. In particular, DC accelerators operate within a tank that is filled with a specific mixture of gases, in order to avoid high-voltage discharges of several million volts.

Today, these machines are no longer built by individual researchers at universities, since they are manufactured by high-tech companies for commercial research and application purposes. Medical

FIGURE 3.9   Left: Cockroft-Walton accelerator with amplifying column and accelerating tube. Voltage in the so-called terminal – the square, metal chamber in the center of the display – can reach as high as 1 million volts and must be well insulated to minimize high-voltage discharges. Right: an open, Van de Graaff-type accelerator with loading belt and accelerator tube. The machine is able to achieve around 1 million volts. Today, higher voltages of up to 20 million volts have been reached, requiring the system to be installed in a high-pressure tank filled with dry gas. Picture collection, Nuclear Science Laboratory, University of Notre Dame.

FIGURE 3.10 Left: classic cyclotron with external magnets that keep the particles on a circular path. The center image shows the so-called 2-D acceleration chamber with a diagonal slit, above which the particles are accelerated to ever higher energies by an electric pulse. The circular path then gradually spirals to a larger radius until the particles can leave the chamber tangentially. On the right: schematic diagram of the 2-D acceleration chamber's setup, including measurement instruments. Courtesy of the Atomic Heritage Foundation and National Museum of American History.

and pharmaceutical industries are the predominant consumers, although electronic chip and other material industries use such machines for ion implantation and material-hardening applications.

The development of particle accelerators in the 1930s was the first breakthrough in the production of radioactive isotopes. In the following decades, more and more exotic isotopes were discovered and measured with these machines.[12] In principle, any isotope can be produced by various nuclear reactions. However, for effective production, reactions with a high-reaction probability are recommended. Various reactions have been identified that lead to the optimal production rate of certain isotopes. In this context, the use of particle accelerators has allowed scientists to systematically determine the strength of various nuclear reactions, as well as the reaction's dependence on accelerated-particle energy.

The probability P of isotope production, within a certain time by a nuclear reaction, depends – in addition to the cross section, as described above – on the number of incident particles N_i per unit time and the number of nuclei in target material N_T per unit area. This can be described as follows (with the standard units for time and area in square brackets):

$$P\left[s^{-1}\right] = N_i\left[s^{-1}\right] \cdot N_T\left[cm^{-2}\right] \cdot \sigma\left[barn = 10^{-24}\,cm^2\right] P\left[s^{-1}\right]$$
$$= N_i\left[s^{-1}\right] \cdot N_T\left[cm^{-2}\right] \cdot \sigma\left[barn = 10^{-24}\,cm^2\right] \tag{3.9}$$

The total number of isotopes produced is determined by the length of irradiation. However, if radioactive isotopes are produced, the decay time must be considered again, since an equilibrium is established between the amount of material that is constantly being produced and the amount that is decaying. This can be described as follows:

$$N(t) = \frac{A(t)}{\lambda} = \frac{P}{\lambda} \cdot \left(1 - e^{-\lambda \cdot t}\right) N(t) = \frac{A(t)}{\lambda} = \frac{P}{\lambda} \cdot \left(1 - e^{-\lambda \cdot t}\right) \tag{3.10}$$

where in the notation already introduced, $A(t)$ is the activity of the produced radioactive material with decay constant λ.

Figure 3.11 shows the behavior for production of two silver isotopes with the same production rate but different decay constants. Usually, the maximum achievable number of radioisotopes is reached after irradiation lasting about five half-lives.

Thus, a high yield of the desired radioisotopes requires a high production rate, which in turn depends on the effective cross section of the chosen production reaction and on the intensity of the accelerator's particle beam. Also, the target material should be isotopically pure: If too many other

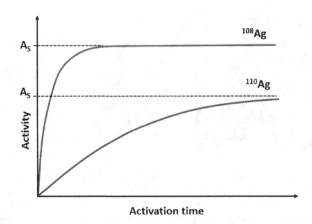

FIGURE 3.11 Activation of radioactive silver isotopes ^{108}Ag and ^{110}Ag, with different half-lives – $T_{1/2}(^{108}Ag)$ = 144 s and $T_{1/2}(^{110}Ag)$ = 24.6 s – achieved at the same production rate. In general, 93% of the maximum yield is achieved after an irradiation period of five half-lives. Longer production runs are therefore considered less efficient. Production of longer-lived ^{108}Ag takes longer and only achieves lower yields, as noted by the decay constant λ in the denominator of the equation.

isotopes including impurities by other elements are present, reactions with these nuclei will also occur; this affects the yield and generates unwanted background activity.

3.1.7 Nuclear Fission and Artificial Production of Radioisotopes in Reactors

So far, we have mainly talked about reactions with charged accelerated particles. The discovery of the free neutron, in 1932 by James Chadwick, opened the possibility of producing isotopes by neutron irradiation. Ernest Rutherford had been thinking about the existence of a neutral nuclear particle for years, but experimental proof was lacking. However, his student Chadwick succeeded when he bombarded a beryllium target with α-particles from a polonium source that he had received from Lise Meitner in Berlin.

Using a Geiger-Müller detector, Chadwick discovered this new type of radiation, which he identified as the neutrons he had been looking for. The nuclear reaction via which the neutrons had been produced was ^{9}Be$(\alpha,n)^{12}$C, the same type of reaction Irène and Frédéric Joliot-Curie had used to produce the first artificial radioactive isotopes. Chadwick had discovered the neutron as the second reaction product.

Chadwick was mainly thinking about medical applications. As neutral particles, neutrons are not affected by electromagnetic force because they have neither positive nor negative charge. Therefore, their interaction with a material is constrained to nuclear scattering and reaction processes, which reduces the probability of absorption. For this reason, neutrons can pass through matter and deeply penetrate biological materials such as the human body. Properly focused, neutrons can help in destroying cancerous tumors (see Chapter 14, Volume 2). To accomplish this, an intense flux of neutrons must be generated.[13]

Neutrons can be generated in an accelerator by nuclear reactions with charged particles. Particle energies are selected to ensure that the effective cross section is as large as possible. This accelerator-based approach is a very successful method of generating neutrons, via various reactions, for the irradiation of different materials.

A second possibility for neutron generation opened after the Second World War, with successful construction of first-generation nuclear reactors. The immediate goal had been to produce energy, from the fission of heavy uranium isotopes (^{235}U or ^{239}Pu), and breed new fuel such as ^{239}Pu for the

US nuclear-weapons program. The fission process itself will produce large quantities of neutron-rich radioactive fission product. This will also generate neutrons in large quantities; these will trigger the chain reaction in an atomic bomb and maintain the neutron flux in a nuclear reactor. In a reactor, the neutrons must be kept at a constant level so as to release the energy in a controlled manner. These reactor neutrons can also be used to irradiate materials.

As mentioned above, the cross section of neutron-induced reactions is different from that of nuclear reactions with charged particles. The greatest reaction probability is achieved with slow neutrons, known as thermal neutrons, because slower neutrons have more time to interact with a nucleus. Neutrons generated by nuclear reactions and fission processes generally have a high kinetic energy, which corresponds to the energy released from the reaction process. Therefore, to achieve a highly effective cross section of neutron-induced reactions, the neutrons must be decelerated to energies that correspond to the thermal energy ($E = kT$) of the environment: T is room temperature in units Kelvin,[14] and k is the so-called Boltzmann constant $k = 8.617 \cdot 10^{-11}$ MeV/Kelvin. This deceleration process is called *thermalization*. Thermalization means that neutrons from a decay process, with a typical kinetic energy of ~5 MeV, will be decelerated to room temperature (20 °C = 293.15K) with energies of $E = 25$ meV.

Thermalization occurs through inelastic scattering of neutrons on light nuclei; with each scattering event, neutrons lose energy. For example, in a game of billiards, when one ball collides with another ball, the biggest effect occurs when both balls have comparable mass. In the same way, a neutron loses about 50% of its energy when it collides with an equally weighted particle, i.e., a light nucleus. Thus, the nuclei of hydrogen atoms in water are the best moderators for fast neutrons. However, even with a highly effective cross section, pure-hydrogen nuclei still capture neutrons, via the $^1_1H_0(n,\gamma)^2_1H_1$ interaction, and form deuterium, thus significantly reducing the number of free neutrons. Lithium, beryllium, and boron are not good moderators because they also have a large neutron-reaction cross section, thereby absorbing the neutrons that one intends to create.

The most used moderator materials are either heavy water (based on the deuteron hydrogen isotope $^2_1H_1$) or carbon (densely packed as graphite); neither ^2H nor ^{12}C has a large neutron-capture cross section.

The question of differentiating these moderators was important in early fission research. German physicists from the Uranium Club, as well as physicists from the Tube Alloy Project in the United Kingdom and the Manhattan Project in the United States, searched for suitable moderators for

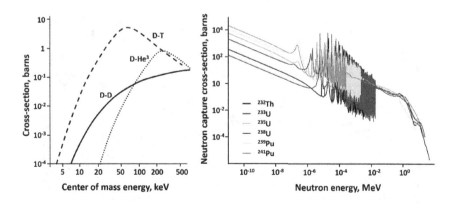

FIGURE 3.12 Left, energy dependence of the cross section for nuclear reactions with charged particles, which decreases exponentially due to the repulsive Coulomb barrier between equally charged particles (http://hyperphysics.phy-astr.gsu.edu/hbase/NucEne/coubar.html). Right, the cross section for neutron-capture reactions at different actinides, which increases with decreasing energy E according to the $1/v$ law ($E = m \cdot v^2$). In the range from 1 eV to 100 keV, the cross section of action is governed by numerous closely spaced resonances. Source: CERN records.

slowing neutrons without absorption by neutron capture. Both deuterium and graphite were independently identified as possible options. Graphite was much cheaper to obtain since deuterium had to first be condensed from normal water by means of elaborate equipment; at that time, such a plant only existed in Norway.

Physicist Walther Bothe was therefore sent, from Berlin to German-occupied Paris where the only cyclotron existed in Europe, to test the possibility of using graphite as a moderator.[15] His experiments showed that graphite was insufficient since it also seemed to absorb neutrons. Although Bothe's interpretation was not entirely correct, based on these results, the Uranium Club focused on the possibility of using deuterium produced in occupied Norway. On the other hand, in the United States, Leó Szilárd (1898–1964) pointed out that impurities in graphite could possibly have influenced the results; he had learned from a manufacturer that, in fact, boron is used in the chemical production of graphite. At Szilárd's suggestion, boron-free graphite was produced, which proved to be a perfect moderator in subsequent experiments.[16]

Thermalized neutrons have a particularly high-effective cross section for fission reactions. If fissile material such as ^{235}U is densely packed, then the neutrons released during the initial fission process can thermalize relatively quickly and be easily captured by neighboring ^{235}U isotopes. This capture process transforms the nucleus into ^{236}U, a more neutron-rich isotope. ^{236}U then decays by fission into two smaller nuclei accompanied by the emission of two to three neutrons. The neutrons released in this process in turn induce new fission processes. This leads to a chain reaction, with the process repeating for as long as the mass of fissile material is large enough to thermalize the neutrons and prevent escape from the critical volume.

The principle behind a fission bomb exploding – or in common parlance, an atomic bomb – is this uncontrollable chain reaction in which the release of fission energy occurs within seconds. The problem physicists faced in the Uranium Club and the Manhattan Project was determining the correct mass at which the chain reaction would begin.

How a chain reaction would in principle proceed is shown, in simplified form, in this advertisement of the American nuclear industry (Figure 3.13). The firing heads symbolize the uranium, from

FIGURE 3.13   Chain reaction illustration: Short matches cause the chain to burn quickly, with each branching corresponding to a nuclear fission. Long matches symbolize the absorption or deceleration of neutrons, allowing for control of the chain reactions. Source: General Electric's promotional brochure.

which neutrons are released during ignition; the length of each match symbolizes the probability of the release causing a neighboring uranium nucleus to ignite or fission. Short matches cause immediate ignition – this is to say, an uncontrolled chain reaction as in the explosion of an atomic bomb. Long matches are neutron absorbers and are intended to cause a controlled chain reaction in the reactor's core.

To avoid an instantaneous chain reaction, the neutron flux must be reduced in a controlled manner. This can be accomplished by embedding movable neutron absorbers such as ^{10}B control rods, in the form of boron-carbide ($^{10}B_4C$), into the fissile material. Conversely, boric acid enriched with ^{10}B can be added to the cooling water around the fuel rods. In either case, neutrons are predominantly absorbed by the $^{10}B(n,2\alpha)^3H$ reaction, producing two ^4He nuclei and a radioactive tritium nucleus; tritium H-3 is captured and used for various applications as will be discussed at a later point (see Chapter 12, Volume 2).

At the reactor, new isotopes are either directly produced by neutron irradiation of material samples – by bringing these samples close to the neutron-producing reactor core – or by guiding the neutrons out of the reactor core to the sample, via so-called neutron guides. The former method is mostly used for small samples, taking advantage of the much higher neutron flux at the reactor core. The second method is mostly used for neutron irradiation of larger objects, with the disadvantage that external neutron flux is less than one percent of the reactor flux. In any case, the production rate depends on the thermal cross section σ_{th} and the neutron flux F_n (with standard units in rectangular brackets),

$$P = \sigma_{th}\left[barn = 10^{-24}\,cm^2\right] \cdot F_n\left[s^{-1}cm^{-2}\right] \tag{3.11}$$

The number and activity of radioisotopes produced by neutron capture again depends on the characteristic half-life, due to the balance between production and decay. As in the case of production by nuclear reactions with charged particles, the optimum irradiation period is about five times the half-life of the radioactive product nucleus.

The neutron flux (Figure 3.14) is determined by the design of the reactor. However, we can divide the flux itself into three components: fast, epithermal, and thermal neutrons. Fast neutrons

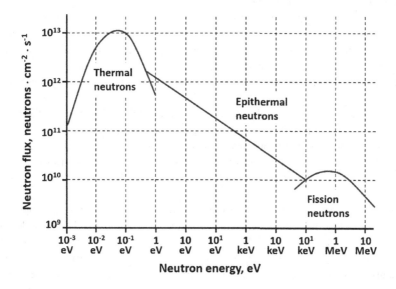

FIGURE 3.14 Typical neutron flux in a reactor as a function of energy, where most of the neutrons have lost their initial high energies due to multiple scattering and have been cooled down to room temperature or thermalized.

FIGURE 3.15 Illustration of the Pile reactor's first test experiment. Developed by Enrico Fermi at the University of Chicago, the Pile was constructed from small uranium blocks that were embedded in a matrix of graphite blocks, which were supposed to thermalize the fission neutrons. When the neutron flux became too large, indium and cadmium rods were inserted to reduce the flux by neutron capture. The flux was measured by the γ-radiation emitted from neutron capture processes at silver. The measuring station can be seen in the background of the illustration. Source: https://www.euronuclear.org/glossary/cp-1/ (November 4, 2023).

are directly produced in the fission process and have not yet lost their energy through multiple scattering on the moderator material. They have energies in the MeV range. A partial energy loss has been suffered by epithermal neutrons, which cover the energy range between 10 eV and 10 keV. The largest flux is composed of thermal neutrons, which have energies in the range of the thermal energy that corresponds to the temperature of the environment. Thermal neutrons have given up all their initial energy to the moderator material by scattering processes. Their flux is more than three orders of magnitude greater than the flux of fast neutrons.

The first research reactor in the United States was the Pile. Built as early as 1942 by Enrico Fermi, the Pile was tested under the bleachers of the former Stagg Field football stadium on the University of Chicago's campus (Figure 3.15).

This first reactor was simply a pile of black graphite blocks with small gray uranium blocks of various shapes placed between them. The control rods were made of cadmium and indium – metals with high neutron-absorption cross sections. An admixture of silver helped to determine the neutron flux, because neutron capture at silver produced γ-radiation that could be quickly and easily measured. If the radiation became too high, the control rods were protracted by hand. This reduced the neutron flux.

The reactor had neither an external concrete wall to absorb the resulting radiation nor a cooling water system to prevent possible overheating. Neutron flux in the reactor core varied between $4 \cdot 10^8$ neutrons/s cm^2 and $4 \cdot 10^{10}$ neutrons/s cm^2, depending on the power of the reactor unit.[17] All in all, it was a risky experiment that would no longer be possible today, due to radiation protection regulations. Spectators at the first demonstration certainly must have received a considerable dose of radiation.[18]

Following the success of the Pile-I reactor, other reactors were quickly developed such as the X-10 reactor at Oak Ridge in Tennessee. Here, a significant amount of ^{238}U was added to ^{235}U fuel to demonstrate that isotopes such as ^{239}Pu could be incubated for the weapons program by neutron capture at ^{238}U. Thermalization of the neutrons has the effect of reducing the capture process at ^{238}U and so can be used to control the breeding process. In 1947, this reactor was removed from the military program and used for civilian purposes, producing the first radioisotopes for medical research.

After the end of the Second World War, multiple countries invested in reactor and application research. It was thought that reactors could produce infinite amounts of energy at reasonable prices, a hope that was soon dashed by the safety industry which was developed at the same time.[19] However, besides reactors for energy production, a whole fleet of smaller reactors were developed that exclusively served scientific purposes. In addition to basic neutron-physics research, focus was on materials-science applications. For example, the structure of new crystalline materials could be investigated and determined in scattering experiments with neutrons. Newly developed neutron guides were implemented to guide neutrons from the reactor core to external measuring stations where experiments were carried out.

In addition to applications in atomic and molecular structure analysis, reactors were also used to produce radioisotopes for nuclear medicine, supporting medical diagnoses and radiation therapy (cf. Chapter 14, Volume 2). Radioisotopes are either produced by neutron irradiation, in which external material samples are activated, or through a process in which fission products are extracted from the reactor core and are sorted according to charge and mass by electromagnetic separators. Today, neutron irradiation is rarely used in tumor treatment; irradiation with charged accelerated particles is preferred, since this results in less radiation damage to the body material adjacent to the tumor. These industrial and medical applications continue today and will be discussed further in Volume 2, Chapters 14, 15, and 16.

NOTES

1. Ernest Rutherford, *Radio-Activity* (Cambridge: Cambridge University Press, 1904).
2. Here one must take particular note: N for number of particles and A for activity are traditional designations that must not be confused with the number N of neutrons in an atomic nucleus or with the atomic weight A.
3. These are mostly quantum mechanical processes that determine the transition probability between the particle configuration of the original nucleus and the configuration of the daughter system.
4. Frequently used size units for physical units: 10^{-15} femto (f); 10^{-12} pico (p); 10^{-9} nano (n); 10^{-6} micro (μ); 10^{-3} milli (m); 10^3 kilo (k); 10^6 mega (M); 10^9 giga (G); 10^{12} tera (T); 10^{15} peta (P). These size definitions will be used by default throughout the rest of the text.
5. Potassium is an important chemical element for the human nervous system, influencing the electrical transmission of information. Likewise, potassium is found in bones and many other body materials. In total, about 0.27% of body material is natural potassium, and of that, 0.021% is the radioactive ^{40}K component.
6. Pascal Froidevaux, François Bochud, Sébastien Baechler, Vincent Castella, Marc Augsburger, Claude Bailat, Katarzyna Michaud, Marietta Straub, Marco Pecchia, Theo M. Jenk, Tanya Uldin, and Patrice Mangin, "^{210}Po poisoning as possible cause of death: forensic investigations and toxicological analysis of the remains of Yasser Arafat," *Forensic Science International* 259 (2016): 1–9.
7. John Harrison, Tim Fell, Rich Leggett, David Lloyd, Matthew Puncher, and Mike Youngman, "The polonium-210 poisoning of Mr. Alexander Litvinenko," *Journal of Radiological Protection* 37, 1 (2017): 266–278.
8. This involves the decay of tiny, biological cell units where no external intervention is possible. Modern genetic engineering may, however, be able to address this decay in the future. In complex biological systems such as humans, the lifespan curve is different. This is explained by the fact that the human body is composed of many biological components, each providing a different probability of misbehavior. The onset of this misbehavior is considered the aging process. All of these aspects, taken together, affect the progression of the aging curve, which is fairly constant but drops off relatively quickly at a certain age. The point at which aging sets in can be postponed by medical care, thereby increasing life expectancy. In birds, however, the lifespan curve corresponds pretty closely to an exponential decay curve, http://de .wikipedia.org/wiki/Lebenserwartung (November 1, 2023).
9. The decay network is predominantly a system of differential equations that must be solved as a function of time. The Bateman equations are simplified solutions of these equations and are based on a mathematical recursion method.

10. Chemical reactions only affect transformations in the electron shell of an atomic nucleus. This does not change the internal structure of the nucleus – does not change the number of protons and neutrons. Therefore, elemental transformation does not result from chemical reaction. Conversely, nuclear reactions, which are much more energetic, cause a change in the proton and neutron configuration in the nucleus itself (as described earlier) and result in transmutation of the original element.

11. In addition to these systems, there are a number of other types of accelerators. These include linear accelerators, in which particles are driven to ever higher energies by high-frequency electrical pulses, and ring-shaped, so-called synchrotron accelerators. The latest technologies target laser accelerators, in which a high-intensity laser pulse accelerates particles to high energies through photon pressure.

12. Michael Thoennessen, *The Discovery of Isotopes: A Complete Compilation* (Heidelberg: Springer, 2016).

13. The neutron flux F_n is defined as the number of neutrons intersecting an aerial cross section per unit time. The typical unit for the neutron flux is $[s^{-1}cm^{-2}]$.

14. 1 Kelvin (K) = –272.15 Celsius (°C).

15. Bothe later developed Germany's first cyclotron in Heidelberg in 1942.

16. The circumstances of this development have been described in detail by Hans Bethe, who was the head of the theory group for the Manhattan Project: Hans Bethe, "The German uranium project," *Physics Today* 53, 7 (2000): 34–36.

17. Cf. Enrico Fermi, "Elementary theory of the chain-reacting pile," *Science* 10 (1947): 27–32.

18. The participants in this first demonstration of a nuclear reactor, as well as their subsequent health and cause of death, are described on the Argonne National Laboratory website: http://www.ne.anl.gov/About/cp1-pioneers/ (November 1, 2023).

19. Jack Devanney, *Why Nuclear Power Has Been a Flop*, https://gordianknotbook.com/ (November 1, 2023).

4 Dosimetry and Biology

4.1 DOSIMETRY AND BIOLOGY

The biological effects of radioactivity can be multifold, some aspects of which are still unknown. As such, the effects of radioactivity remain a matter for heated debate since human biology and chemistry are not as easily quantifiable as the physics processes. In addition, the processes themselves and their correlations are extremely difficult to identify, measure, and quantify. We again recognize that effects depend on the type of radiation, since α-, β-, and γ-radiation are fundamentally different in nature, each having very different effects with varying levels of danger for complex biological systems.

A standard-based framework, developed through experimentation, might be able to quantify the impact of radiation but only with considerable uncertainty. This is because the effects of radiation on biological systems are not absolute and only understood statistically. These assessments consider billions of individual processes occurring between radiation particles and the molecules that comprise the biological organism. Further, the processes themselves depend on the effectiveness and particularity of enzyme-controlled repair mechanisms within the biological system, as these react to and mitigate possible damage.

This chapter focuses on four important aspects of radiation's effect. First, we will consider radiation absorption and depth of penetration with regard to different materials, including soft body tissue. This concerns both the body's protection against radiation and the effect of different types of radiation on the human body. Due to their physical nature, different types of radiation vary in how they impact the human body as a biological system. Therefore, different types or levels of radiation protection are required.

The second question concerns dosimetry. This is the quantification of radiation effects, the probability that a certain dose of radiation will have a biological impact on the human body. Dosimetry characterizes a series of standardized terms that are intended to make radiation exposure and the resulting damage quantifiable, providing the basis for legal limits. The dose corresponds to the amount of energy which is transferred by radiation to a body of a certain mass.[1] Legal exposure limits are, however, different for the general population, radiation workers, and radiation accidents. The decision to vary exposure limits has brought the scientific quality of the assessment into question.

Radiation protection parameters are usually based on the LNT method (linear no threshold), which assumes a linear relationship between radiation dose and radiation damage. This is particularly important to consider when there is low-level radiation exposure for which no statistically reliable data exists regarding radiation effects. In this range, the LNT method provides a way of extrapolating data from high-exposure measurements to provide probabilities for this unknown range. As such, the method is not entirely undisputed since the consequences of low-level radiation, as well as the human body's possible repair mechanisms at low levels, are unknown. Nevertheless, the LNT approach is currently widely used to determine permissible limit values.[2]

However, there are differences in nationally defined limits: In Europe, a legally permitted annual-exposure limit[3] for occupationally exposed persons is 20 mSv,[4] while in the United States the limit is 50 mSv.[5] These limits have less to do with Europeans being more sensitive to radiation and more to do with the relative uncertainty of scientists regarding the actual effects of low-level exposure. These uncertainties lead to different interpretations and threshold values, whereby legislators are mostly trying to keep parameters within safe or politically opportune limits.

DOI: 10.1201/9781003435907-4

It is worth noting that the permissible dose for occupationally exposed people is usually higher than for the general population. As determined by the NRC, the radiation exposure limit for the general public is 1 mSv/year, as compared to 20mSv/year for workers. This distinction in acceptable radiation exposure is not made because radiation workers can build up a tolerance to radiation and therefore endure more exposure. Rather, the limit value is set so high because radiation workers would reach the legal-limit values established for the normal population after only two to three months and would therefore be absent from work for the rest of the year. The low limit is therefore regarded as economically irresponsible when considering the radiation worker population.

The third part of this chapter deals with the direct chemo-biological effects of radiation and the possible chemical consequences of radiation. The latter are so-called stochastic consequences because their occurrence is not absolutely certain and can only be predicted with a certain probability. When considering chemo-biological effects, a distinction must be made between short-term irradiation with high doses and long-term irradiation with relatively low but constant doses.

In the case of high-dose irradiation, known cases clearly show harmful effects on the biological system, which can lead to death. To determine stochastic data on the late effects of lower-dose radiation, it has been important to study surviving victims of major radiation accidents for possible health effects. However, the small number of such radiation accidents makes reliable statistical analysis difficult; the number of radiation victims is too small, and the available observation period is too short to allow for determining and statistically evaluating long-term effects. In addition, there is often little or no reliable data on the dose and duration of exposure, resulting in serious obstacles to accurate analyses. This is particularly true in cases where the dose is low or below the LNT thresholds. Here, despite intensive research, the biological effects are still largely open to interpretation and debate.

However, the number of studies done on groups who have been exposed to increased radiation is growing. The classic and probably best-known example, of a more precise statistical analysis of the short-term and long-term effects of radiation exposure, is the continuous investigation and monitoring of radiation victims of the Hiroshima and Nagasaki bombings.[6] Over decades, up to 120,000 victims have been continuously examined for long-term damage as well as cause of death. Difficulties arise when determining the absolute dose of radiation received by each person. This information relies not only on exact knowledge regarding distance from the explosion but also on whether the victim was in an open area or was protected by buildings or other radiation-absorbing structures. Also, the exact gamma flux and especially the neutron flux of the radiation have had to be determined by indirect methods, and this information has been corrected several times as these methods have been continuously refined.

In two further cases, more precise statistical analysis of stochastic effects, involving high-radiation exposure of larger groups, would be possible but sufficient information, including necessary existing data, has not yet been released. This information concerns US soldiers who actively participated in the nuclear weapons testing program.[7] Troops were positioned as close as possible to an atomic bomb explosion in order to storm the blasting area within the framework of an infantry exercise.[8] While several hundred thousand soldiers took part in these exercises,[9] a precise statistical evaluation requires data about duration of stay and exact location of respective units in the blasting area – information that has not been made available.

The second case concerns the so-called liquidators who were deployed to clean up the Chernobyl reactor building.[10] The estimated number of liquidators, who worked over a long period of time, varies between 50,000 and 500,000; documents from a systematic analysis of the consequences of irradiation are only available in isolated reports.

In both cases (Figure 4.1), the challenge for long-term analysis is to determine the actual dose rate received and its duration as well as obtaining reliable medical data on individuals over the years after exposure. This is not fully possible for either US soldiers or Russian-Ukrainian liquidators. Data sets have either not been compiled, have been lost, or are declared classified. This lack of openness is of course an ideal breeding ground for conspiracy theories about horribly disfigured Chernobyl

FIGURE 4.1   Left, US infantry at the nuclear-bomb test in Yucca Flat, Nevada, in November 1951 (Atomic Heritage Foundation). Right, Russian liquidators cleaning the reactor complex in Chernobyl, May 1986. Source: https://rarehistoricalphotos.com/chernobyl-disaster-pictures-1986/ (May 23, 2023).

victims whose sad but mostly unproven fate is the subject of many discussions on the Internet. The horror-film industry also makes use of these opportunities, as in the 2012 film *Chernobyl Diaries* which dramatically depicts the dangers of adventure holidays in the radiation region.

A third case is the Fukushima catastrophe of 2011 caused by an earthquake-triggered tsunami that flooded the Fukushima Daiichi reactor complex, paralyzing the electrical control system and, as a result, the reactor's cooling system. This led to over-heating of the reactor's core with subsequent explosions, which released large amounts of radiation.[11] Multiple reports on health consequences for the general population have been published but, due to the limited observation period of a few years, it is not yet possible to make any statements on possible long-term effects.[12]

Due to the lack of data on human exposure, statistical statements regarding these major radiation accidents are usually supplemented with results from long-term irradiation experiments on rats and mice. Yet, these studies also typically focus on rather large exposure doses, so that the long-term impact of low-dose exposure remains uncertain.

The effects of long-term irradiation at low doses below the legal thresholds are therefore the subject of intense discussion. These discussions primarily focus on the validity of the LNT hypothesis, which assumes that the radiation-induced damage increases linearly with the radiation dose without any deviation toward low and unexplored dose values. This topic is extremely controversial since the experimental data for such statements is extremely difficult to obtain and largely based on the statistical analysis of group studies with many uncertainty factors. Especially in the low-dose range, the data allows for alternative extrapolations which could lead to considerably different interpretations regarding possible effects of low-dose irradiation.

Edward Teller (1908–2003), father of the hydrogen bomb, was also the founder and long-time director of the Lawrence Livermore National Laboratory. He swore to the health benefits of long-term low-dose irradiation, in contrast to the LNT hypothesis.[13] He thus advocated the thesis of radiation hormesis, which suggests that low-dose radiation improves the enzymatic mechanism for cell repair and can contribute to prolonged life. The fact that Teller lived to the ripe age of ninety-five seems to underline his argument and yet, such a generally valid statement requires the careful investigation of control groups who have been exposed to radiation for long-term medical or occupational reasons.

A summary of such studies has been produced since 1955, as a so-called BEIR (Biological Effects of Ionizing Radiation) report, by the National Academy of Sciences.[14] This report supports

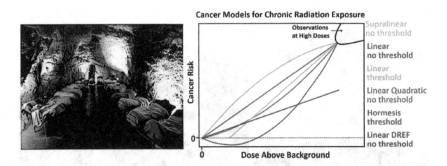

FIGURE 4.2 The claim of radiation hormesis, the health-promoting effect of low radiation doses, remains controversial. However, patients of radiation spas such as Bad Bleiberg in Austria (left) are convinced of the healing effect of such radiation; in many treatment cases, it seems to work with great success. Statistical analysis of the effect of such irradiation is difficult. Curves shown on the right reflect the different models that are currently used to assess the risk of irradiation. Radiation experts usually use the conservative supralinear model (A) enhancing the risk, or the linear LNT model (B), but other models such as the quadratic linear model (C) and the hormesis model (D) cannot be excluded, due to statistical uncertainties. The lack of data in this low-exposure range simply tells us that we do not know how to extrapolate the prediction of radiation damage into the range of daily exposure. This is the fundamental question: Are low levels of radiation exposure dangerous? Source: https://sturgeonshouse.ipbhost.com/topic/342-lets-talk-about-radiation-exposure-thresholds/ (November 4, 2023). Jerry M. Cuttler, "Application of low doses of ionizing radiation in medical therapies," XVIII Polish Radiation Research Society Symposium (right).

the thesis that the risk of long-term health effects increases linearly with dose accumulation over time. With these arguments, there do not seem to be any critical-threshold values according to which radiation below a certain level is beneficial to health, while radiation above such a level represents a health risk.

Again, these arguments correspond to the LNT curve, which is generally applied based on results summarized in the BEIR reports. However, uncertainties regarding various data points also allow for a so-called quadratic-linear interpretation, which shows no direct, harmful long-term effects at low doses but registers a stronger linear increase with higher doses. Such curves only illustrate health damage; health benefits, as propagated by Edward Teller, are difficult to register and prove. Possible benefits are therefore not taken into consideration within radiation protection guidelines. For these reasons, there are several objections, among radiation researchers, to the linear extrapolation of the LNT curve to low dose rates.[15] A recent re-evaluation of the mortality data for nuclear bomb survivors indeed "shows that the data no longer support the LNT model but are consistent with a radiation hormesis model when a correction is applied for a likely bias in the baseline cancer mortality rate."[16]

In spas such as Bad Gastein in Austria (Figure 4.2), where radon therapy is performed on a large scale, the annual whole-body dose is 1.80 mGy with 1.36 mGy affecting the lungs.[17] Since radon is an α-emitter, the equivalent dose is 0.04 Sv and 0.03 Sv,[18] respectively. Here and in other radon spas, comparing potential negative and positive long-term health effects would be possible with careful investigation. However, previous studies have yielded largely controversial statements;[19] therefore, radiation hormesis and the LNT hypothesis remain controversial[20] (for further discussion, see Chapter 8 and Chapter 14, Volume 2).

4.1.1 Absorption and Depth Effect of Radiation

Four types of ionizing radiation[21] – α, β, γ, and neutrons[22] – are fundamentally different from each other in their nature, effect, and properties. This difference influences in particular their interaction with matter, their penetration depth, and their effect on the chemical structure of material. Penetration depth is inversely proportional to the absorption of radiation, such that interaction is determined by the scattering or Coulomb repulsion as radiation interacts with a material's atoms and electrons. These are, however, statistical-probability processes, since each of the millions to billions of incident particles[23] travel through the material along different paths and are scattered in different directions, depending on the different processes and forces involved.

Nevertheless, it is possible to formulate a general law that is analogous to the decay law. This general law describes the absorption, and therefore the depth of penetration, of each type of particle; this is achieved by introducing a probability for absorption, the so-called absorption coefficient μ. .

$$I(d) = I(0) \cdot e^{-\mu \cdot d} \tag{4.1}$$

According to this law of absorption, the initial intensity of the radiation $I(0)$ decreases exponentially with the absorber material's thickness d. The absorption coefficient μ μ depends on the type of radiation and the probability that it will interact (cross section σ_{abs}) with the material's components. The absorption coefficient μ also depends on the density ρ of the material, so that the absorption coefficient can be described by the following equation (with standard units in square brackets),

$$\mu \left[cm^{-1} \right] = \rho \left[g / cm^3 \right] \cdot \frac{N_A}{A} \cdot \sigma_{abs} \left[barn = 10^{-24} cm^2 \right] \mu \left[cm^{-1} \right]$$

$$= \rho \left[g / cm^3 \right] \cdot \frac{N_A}{A} \cdot \sigma_{abs} \left[barn = 10^{-24} cm^2 \right] \tag{4.2}$$

where N_A is the Avogadro constant, $N_A = 6.022 \cdot 10^{23}$: the number of atoms N in material with a mass of A grams, where A is the atomic mass or molecular weight. Thus, the law of absorption states: The intensity of the radiation decreases exponentially with the density of the absorbing medium. However, intensity also depends on the probability of the interaction, which can vary greatly depending on the type of radiation.

Probability is again formulated as a cross section that depends on the energy of the particles and the type of interaction. The cross section is usually determined by measurement. Similar to the decay law, the half-value thickness $d_{1/2}$ of an absorber can be defined as the thickness at which 50% of the radiation is absorbed: $d_{1/2} = ln2 / \mu$. The range, on the other hand, is defined as $R = 1/ \mu$ an inverse to the absorption probability.

Figure 4.3 demonstrates the range of different radiation particles in a plastic medium with the chemical consistency of biological material. The low mass of the radiation particles and their interaction with the material's electrons, being of the same size, cause quick and wide scattering of the electrons into a broad angle range. As a result, the initially sharp electron beam fans out even at shallow depths.

Due to their relatively low energy, electrons have only a short range. This is why β-radiation essentially affects only surface materials. Protons, on the other hand, have a much larger mass and are scattered less. Their penetration range is therefore many times greater than that of electrons. All charged particles do, however, lose energy by electromagnetic interaction along this path.

This energy loss is not linear. Heavy particles experience maximum energy loss at the end of their trajectory, when they have already reached relatively low energies. At that time, the beam expands like a club as seen on the far left of Figure 4.3. The chart on the right side of the figure shows the energy loss of a beam with typical energies as a function of penetration depth. This curve is called the Bragg curve, which demonstrates that maximum energy loss is reached at the greatest penetration depth and then drops abruptly.

This behavior is described theoretically by the so-called Bethe-Bloch formula, which describes interaction between the Coulomb fields of the electrons and nuclei in the material and the incoming charged particles causing scattering and energy loss; the latter reaches its maximum when the

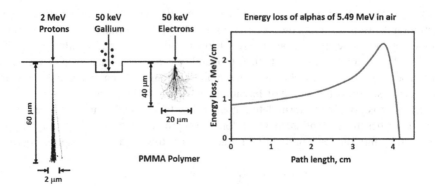

FIGURE 4.3 Simulation of particle absorption in plastic. The left figure shows that the range and extension of the particle beam, by scattering interaction, is a function of particle type and energy. The figure on the right shows energy loss of particles in air, with typical decay energy; the typical path length or range is around 4 cm. Source: https://en.wikipedia.org/wiki/Bragg_peak (November 4, 2023).

particles move only at very low energy at the end of their trajectory. The equation applies to heavy, positively charged particles and not to negatively charged electrons, which are, as we have just discussed, subject to different behavior due to their light mass and negative charge. With this formula, a complex interaction, between a positively charged incident particle and the positive and negative charge carriers in the material, is reflected. This formula was developed in the early 1930s by the German-American theorist Hans Bethe (1906–2005) and improved by his collaborator Felix Bloch (1905–1983).

Figure 4.4 demonstrates the thickness of different materials with different densities necessary for complete deceleration of α-, β-, and γ-radiation. The probability of heavy charged particles decelerating is very high due to the Coulomb interaction. This effect causes a very short path length for massive helium nuclei or α-particles that are being very swiftly stopped. Less massive particles such as deuterons and protons can penetrate deeper into materials. Electrons or β-particles are also rapidly decelerated, but this is more due to the many scattering events that cause rapid broadening of the beam. Also, neutrons distribute their initial energy through multiple scattering events.

Figure 4.4 also schematically shows possible absorption processes for different particle types, whereby the cross section determines the respective probability of these processes. The higher the cross section, the higher the respective absorption probabilities and the lower the range of radiation in the materials.

Charge-free particles such as neutrons are not subject to the Coulomb force, although other processes slow them down and determine the range. Gamma radiation and, to a lesser extent, X-rays are extremely high-energy electromagnetic waves. These can be described as photons, according to the quantum-mechanical wave-particle dualism. Depending on their energy, photons cause various processes in which they lose energy, releasing it into the material. At lower energies – between 100 keV and 10 MeV – the probability of absorption is dominated by the so-called Compton process. Practically speaking, this corresponds to the scattering and deflection of incident photons on the electrons in the material. In this scattering process, photons pass part of their initial energy $E = h \cdot v$ to the electrons, which corresponds to a reduction in frequency or an increase in the wavelength of the photons. This scattering process, known as Compton scattering, is named after Arthur Compton (1892–1962) who first observed and interpreted this effect in 1922.

Photons of higher energy increasingly lose their energy due to pair formation. These photons, as electromagnetic wave packets, lose an energy of 1,022 keV. This loss corresponds to the energy necessary to form an electron-positron, which is to say a particle-antiparticle pair.

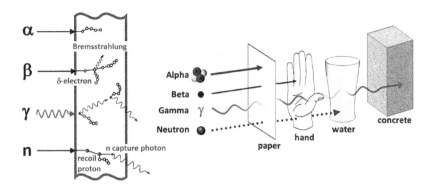

FIGURE 4.4 Interaction and absorption behaviors of alpha, beta, gamma, Bremsstrahlung X-rays, and neutron radiation in different materials.

In 1933, Irène Joliot-Curie and Frédéric Joliot proved this effect experimentally with the first measurement of the pair-formation process. In a similar way, proton-antiproton pairing may occur but, since the mass of these particles is approximately 2,000 times larger than that of the electrons, such a process requires 2,000 times the photon energy for this to occur. Since the absorption probability of photons increases in high-density materials, the influence of both the Compton and pair formation processes is maximized. For this reason, lead is mostly used as an absorption material for γ- or X-ray photons (Figure 3.7).

Neutrons have a low probability of interaction with matter. Their energy loss is mainly associated with the probability of collision with other particles in the material. Impact with considerably lower-mass electrons does not change the direction or energy of incident neutrons. Impact with massive particles causes a change of direction but has only a small influence on the energy of the neutron since it is redirected during impact much like an elastic rubber ball. This corresponds to elastic scattering, which takes place without the scattered particle losing energy.

Neutron energy loss is greatest through impact processes with particles of comparable mass. In this process, inelastic scattering takes place in which up to 50% of the neutron's original kinetic energy can be transferred to its impact partner. The best impact partners are light nuclei with mass $A = 1–12$, i.e., hydrogen, lithium, beryllium, boron, and carbon (see Chapter 2). Therefore, light materials such as water or paraffin are mostly used for neutron deceleration or moderation. In addition to inelastic impacts, neutrons can also cause direct nuclear reactions that completely absorb neutrons, thereby generating protons or different types of radiation such as γ- and α-radiation. This process is very effective in materials such as boron and cadmium. Moderators such as water and paraffin are often mixed with boron to absorb neutrons.

The deceleration behavior of radiation is of great practical importance in medical applications such as the irradiation of small tumors. The energy of the beam is chosen so that the particles have their maximum energy loss as close as possible to the location of the tumor, according to the Bragg curve, and thus deposit the maximum dose. Range and energy deposition, for the different types of radiation, can be precisely calculated by means of the scattering processes and the respective effective cross section.

The graph in Figure 4.5 shows the differing energy ranges for different types of radiation in water. The absorbing water corresponds to biological material, which consists of 70% water. The energy of light or weakly interacting particles is deposited near the surface of the material. According to the Bragg curve, heavily charged particles initially have only a small, gradual energy loss, which increases rapidly at lower energies and leads to a complete stop. This means that most of the initial energy can be accurately deposited as a dose, at a particular depth in the body.

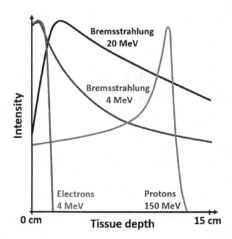

FIGURE 4.5   Range and dose deposition of different types of radiation with typical energies in biological tissue.

4.1.2   DOSIMETRY

Dosimetry quantifies the extent of radiation exposure so that we can investigate and predict its effects on various materials. Of particular interest, of course, is its effect on human organs, including possible chemical and biological consequences. The effects of radiation are primarily due to ionization. This means that the electromagnetic or scattering interaction, between radiation and material, transfers enough energy to remove electrons from their atomic bonds. What remains is a positively charged ion that is chemically very reactive and can therefore form new compounds. The radiation also transmits sufficient energy to molecules to break up their molecular bonds, leaving behind chemically active, differently charged molecular radicals. Through these processes, radiation transfers energy to the material.

The dose identifies the quantity of radiation required to cause ionizing effects in matter. Various units have been defined for this purpose, the best known of which is the dose energy or simply, dose D. This corresponds to the amount of energy $\Delta \Delta E$ transmitted by the radiation to the material. The dose D is quantified as energy per mass of material m.

$$D[Gy] = \frac{\Delta E[J]}{m[kg]} D[Gy] = \frac{\Delta E[J]}{m[kg]} \tag{4.3}$$

The equation indicates the relationship between the different parameters, with regard to the international units commonly used today. The unit for the dose is gray $[Gy]$, which is defined as energy in joules per mass in kilograms $1Gy = 1J/1kg$.[24] However, the energy of radiation is traditionally expressed in electron volts (eV), not in joules. Therefore, one usually has to convert the energy units: $1 \text{ eV} = 1.6 \cdot 10^{-19}$ J, where the conversion factor $1.6 \cdot 10^{-19}$ is the electrical elementary charge in Coulomb units.

For example, a single particle of 5 MeV energy being fully absorbed in a body part, such as a hand with a mass of 1 kg, deposits a radiation dose of $8 \cdot 10^{-13}$ Gy. While a single particle won't have much of an effect, a radioactive source with an activity of 1 MBq or in classic curie units, 27 Ci – having a decay rate of 1 million or 10^6 decays per second – increases the total dose substantially. The total dose rate for the hand will be $8 \cdot 10^{-7}$ Gy/s or 0.8 μGy/s. If one keeps the source in the hand for one hour, the total dose will be about 3 mGy. This may cause a reddening like sunburn because the radiation would not penetrate the skin. However, if such a source were to enter the body, it would have a dangerous effect because radiation can destroy thin mucous membranes; these effects are explained later in more detail. If the radiation intensity and energy are known, the dose rate can easily be calculated. The period of irradiation results in the total dose.

The definition of *dose* is independent of the radiation's effect. Therefore, dose D only indicates the amount of energy deposited in a body and says nothing about the effect of the radiation. There are considerable differences in radiation impact, which in turn depend on the type of interaction between the radiation and the material or the ionization probability of the radiation and the chemical consequences. For this reason, the dose is multiplied by a weighting factor Q, which corresponds to this ionization probability. The result is the equivalent dose H, with $H = Q \cdot D$ which reflects the biological effect of the respective radiation type. The weighting factor Q is defined as a function of the linear energy transfer (LET). LET expresses how much energy the radiation deposits into the material over a distance unit (cm). This, in turn, approximately corresponds to the amount of energy that the radiation will lose in matter. It is important to note that LET only covers energy loss by ionization – i.e., interaction with electrons – while the Bethe-Bloch formalism also considers interaction with the positively charged atomic nuclei. Table 4.1 lists the approximate Q-factors for the most common types of radiation confronting the population.

The physical unit for the equivalent dose is the sievert (Sv), which is the international standard unit[25] in which the biological impact of radiation exposure and legal radiation limits are specified today.

If one goes back to the example of a 1 MBq source, an hourly dose of $D = 3\mathrm{mGy}$ with weighting factor $Q = 20$ results in an equivalent dose of $H = 60\mathrm{mSv}$ or $H = 0.06\mathrm{Sv}$, which can cause considerable damage in the body. This equivalent-dose value corresponds to three times the legally permitted annual dose in the United States.

Here, however, one must differentiate. In medical dosimetry, one distinguishes between the whole-body dose and the dose as it pertains to specific organs. This is because some organs are more chemically sensitive to radiation than others. To take this into account, various organ-specific weighting factors w_O are introduced. These are standardized in such a way that the sum of the weighted organ doses H_O results in the effective whole-body dose:[26]

$$H = \sum_O w_O \cdot H_O \tag{4.4}$$

Relatively insensitive organs such as the skin, salivary glands, and brain have a small weighting factor of $w_O = 0.01$, while sensitive organs such as those of the digestive system and the lungs have a tenfold-higher weighting factor of $w_O = 0.12$ (Table 4.2).

TABLE 4.1

Weighting Factors for Different Types of Radiation, for Calculating the Equivalent Dose

Gammas & Photons		Q=1
Beta particles		Q=1
Neutrons	E<10keV	Q=5
Neutrons	E>10keV	Q=15
Protons		Q=5
Alpha particles		Q=20

Source: https://www.nrc.gov/reading-rm/doc-collections/cfr/ part020/part020-1003.html (November 4, 2023).

TABLE 4.2

Weighting Factors for Calculating the Effective Dose According to the Various Dose Determinations of the International Commission on Radiological Protection (ICRP). Changes in the Various Weighting Factors, between 1990 and 2007, Are Due to the Increasing Accuracy of Radiation Medicine's Results

Organs and Tissues	ICRP 60 (1990)	ICRP 103 (2007)
Gonads	0.20	0.08
Bone marrow (red)	0.12	0.12
Colon	0.12	0.12
Lung	0.12	0.12
Stomach	0.12	0.12
Bladder	0.05	0.04
Chest	0.05	0.12
Liver	0.05	0.04
Esophagus	0.05	0.04
Thyroid gland	0.05	0.04
Skin	0.01	0.01
Bone surface	0.01	0.01
Salivary glands	not defined	0.01
Brain	not defined	0.01
Other organs and tissues	0.05	0.12
Total	1.00	1.00

Source: https://www.nrc.gov/reading-rm/doc-collections/cfr/part020/part020-1003.html (November 4, 2023).

The effective dose is thus defined as the sum of the weighted organ doses and is regarded as the measure of radiation risk from a particular radiation source. Radiation dose without a more precise designation usually refers to the effective dose. The details of legal limits, as defined by dosimetry and the statistics for radiation exposure of a person, will not be discussed here; the relevant technical literature given at the end of the chapter provides information on this.

Everyone is exposed to radiation on a daily basis. However, respective doses differ considerably. Variations depend on local conditions, including where and how people live as well as where they work and what they do. As shown in Figure 4.6, during our daily lives we are all exposed to many natural radiation sources. This primarily includes radiogenic sources from radioactive materials present in our natural environment. Cosmogenic sources also contribute as we experience radiation from the sun as well as high-energy cosmic radiation from stellar explosions in space. The intensity of these explosions does, however, rapidly decrease with increasing energy and is eventually only detectable using complex measuring methods.

In addition to natural exposure, humans are increasingly exposed to anthropogenic (man-made) radiation sources. These include radiation from radioactive material that has been deposited into our environment from nuclear weapons tests. This radiation has been distributed globally by air currents in the stratosphere and even today is reflected in our environment. Accidents in nuclear power plants will impact our environment with radioactive material being distributed by atmosphere and hydrosphere. The steady growth in nuclear medical applications is also considered to have a growing impact. In the last thirty years, nuclear medicine methods and applications have nearly doubled mankind's average radiation exposure, as shown in Figure 4.7.[27]

Currently, the average effective dose per year from cosmogenic, radiogenic, and anthropogenic sources averages approximately 6.2 mSv for any individual. Of these, about 3.1 mSv per year are of

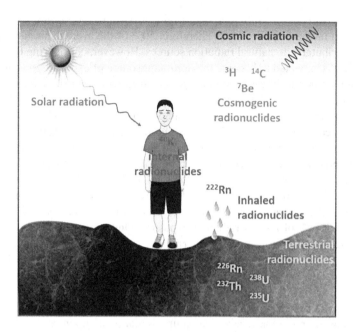

FIGURE 4.6 There are three natural radiation sources that contribute to human exposure. Cosmogenic sources are the sun and the explosion of distant stars. Radiogenic sources are long-lived radioactive isotopes in rock and earth. Within our bodies, radioisotopes are chemically integrated into the molecular structure of the body.

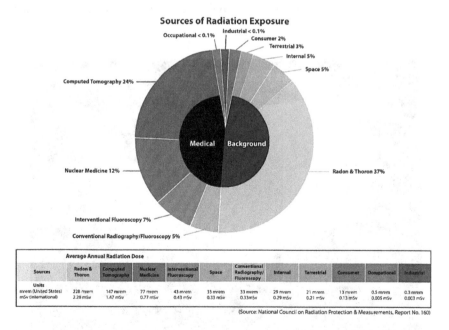

FIGURE 4.7 Percentage of various natural and anthropogenic radiation sources contributing to the average radiation exposure of 6.2 mSv in the United States. The share of natural radiogenic or cosmogenic radiation sources is currently less than 50% and is rapidly decreasing due to the rapid increase in radiation-related diagnostic and therapeutic nuclear medicine methods, which have become the main anthropogenic radiation sources for the average human's radiation exposure. Source: https://www.epa.gov/radiation/radiation-sources -and-doses (May 23, 2023).

natural origin; approximately 3.1 mSv per year originate from anthropogenic sources, mainly ionizing radiation from medical applications. Therefore, 50% of the average human's radiation exposure is currently attributable to natural radiation sources. However, considering the rapid growth in nuclear medicine, it is expected to become the dominant source of exposure in some near future.

It should be noted that the average for anthropogenic-radiation exposure, due to medical examinations in the first decade of the 21st century, rose from 20% to more than 40% of the total annual dose. Of course, this dose is not evenly distributed among the population. With the possible exception of use for dental procedures,[28] the average dose is mainly factored from radiation exposure involving primarily elder patients, for cardiovascular and tumor diagnostics and for cancer treatment. Computer tomography and nuclear-medicine examinations contribute up to 75.4% of the cumulative effective dose of elderly people, while 81.8% of the total medical-radiation dose is applied in outpatient examinations.[29] This topic will be discussed in Chapter 14, Volume 2 in more detail.

Contrary to general expectations, other anthropogenic radiation sources account for only a very small share, less than 2% of annual radiation exposure. One third of this – less than 1% of the total anthropogenic share – comes from the Chernobyl reactor disaster, with another third from the nuclear-weapons testing program and operation of the reactor program, while the last third is from normal household operations.[30] These are, of course, only average values that may fluctuate with location.

The geographical dependence of radiation exposure depends on a number of conditions. One is the amount of cosmic radiation received, a natural radiation component that currently accounts for 8% of total radiation exposure. This corresponds to an effective average dose of 0.33 mSv per year. Cosmic radiation comes mainly from the sun. Additional high-energy components are due to radiation generated in distant supernova explosions in our galaxy as we will discuss in Chapter 6.

Most cosmic rays are absorbed by the Earth's atmosphere,[31] so radiation exposure increases with altitude as shown in Figure 4.8. With every 1,500 vertical meters, the dose approximately doubles. Therefore, mountain dwellers receive about two to three times more cosmic radiation than coastal

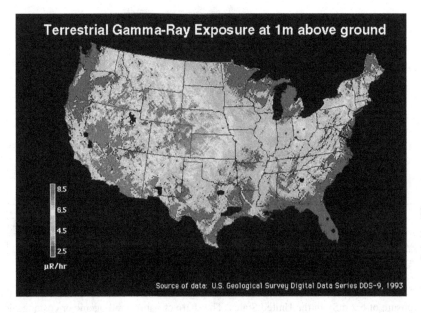

FIGURE 4.8 Hourly exposure to radiogenic γ-radiation in the United States at a 1 m distance from the ground. This activity is mainly due to the radiation associated with the decay of ^{40}K in the ground and radon diffusing out of the ground. Activity is largely distributed along the granite-containing southwestern mountain ranges. Source of data: U.S. Geological Survey Digital Data Series DDS-9, 1993.

dwellers. At a normal flight altitude of 10 km, the cosmic dose rate, for personnel and passengers, is up to ten times higher than at home in the lowlands.[32] An eight-hour flight means an additional exposure of 37 µSv – this is comparable to the annual average exposure from nuclear power and nuclear power plant accidents.

Long-lived radioactive elements form natural decay chains in the earth's rock, and these also contribute to natural radiation exposure. Because of their longevity, these elements are often older than our Earth and have their origins in past star systems, from whose dust our solar system once originated. Apart from the long-lived potassium isotope ^{40}K, these elements mostly belong to the group known chemically as actinides, which form mineralogical concentrations in certain types of rock such as granite. Therefore, the annual radiogenic-radiation distribution for the mountain states in the United States is about 0.07 µSv per hour, which is about 0.6 mSv per year. This dose is about three times higher than that of inhabitants residing on southeastern or northwestern coastal plains, who are on average exposed to 0.2 mSv of external radiogenic radiation per year.

From the decay of long-lived actinide isotopes, the noble gas radon develops as part of the decay chain. Due to its volatility, radon easily enters the atmosphere but typically remains at ground level because it is heavy. It can, however, accumulate in rooms without sufficient ventilation, becoming a strong radiation source in such environments.[33] When inhaled in large quantities, radon represents a potential danger for the lungs, but more so because from the lungs it may enter the bloodstream through gas exchange processes. While radon itself is exhaled rather quickly, its decay products may remain in the lungs and enter the bloodstream. Radon and the other members of the natural decay series account for more than 40% of natural radiation exposure.

Overall, the effective dose from natural sources for humans in the United States is about 3.1 mSv per year.[34] However, there are considerable regional variations, which range between 1 and 5 mSv per year. Maximum values result for mountainous granite-containing regions of the American Southwest. In Europe, the average values are comparable. In certain areas, however, the dose can be estimated at about 10 mSv per year, either due to cosmic radiation or uraniferous rock. The extent of such variations will be discussed in a later chapter.

4.1.3 Radiation Measurements and Detectors

An important goal of dosimetry is reliable measurement of the radiation intensity and energy that are deposited in organic and inorganic materials. Our eyes, ears, nose, and skin are the means we have for assessing our environment. However, since radiation cannot be felt, heard, or seen, these systems are relatively insensitive to the detection of radiation. Although they are sensitive to the effects of radiation – the light flash and subsequent thunder of an atomic bomb explosion – we do not possess an instantaneous radiation-exposure sensing organ.

Therefore, radiation intensity can only be measured through secondary effects, which is to say through the interaction of radiation with matter. This interaction usually includes an energy transfer, from the radiation source to the surrounding material, by scattering or absorption (see Section 4.1.1). Energy transfer leads to chemical reactions or also to atomic and molecular excitation and ionization processes, which can deliver measurable signals.

The classic example of chemical reaction is the blackening of a photographic plate by uranium-crystal radiation, which Henri Becquerel discovered by chance in 1896. Becquerel interpreted the blackening to be a result of unknown radiation, but neither the energy content nor the intensity of the radiation could be determined (Figure 4.9).

Special photo emulsions were developed to improve the efficiency of chemical processes and also to determine the type of radiation. In particular, magnetic spectrometers used photographic plates in the focal plane to determine the paths of charged particles deflected in the magnetic field. Particle traces in the photographic plate were then examined under a microscope, where particles could be identified from the length and extension of their traces. This method was used until electronic detection methods improved in the 1980s.

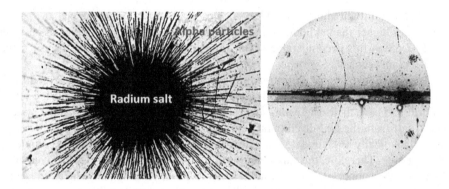

FIGURE 4.9 Photographic evidence of α-radiation from a radium source. The α-particles emitted from the source are evidenced in the blackening of the photographic emulsion. The image demonstrates that the particles have only a certain range and are then stopped in the material due to energy loss, https://www.sci-encephoto.com/media/1242/view/radioactive-emission-from-radium (November 2, 2023). The photo on the right was taken through the window of a cloud chamber by Carl Anderson in 1933. It shows the upward curved orbit of a positron together with the reverse downward curved orbit of an electron, which were generated by the pairing process at high-energy cosmic rays; this is regarded as the first direct experimental proof of the existence of anti-matter. Source: https://www.aps.org/publications/apsnews/200408/history.cfm (November 2, 2023). Carl D. Anderson, *Physical Review* 43 (1933): 491.

In addition, particle physics often used cloud chambers, which are chambers filled with a supersaturated air-alcohol mixture. When high-energy particles enter this area, these particles ionize individual atoms in the gas mixture. The ions serve as condensation nuclei so that very fine mist droplets form that make the path of the particle easily visible. In turn, this allows physicists to directly observe the influence of electric and magnetic fields. This does, however, require suitable gas-pressure conditions.

For the first time, in 1933, Carl Anderson used such a chamber to detect the antiparticle of electrons, the positron (see Chapter 1). The positron and electron were produced by high-energy cosmic rays in the gas mixture and observed, after the pair-production event, by their curved trajectories (Figure 4.9). Although photographic plates and cloud chambers have largely disappeared from research, they are still used to demonstrate the existence of cosmic rays.

In the case of excitation processes, atoms or molecules are transformed into a higher energy state by radiation energy. Such states correspond to rotational or vibrational modes but also other quantum configurations. De-excitation, through transition from the excited state to the ground state of a nucleus, releases energy as electromagnetic radiation, in this case as γ-radiation. This γ-pulse is converted by scintillator material into light that can be measured or also converted into an electric signal or pulse that can be electronically recorded. The number of pulses correlates with the intensity of the radiation and the pulse height with the energy of the incident radiation particle or photon.

In ionization processes, radiation releases electrons from the material. This can generate an electric pulse, which can also be recorded. The pulse height is proportional to the amount of charge carriers that are released, which in turn is proportional to the energy that the radiation event has deposited in the material. For radiation detectors, the mode of action is based on these two possibilities. The release of light is referred to as scintillation, the release of electrons as ionization.

In scintillators, light can be emitted over a wide wavelength range from infrared to ultraviolet; the frequency or frequency range depends on the properties of the scintillator material itself. Scintillators are usually salt crystals of high purity since internal contamination or impurities in the crystal can lead to the absorption of light. In 1908, Ernest Rutherford and Hans Geiger discovered that zinc sulfate (ZnS) emits light flashes when irradiated with α-radiation. Rutherford and his students Marsden and Geiger then sat in a darkened room for days watching and counting flashes of light, in order to determine the activity of the radiation source from the number of individual flashes. The intensity, or radioactivity, of a radioactive sample was determined by the number of light flashes within a fixed time range. This was the birth of the scintillator.

The photoelectron multiplier, or photomultiplier for short, replaced this laborious and eye-taxing procedure. This instrument is based on the photoelectric effect whereby the incidence of light on a material causes electrons to be emitted by energy transfer from the photons. This emission generates an electric current.[35] This effect is used today with every light sensor.

In photomultipliers, the photons, produced by radiation in the scintillator, hit a photoelectric layer – the so-called photocathode. This consists of a material from which electrons can easily be removed; these are usually thin layers of various metal alloys with a strong alkali component, which are optimized for the respective applications.

The released electrons are accelerated, in an electric field of about 100 V, toward so-called dynodes. These dynodes are electrodes usually made of BeO or MgO; upon impact each electron creates multiple secondary electrons. With each dynode, the number of electrons multiplies in a cascade, resulting in an electrical pulse that can easily be measured with an appropriate resistance circuit (Figure 4.10).

The height of the electrical pulse corresponds to the energy of the original radioactive radiation. The number of electrical pulses is again proportional to the number of photons originally generated and thus depends on the activity of the radioactive source. There are device-specific factors that determine this relation, such as the type of scintillator, the connection between scintillator and photomultiplier, the composition of photocathode and dynode, and the number of dynodes. Therefore, each scintillation detector must be calibrated with radioactive sources of well-known activity to obtain reliable data.

Today there are many commercially produced scintillators for different applications. For the most part, their materials have been optimized so that they are effective for various types of radiation. This often relates to the internal structure of the material and how the radiant energy is absorbed by the material and converted into light. Besides salt crystals, such as NaI (sodium iodide) and BaF_2 (barium fluoride), the most commonly used scintillator detectors for measuring gamma radiation are BGO ($Bi_4Ge_3O_{12}$ – bismuth germanate) and $LaBr_3$ (lanthanum bromide) or also Br_3Ce (cerium-bromide) crystals.

Some of these scintillator types do, however, have a high inherent activity, since long-lived radioactive barium, bismuth, lanthanum, and cerium isotopes are inherently built into the crystal, due to their natural isotope distribution. In general, scintillator detectors are used to measure gamma radiation and neutrons. For neutrons, the scintillator material is primarily an organic liquid since neutrons usually release their energy by scattering to light particles.

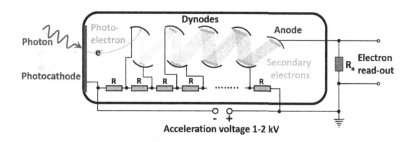

FIGURE 4.10 Principle of a photomultiplier. The photons generated in the scintillator hit a photo layer where they release electrons by the photoelectric effect. Electrons are then accelerated by a resistance circuit and multiplied by dynodes, on which secondary electrons are generated until a readable electrical signal is obtained.

In many spectroscopic applications for detecting or counting radiation, semiconductor counters play an important role. Semiconductor detectors are based on the special electrical properties of semiconducting materials. Silicon or germanium are typically used since they detect particle or gamma radiation. Here, the radiation excites electrons, from the electronic band structure of the semiconducting material, so that they migrate under electrical voltage as free-charge carriers. These electrons are thus carried to specially attached electrodes made of metal, where they are picked up and amplified as an electronic signal.

Silicon-based semiconductors are mainly used for measuring particle and X-ray radiation, while germanium detectors are preferred in gamma spectroscopy. The great advantage over scintillators is the almost one-hundred-times-better energy resolution of the detector signal, by which even small energy differences in the various radiation components can be measured. Due to their high cost, semiconductor detectors are primarily used for spectroscopic applications in nuclear physics and particle physics research. These detectors are, however, increasingly used in industrial applications requiring high energy resolution. Applications range from materials and electronics research to oil exploration and art analysis, as will be discussed in detail in later chapters.

The silicon meters shown in Figure 4.11 typically consist of thin silicon layers. Layer thickness depends on the application and is usually dimensioned to stop particle radiation, in order to receive the full energy signal. In so-called telescope meters, very thin detectors are mounted in front of a thicker silicon layer so that the first, thin detector layer measures only an energy loss while the second layer measures the full energy of the particles. Since energy loss is different for different particles – for example, protons and alphas – signal strength can be used to identify the radiation particle type.

For measuring electromagnetic radiation, semiconductor detectors are often cooled with liquid nitrogen, which can suppress thermal fluctuations in the detector material. This improves resolution and is particularly necessary when measuring gamma and X-ray radiation.

Germanium detectors as shown in Figure 4.12 are mostly used with special crystal configurations at larger volumes. Specifically, since γ-radiation has a higher range than particle radiation, it must be absorbed in crystal so as to generate the full energy signal.

Despite numerous innovations in semiconductor detectors, many particle-radiation detectors are still based on the ionization-chamber principle because of their high detection efficiency. These are mainly used to measure charged particles as they occur in β- or α-radiation (Figure 4.13). In a gas volume, a charged particle interacts with gas particles, ionizing them along its path. The initial

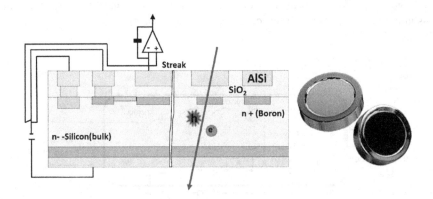

FIGURE 4.11   Electronic structure of a silicon counter. Incident radiation generates free electrons that were previously bound in the silicon-crystal bond. These migrate to the positive electrodes and are picked up as a signal by an electronic circuit. On the right is a typical silicon detector. Today such detectors can be obtained commercially for a few thousand dollars, but the electronics required for amplifying and processing the data to digitized signals often come at higher costs. Source: https://www.ortec-online.com/products/radiation -detectors/silicon-charged-particle-radiation-detectors.

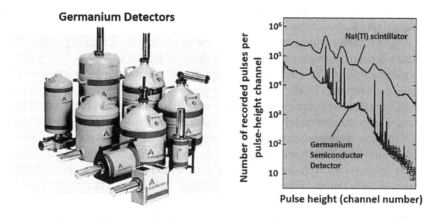

FIGURE 4.12 On the left, germanium detectors mounted, in different configurations, on liquid-nitrogen containers for cooling. Crystals are located in the aluminum-clad cylindrical containers. These detectors cost tens of thousands of dollars depending on the size. On the right is a diagram comparing the signal resolution of a sodium-iodide spectrum (above) and a germanium counter (below). The better resolution of the germanium detector allows for analysis of adjacent lines. Source: https://www.britannica.com/science/active-detector (May 25, 2023).

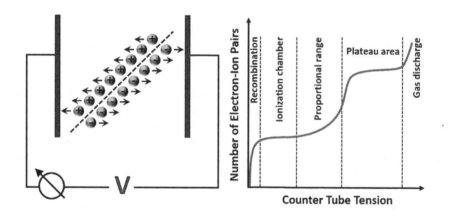

FIGURE 4.13 On the left is a simplified scheme for the operation of an ionization chamber that measures particle radiation. Incident particles ionize the detector gas and the applied voltage accelerates the negatively charged electrons to the positive pole (anode) and the positive ions to the negative pole (cathode), where they trigger an electrical signal or pulse. On the right, the correlation between the applied voltage and the pulse height at the measuring instrument. At low voltages, many of the electrons and positive ions recombine. At higher voltages, in the ionization range, all electrons reach the anode; the pulse height corresponds to the original energy of the incoming particle. In the proportional range, the voltage is so high that secondary electrons are generated in the gas by the accelerated electrons. The higher the voltage, the higher the signal, so that the intensity of the incident radiation can be determined. In the plateau range, the instrument has reached saturation.

radiation energy is transferred to the gas and the incident particle is stopped in the detector. The resulting positive (ions) and negative (electrons) charge carriers are separated by an external electrical field and collected at two electrodes. This in turn generates an electric signal that can either be read as a current or registered as an electrical pulse. Current readings are mainly used in dosimetry. The measured current intensity can be correlated with the activity of a source or with the average radiation dose. Electrical-pulse registration is usually limited to low radiation activity, since individual particles trigger a measurable electrical pulse.

The pulse height is directly proportional to the number of ion-electron pairs produced by ionization. This in turn depends strongly on the applied voltage or kinetic energy that the charge carriers absorb in the electrical field. The number of charge carriers, determined by the energy of the original radiation particle, depends on the acceleration voltage (Figure 4.13).

There are three typical application areas for an ionization chamber. In the lower voltage range, electrons released by ionization receive sufficient energy to be extracted from the ion range; at the electrode, each of these electrons can be measured and a complete count is possible. This is the saturation range in which most ion chambers operate. At higher voltages, the released charge carriers generate further electrons through secondary ionization, which results in a proportional increase in the signal height. This is the so-called proportional range. In these two ranges, the proportionality between pulse height and radiation energy is maintained.

At still higher voltages, the so-called Geiger-Müller range is reached. Here the correlation between radiation energy and pulse height is lost. In this counting range, only the number of radiation particles is measured – the activity of the source and thus the radiation dose are measured but not the energy of the radiation. This is why the Geiger-Müller counter has become the preferred measuring device for quickly determining the presence of radioactivity within a locality as well as the potential radiation hazard.

4.1.4 Chemical and Biological Effects

In this section, the possible impact of radiation on biological materials is presented in more detail. This topic is very broad and fiercely discussed due to the challenges involved when interpreting the possible impacts of low dose exposures. These interpretations are traditionally based on the linear extrapolation of the LNT theory as summarized earlier. Here, I present a broader discussion based on the available data and the possible interpretations found in the current scientific literature.

The potential chemical and biological effects of radiation are among people's greatest fears. This includes fear of cancer and of possible genetic changes that could result in dreadful deformities in subsequent generations. However, when the causes of cancer are statistically analyzed, other causes – such as the chemical reactions caused by smoking, malnutrition, and obesity, as well as air pollution – play a much greater role than natural and artificial radiation exposure. Thus, radiation exposure is estimated to cause around 2–5% of cancer cases.[36] It should be noted, however, that smoking, which is one of the main causes of cancer at 20%, might be directly correlated with increased radon radiation in the lungs as frequently suggested.

Tobacco plants are considered effective absorbers of the radon gas emitted from fertilized ground.[37] However, due to the short half-life of ^{222}Rn, $T_{1/2} = 3.8\ d$, most of the radon has already decayed during the curing, fermenting, and storage time for tobacco, which may last from several weeks to years depending on the desired quality. Only the remaining traces of radioactive noble gas ^{222}Rn are eventually released when smoke, from burning the dried tobacco leaves, is inhaled into the lungs. The ^{210}Po with a half-life of $T_{1/2} = 138.4\ d$ and the long-lived decay product, β-emitter ^{210}Pb with a half-life of $T_{1/2} = 22\ y$ will also be released.

There is no question that high doses of any kind of radiation are extremely harmful if not fatal. The dispute is about the existence or absence of a low-radiation dose threshold. While some maintain that such acceptable exposure limits exist, others claim that radiation is a potential health hazard even at the lowest doses. It is therefore necessary to investigate whether radiation can in principle be dangerous even in very small doses. These kinds of tests are challenging due to the lack of sufficient observational evidence as well as the inherent uncertainties associated with the inability to exclude other hazards that may be present besides low-dose radiation.

The view that radiation is harmful at any dose is held by many, including the Australian anti-nuclear activist Helen Caldicott (*1938).[38] Alternatively, there are many others who maintain that the perspective of physician and alchemist Paracelsus (1493–1541) applies to radiation: "Nothing is without poison. Only the dose determines whether a substance is a poison."[39] Before we turn to this question, however, the principal effects of radiation and its chemical reaction with biological systems should be discussed.

4.1.4.1 Ionizing Radiation and Chemical Effects

Radiation induces chemical reactions that in turn cause biological molecules to modify. As with nuclear reactions, chemical reactions are considered exothermic when energy is released and endothermic when energy is required to initiate the process of transforming a previously stable chemical system.

Radiation pumps a significant amount of energy into a molecular system and this can cause drastic changes. The impact of direct α-particle radiation can destroy molecules directly on impact, whereas γ- and β-radiation can be responsible for ionization processes. The classic example is the impact of radiation on water molecules, which is of particular interest since more than 70% of the human body consists of water.

When radiation hits water or body material, the radioactive particle deposits energy through collisions or other interaction processes along its path. This energy is transferred to the molecules of the transport medium, i.e., to the water molecules, via excitation or ionization processes as described earlier. The ionization process produces positively charged water molecules H_2O^+. Within less than picoseconds, neighboring water molecules react chemically, forming a so-called hydronium ion and a hydroxyl radical: $H_2O^+ + H_2O \Rightarrow H_3O^+ + OH$. Also along the particle path, lower-energy transfer processes lead to highly excited water molecules H_2O^*, which split directly into hydrogen and OH radicals: $H_2O^* \Rightarrow H + OH$. In both cases, OH radicals are formed that are extremely chemically active and can easily attach themselves to other molecules present in the body.

While ionization processes take place along the particle path, the largest energy transfer occurs toward the end of the path. This then is where the number of excited and ionized particles increases. The light particles, such as electrons or β-radiation, have an irregular trajectory and so can be directed in other directions by direct scattering processes with massive particles or by Coulomb interaction with local charge carriers. On the other hand, massive particles, such as α-particles or protons, follow their original direction of incidence and can hardly ever be deflected from their path. These particles are, however, stopped within one picosecond (Figure 4.9), while resulting ionized atoms and residual molecules gradually diffuse away from the original radiation path. And so, within nano- to microseconds the latter have spread over a radius of up to one micrometer.

Simulations in Figure 4.14 show that the number of reactive particles decreases rapidly. During this time period, rapid chemical reactions as described above take place between the ionized and excited residual molecules and the surrounding medium. This means that the radiation-induced chemical processes take place within a microsecond, in a volume with a 10 to 100 micrometer radius, along the particle path. The biological after-effects then depend on which and how many molecules have been affected. Although only a few reactions lead to after-effects, these effects can emerge within days or decades depending on the dose.

The more intense the radiation, the greater the chance of radiation damage. Damage is essentially caused by the chemical reactions described, which modify the chemical structure of the affected molecules. These chemical modifications have biological consequences and, at high doses, can lead to radiation sickness if not death due to the alteration or damage to critical organ cells.

As an example, Table 4.3 shows the approximate yield for the production and release of chemical radicals in one gram of body tissue, at an irradiation dose of 10 µGy. This dose corresponds to an energy of 0.624 MeV per gram deposited in the tissue. In the second column, the table shows the yield of the respective radicals per 100 eV energy loss, and the respective total yield is noted in the third column. This demonstrates that at 10 µGy more than six billion free water radicals are produced in one gram of body material. At a dose of approximately 1 mGy,[40] this is almost a trillion

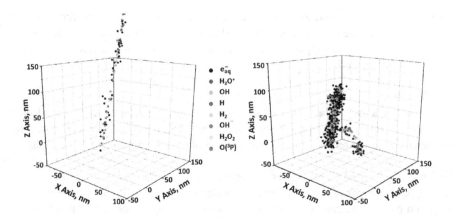

FIGURE 4.14 Simulation of the trajectory and ionizing effects for 10 MeV protons (left), which are considered typical for cosmic radiation, and 5 MeV α-radiation (right), which corresponds to the typical decay energy for α-particles from the natural decay series. The time period of the simulation corresponds to one picosecond. Courtesy Jay LaVerne, University of Notre Dame.

TABLE 4.3

Typical Yield of the Most Important Chemical Radicals by the Ionizing Effect of Radiation in One Gram of Body Tissue

Radical	Radicals/100eV	Radicals/1μGy
H*	3.2	$2.0 \cdot 10^9$
OH*	2.7	$1.7 \cdot 10^9$
e⁻	2.7	$2.0 \cdot 10^9$
H_2	0.45	$2.8 \cdot 10^8$
H_2O_2	0.7	$4.4 \cdot 10^8$
Average	1.95	$1.3 \cdot 10^9$

Source: Courtesy Jay LaVerne, University of Notre Dame.

---

free chemical radicals per gram that can trigger chemical reactions with body material and possibly lead to biological consequences.

A summary of the most important of the possible bio-chemical consequences is presented here because this information is necessary when considering the effects of radiation-induced processes on biological systems. Therefore, we will examine both the possible primary and secondary consequences of radioactive irradiation.

4.1.4.2   Basic Biological-Chemical Processes

Biological effects can be caused by either direct, primary radiation interaction with cell molecules or by the indirect, secondary chemical reactions of aggressive hydroxyl molecules with cell material. Cells are both the function and information carriers of biological systems and are therefore particularly sensitive to changes caused by radiation-related or other chemical reactions. While the primary reactions of radiation-induced ionization occur within picoseconds, secondary biochemical reactions take place over a much longer period, up to one second.

The cell is the basic component of living biological systems. Depending on origin and function, cells have different sizes, shapes, and structures. The oldest living organisms consist of single cells, while younger organisms such as plants and animals consist of multiple cells. In these multicellular life forms, subgroups of cells perform various functions. These groups depend on each other and exchange information via a complex neuronal network.

Cells also have a complex internal structure of proteins, enzymes, and other macromolecules such as DNA and RNA,[41] which mainly contain cell information and determine cell function. Radiation damage to the DNA can lead to cell death or mutation, with mutation changing the cell's long-term mode of action.

Biologists distinguish between two classes of cells: prokaryotes and eukaryotes. Prokaryotes do not have a cell nucleus, so the DNA is freely located in the cell's plasma and consists of a single double-stranded densely structured DNA molecule. On the other hand, eukaryotes are considerably larger cells characterized by a solid cell nucleus, which is enclosed in a membrane containing the DNA. Most known multicellular organisms are based on eukaryotes, including plants, animals, and humans.

DNA molecules have the well-known double helix structure, whose two strands consist of alternating deoxyribose phosphoric acid subunits, the so-called nucleotides. Each of these deoxyribose units (sugars) is bound to one of the four nucleobases: adenine, thymine, guanine, and cytosine (A, T, G, C). These bases are in pairs so that A is located in one strand opposite T while in the other strand, G sits opposite C. In this way, A-T and G-C pairs are formed. Hydrogen bonds hold the pairs together.

Cell or genetic information is contained in these base-pair sequences along the two DNA strands; these then are the genes. The human genome – the entirety of human genetic information – contains about 20,000 to 30,000 genes distributed over 3 billion DNA bases ($3 \cdot 10^9$). Genes and proteins within a cell form the chromosomes. Each human cell contains twenty-three pairs of chromosomes, which comprise the genetic information on cell function. The human body contains billions of cells whose function depends on their DNA structure and information.

Cell division is an important mechanism for the development of the human body, for the function as well as the lifespan of the cells. The success of the division processes depends on the correct transfer of information through the DNA. This is why radiation damage that occurs during often-protracted cell division processes can be particularly dangerous.

Cell division often takes place differently depending on the cell type. Somatic cells divide in two phases, in order to maintain bodily function: In the synthesis phase, division is prepared by copying the cell's internal DNA information and building up two chromosome copies; in the mitosis phase, the cell divides, and the chromosomes are distributed to two daughter cells. The division of hereditary cells, the so-called meiosis, is somewhat more complicated because a cell divides into four daughter cells with mixed genetic information.

During the synthesis phase, the DNA also duplicates: The original double strand separates into its individual strands; each individual strand serves as a template for the construction of a second strand. The base sequence of the original DNA strand is thus replicated in the sequence of the new, opposite strand.

Errors can occur during this copying process. One example is the so-called polymerase, in which the wrong base appears in the opposite strand and so induces a wrong base pair. This could result, for example, in G-C instead of A-T pairing. In this case, a new DNA strand is formed that changes the DNA information distributed by its carrier, the ribonucleic acid RNA. These processes are called *mutations*.

Mutations are changes in DNA that are retained even after division and thus pass on the modified genetic information. Such mutations are normally regarded as random errors, which can occur when the information of the original genetic-base sequence is being copied. However, such genome errors can also occur because of direct or indirect radiation or due to faulty DNA repair. The extent to which mutations occur stochastically (randomly) or are induced by external processes, such as chemical environmental influences or natural radiation exposure, will be discussed in more detail in Chapters 7 and 8.

In the case of mutations, a distinction is made between changes in somatic cells and in germ cells. Somatic mutations alter individual body cells. These mutations remain in place during further division and so can change cell function. However, somatic mutations are not part of the genetic material and are not passed on to the next generation. Somatic mutations can cause malignant changes in the cells and form tumors. If, however, the mutation affects germ cells – sperm or oocytes – the respective change can be transferred to the next generation. Germ cell mutations are particularly important for the evolution of biological systems. These mutations can develop new cell properties that affect the characteristics and capacity of subsequent generations.

Mutations can have negative as well as positive consequences. For example, mutations can lead to the death of an individual or the extinction of the species, in either a shorter or longer sequence. But mutations can also improve adaptation to living conditions or even develop a new species. This development has been intensively discussed in recent years, particularly in connection with the identification of biological units including cells as self-organizing systems[42] (cf. Chapter 7).

4.1.4.3   Biochemical Effects of Radiation

DNA, with its double-helix structure, is an extraordinarily complex biomolecule and as such, radioactive radiation can lead to various forms of damage. For example, DNA strands can be destroyed by direct interaction with radiation or there can be a change in the four base molecules: adenine, thymine, guanine, and cytosine. In the case of direct interaction, fractures can easily be repaired by enzymes, if they are single-strand fractures. In the case of double-strand breaks, there is the possibility of incorrect repair with information modification or losses in the DNA. Chemical changes, in the individual bases, may cause mismatches and wrong connections, base losses, or damage to the deoxyribose and phosphoric acid compounds. Chemical changes also entail the risk of breaks along the DNA strands and thus incorrect reassembling during subsequent enzyme repair. Several single breaks in the DNA strand can easily lead to double breaks if the damaged sites are close together. This can be expected at high radiation intensities.

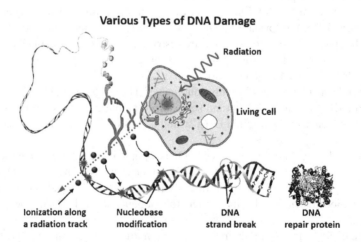

Various Types of DNA Damage

Radiation

Living Cell

| Ionization along a radiation track | Nucleobase modification | DNA strand break | DNA repair protein |

FIGURE 4.15   Effects of direct and indirect radiation on a DNA molecule series. Base sugar or phosphate bonds in the DNA strands, as well as base-pair stabilizing hydrogen compounds, can be destroyed in primary reactions. Secondary chemical reactions by free radicals can also cause destruction or modification. Such damage is usually repaired by cell enzymes, but the quality of the repair depends on the remaining information. Source: https://asrc.jaea.go.jp/soshiki/gr/eng/mysite6/index.html (May 23, 2023).

In addition to these primary processes, radiation-induced break-up of the body's water molecules results in the release of electrons, hydrogen ions, and hydroxide peroxide radicals (cf. Table 43), which then themselves initiate chemical reactions with the RNA and DNA molecules, potentially causing far-reaching changes. Secondary effects are the chemical reactions of these hydrogen radicals with DNA molecules. The difference in how energy is transferred, in these primary and secondary processes, causes the different correlations between the deposited dose and the potentially damaging effects (see Figure 4.15) as will be discussed in this section in more detail.

Direct interaction, between radiation and a DNA molecule, can destroy individual DNA strands and can also cause double-strand breaks. These damaging direct processes are characterized by an exponential relationship between dose and damage. If the initial number of undamaged molecules is M_0, it decreases exponentially with radiation dose D. The factor k in the exponent determines the efficiency of the radiation effect, which depends on the type of radiation as well as on the sensitivity of the respective organ:

$$M(D) = M_0 \cdot e^{-k \cdot D} \tag{4.5}$$

$M(D)$ is the number of remaining, undamaged molecules. Thus, the equation states that a small dose has only a small effect, although this effect can vary from organ to organ. This relation is based on the so-called target theory, according to which the probability of radiation interaction depends directly on the size of the molecule concerned. If an average energy of 75 eV is delivered to a molecule, there is a direct correlation between the efficiency factor k and the molecular weight of the affected molecule W_t, which in turn directly depends on the molecule size:

$$k = 1.37 \cdot 10^{-12} \cdot W_t \tag{4.6}$$

This means that the probability of primary radiation damage directly depends on the size of the molecule involved. DNA molecules are among the largest molecular components of a cell and are therefore very sensitive to primary radiation-induced damage.

After radiation-induced break-up of water molecules in the body, secondary indirect processes between radicals and the molecules in surrounding body tissue occur along the radiation path, as shown in Figure 4.14. As depicted, different kinds of radiation result in large differences in chemical impact. The most damaging of such processes would be chemical reactions of OH^+ radicals with the DNA or RNA molecules. Aggressive OH^+ radicals are chemically much more effective than free hydrogen ions or electrons and are therefore very quickly absorbed and chemically reintegrated, so that their average range is only 3.5 nm. This range is considerably smaller than typical cell sizes, which are in the micrometer range. Therefore, primary damage to a cell's water molecules has a direct effect on the cell's DNA and RNA molecules.

Secondary reactions include a broad range of chemical processes. One component is extraction processes that remove hydrogen atoms from the DNA or RNA molecule; in turn, these hydrogen atoms form hydrogen gas and water molecules. In dissociation processes, smaller molecule segments are split off, leading to structural changes in the molecules. In contrast, radicals such as OH^+ can also be incorporated into DNA sequences, also causing a change in the DNA structure. In addition, there is the already mentioned exchange of basic sequences leading to changes in DNA information. Further effects include interactions leading to single or double fractures in the DNA chain. In addition, transfer of a radical, from the base into the DNA chain, leads to loss of base function and can lead to the DNA strand breaking. As in the case of primary-induced breaks, this can easily lead to double-strand breaks within a single strand if there are several breaks close to each other.

Single-strand breaks are easy to enzymatically repair because the genetic function is retained and can be determined by copying the base sequence in the second strand. However, in the case of double-strand breaks, the process is no longer so simple; incorrect copies and associated changes in the information structure can follow. The radiation's dose-related mutation rate therefore depends not only on the dose but also on the efficiency of the repair mechanism in the respective cells. In the following, the function of the biochemical repair mechanisms will be presented in more detail.

4.1.4.4 Biochemical Repair Mechanisms

An internal, enzymatic mechanism detects damage to DNA and its components. Even in incidents of radiation-induced damage, this mechanism autonomously controls and repairs damage using the cell's own enzymes. These repair mechanisms may have arisen during evolution due to constant radiation exposure from natural sources, which initially was substantially higher than it is today. This repair mechanism is regarded as part of cellular development into self-regulating biological systems.

In addition to the enzymes required for these processes, cell plasma also contains substances that can bind and neutralize aggressive chemical radicals. However, the efficiency of these repair and protection mechanisms depends on the type of radiation and the extent of the damage. Within individual cells, the concentration of these repair enzymes and protective substances also represents an important factor in terms of repair efficiency.

Further critical parameters for an effective repair mechanism are determined by the cell's internal conditions. These include the respective phases of the cell's cycle as well as the energy and oxygen content of the cell. The faster the cycle, the less time available for the enzymes to repair. This explains the high susceptibility of blood cells,[43] since they have a short lifespan and are therefore subject to a rapid regeneration cycle.

Immediate cell repair takes place on the directly affected molecule when damage is locally confined and limited to only the DNA bases. With major damage, damaged segments are enzymatically removed from the DNA strand and re-synthesized. The template for this copy is the opposite, second strand – unless it has also been affected. If both strands are damaged, repair takes place after the DNA has been replicated. Chromatin, the information carrier, serves as a template for this repair. If chromatin is not available, the repair is carried out without a template. This often leads to genetic damage since the associated information is lost. The efficiency of DNA repair therefore depends to a large extent on the integrity of the cell material, since its information is necessary for the reliable transfer of information to the enzymes.

Cell sensitivity to ionizing radiation also depends on several other cell-specific physical and chemical parameters. Some cells have special protective chemical substances that intercept aggressive radicals, reducing risk. Other cells require oxygen exchange for repair. Their cell fluid contains free oxygen molecules O_2. Stimulated or ionized by radiation, these molecules form particularly aggressive radicals, such as perhydroxyl OH^- and peroxide O_2^{2-}, which dramatically increase the likelihood of secondary chemical reactions thereby reducing the amount of available oxygen for repair. Radiation biology speaks of this as the oxygen effect.

An additional critical aspect for the impact of the radiation effect on a cell is its cycle phase and the time of the radiation exposure. Cells are particularly sensitive to radiation in the mitosis or cell division phase. During this phase, the radiation effect often leads to cell death, which primarily causes changes in and destruction of the DNA material. These materials can no longer be repaired within the short time that remains in the cycle and so they replicate incorrectly with the next cell division. This can occur even at relatively low radiation doses.

Although the cell is less sensitive in the phase between cell divisions, massive disturbances in chemical metabolism and enzyme formation can still occur, including drastic changes or cell death at high doses. Due to this time-dependent sensitivity regarding cell formation, cell tissue with a particularly high regenerative capacity is far more sensitive to radiation than less cell-active body substances. In humans, this particularly affects the hematopoietic and lymphatic systems as well as the mucous membranes, where the first consequences of high radiation doses occur in the form of radiation sickness.

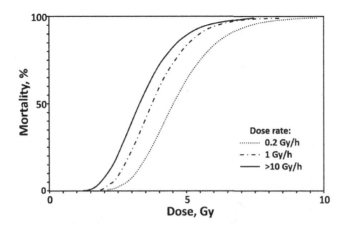

FIGURE 4.16 Mortality of body cells at three different dose rates, as a function of the whole-body dose. At high short-term doses, the 75% threshold is 4 Gy, but this shifts to 5.5 Gy at low dose rates. With a dose of more than 6 Gy, the chances of survival are practically zero. Source: https://www.wikipedia/commons/f/fb/Death_by_haematopoietic_syndrome_of_radiation_sickness-_influence_of_dose_rate.png (November 2, 2023).

The effects of radiation dose therefore depend on the time conditions of exposure. This causes a difference in radiation damage occurring at the same dose, which depends on whether exposure is spread over a short period of time or a long period of time – although the result is primarily explained by the efficiency of the cell-specific repair mechanism. All cells are affected by short-term high doses. At low long-term doses, cells with insufficient repair or repopulation possibilities show a higher sensitivity than cells with a rapid repair mechanism. This is because rapid-repair mechanisms have enough time to repair and regenerate cells between individual damage events. In general, radiation is more easily tolerated if the dose rate is low, as in when the radiation dose is distributed over a long period of time (cf. Figure 4.16).

The average cell mortality also depends on the exposure rate of a cell; the slower the exposure takes place, the higher the probability of successful cell repair. The 50% probability for survival occurs between 3.5 and 4.5 Gy. At higher doses, the chance of cell survival becomes increasingly low.

An interesting effect is the so-called bystander effect. This is observed at low radiation doses when cell damage occurs due to external processes, such as from direct radiation or other mutagenic substances. In these circumstances, not only is the affected cell damaged, possibly showing signs of mutation, but neighboring cells – the so-called bystanders – also start to prepare for defense and repair processes. This is interpreted as a sign for the existence of communication processes between the cells.

The carrier of this intercellular information exchange is not known, and the bystander effect does not occur with every type of cell material. To varying degrees, the exchange depends on radiation dose as well as cell type and function. In general, however, it can be said that this process is predominantly observed at low radiation doses.

In the following, radiation damage and effects occurring at relatively high doses will be presented.

4.1.5 Radiation Damage, Radiation Sickness, and Radiation Death

Discussions about possible radiation damage focus on several questions: What are the details regarding how high doses affect body function? At increased doses, what are the immediate and stochastic long-term effects? Are there threshold values below which no radiation damage can be detected or

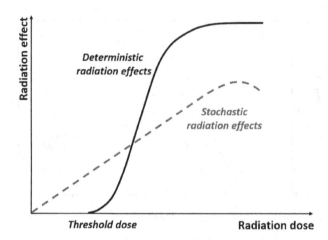

FIGURE 4.17   The effects of radiation schematically shown as a function of dose. In the case of short-term or deterministic damage, a threshold dose is assumed - below this, nothing can be detected. In comparison, a linear dependence on the dose is assumed for long-term or stochastic effects.

is there always a risk of possible long-term effects that can only be detected via long-term statistical studies of radiation victims?

A distinction is made between two classes of radiation damage, the so-called deterministic consequences and the stochastic radiation consequences. Deterministic effects occur primarily at high doses and reflect massive cell-function failure. The entire body's ability to function is significantly impaired, eventually causing the entire system to fail. Stochastic radiation effects, on the other hand, are the late effects of radiation that can occur with a certain probability due to somatic mutations. Figure 4.17 shows the relative probability for the occurrence of these effects in a simplified graph form.

As a result of natural radiation exposure in our environment, a balance between cell death and renewal has been established through repair or regeneration for all biological systems. A high radiation dose increases the possibility of cell death and shifts this balance. There is, however, a considerable difference in the effect when considering individual organs and their function; this effect depends on the physical-chemical characteristics of the respective cell structures, as described above.

Deterministic radiation damage represents early damage occurring within a few days to months after radiation exposure, while stochastic radiation consequences occur later. The short-term effects of deterministic radiation damage are well known and have been sufficiently studied. Especially American and Japanese physicians have been involved in studies examining victims of the atomic bombs dropped on Hiroshima and Nagasaki.[44] In the following decades, radiation-exposed military units, radiation workers, and those exposed due to radiation accidents in the civilian sector - within the framework of the US and Soviet weapons programs – have also been included in this category for study.[45] In addition, analyses and animal experiments have been added to the information pool.

In contrast, stochastic radiation consequences only occur after a longer time period. This is because they are based on biological processes that have a long incubation period. These effects are often linked to personal preconditions such as age, health, and living conditions, making general analysis and prediction difficult. Therefore, in the following, these two types of radiation-related damage will be considered separately.

4.1.5.1 Deterministic Radiation Damage

Deterministic radiation damage, often referred to as acute radiation damage, occurs within a short time after exposure to high doses of radiation. At high radiation-dose rates – i.e., high doses over a short period of time – tissue with high cell-regeneration rates is particularly affected. This includes the cells forming skin, the blood system, and the mucous membranes of the digestive or gastrointestinal system.

Late damage occurs after a few months to years. Cell systems with a lower renewal rate are part of this group, and damage is mostly associated with certain organs losing the ability to function. This includes the failure of glands, such as the salivary glands and sterility of the gonads. Another consequence is the formation of cataracts on eye lenses as well as hardening of other tissue units.

Radiation's influence on the skin shows symptoms similar to severe sunburn caused by UV radiation. However, there is a greater, deeper effect and the healing phase lasts longer. Low doses cause reddening of the skin, so-called erythema, which can lead to the formation of burn blisters (see Figure 4.18). As with prolonged exposure to UV light, skin is altered by prolonged low-dose irradiation. The tissue develops a parchment-like tanned appearance with carcinoma formations in particularly affected areas. In the extreme case of radiation ulcers, which spread over larger areas of the skin depending on the dose, only amputation of the affected limbs addresses the damage.

The best-known case of deterministic radiation damage is Marie Curie. As described earlier, her hands accumulated a considerable dose from the α-decay of radium, together with other long-lived substances, as a result of her work. Another case is that of radiologist Mihran Kassabian (1870–1910) who was unprotected as he worked with X-rays over many years. Kassabian's hands suffered severe burns due to daily high-dose exposure. It was presumably scientific curiosity that drove him to carefully document the radiation damage to his hands by photographing his skin's changes. One of his late-phase images (Figure 4.18) shows the destroyed skin and two finger-amputation sites.

Radiation poses a considerable risk to human blood and the hematopoietic system. Blood consists of white and red blood cells and blood plasma. White blood cells are known for leukocytes, which are important for the immune system, while thrombocytes or blood platelets ensure blood

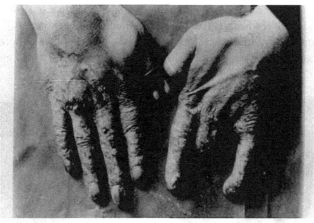

FIGURE 4.18   Left: Clear evidence of erythema in a Japanese radiation victim. This thirty-year-old man was 3.8 km away from the center of the Nagasaki bomb explosion. The picture was taken two months after the bombing. Source: https://www.archives.gov/news/topics/hiroshima-nagasaki-75 (November 2, 2023); Right: Mihran Kassabian's hands as an example of radiation-related long-term skin damage after years of handling radiological materials without protection. Source: https://www.wikipedia.org/wiki/Mihran_Kassabian#/media/File:Röntgen_rays_and_electro-therapeutics_-_with_chapters_on_radium_and_phototherapy_(1910)_(14755919334).jpg (November 2, 2023).

coagulation. Red blood cells or erythrocytes are essentially responsible for transporting oxygen from the lungs to the other organs of the body.

The most radiosensitive cells are lymphocytes, which are leukocytes. An average of 35 billion leukocytes are found in the blood of a normal adult, although there are considerable fluctuations. Thirty percent of lymphocytes have a lifespan between hours and months, depending on the type; thus, the regeneration phase is very short for many of these cells. This explains their sensitivity to radiation exposure.

Erythrocytes, on the other hand, are somewhat less at risk, as their lifespan is 120 days. However, 0.8% of the approximately 25 trillion erythrocytes in an adult's body are renewed every day, at a rate of 160 million erythrocytes per minute. This high regeneration rate means that there is not sufficient time for enzymatic processes to repair after radiation damage. This in turn greatly contributes to the blood system's sensitivity to radiation.

Blood stem cells or hemocytoblasts are found in red bone marrow and are responsible for the formation of most blood cells, ensuring constant reconstruction of the blood. These blood cells regenerate continuously and are therefore also extremely sensitive to radiation damage. In addition, although platelets or thrombocytes are not cells, their chemical structure does make them vulnerable to radiation. Thrombocytes remain in the blood for five to eleven days before being broken down in the liver and spleen.

Even a relatively low dose of radiation, 0.25 to 1 Gy, can lead to changes in the lymphocyte count, with internal or external bleeding due to a lack of platelets. These hematological symptoms are the dominant effect in whole-body radiation doses up to 10 Gy. With blood-count changes, the immune system's defense leads to point bleeding in the entire body, so-called purpura (Figure 4.19). The lack of red blood cells is, on the other hand, reflected in the resulting anemia.

The gastrointestinal tract, i.e., the digestive system, is also in grave danger at high radiation doses. This is particularly true for the mucous membranes of the intestinal system, whose cells have a high regeneration rate of three to five days. At higher doses between 10 and 50 Gy, the destruction of these mucous membranes in the gastrointestinal tract will start instantaneously, destroying the ability of the membranes to absorb water and important electrolytes without which the digestive process will fail to function. The associated symptoms of the most severe gastrointestinal disorders are loss of appetite, nausea, vomiting, and severe diarrhea with significant fluid loss. Destruction of the epithelium, the cell layer protecting the inner-intestinal mucosa from the stomach's digestive juices, leads to ulcer formation and self-digestion of the intestinal walls.

Radiation doses of more than 50 Gy lead to the decomposition of proteins and, thus, to slow self-poisoning and circulatory failure up to death. This dose range also attacks long-lived nerve cells.

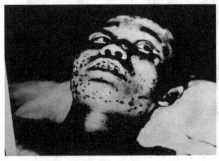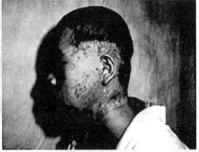

FIGURE 4.19   Examples of purpura on Japanese bomb victims of Hiroshima. The twenty-one-year-old man on the left was about one and a half kilometers from the center of the bomb explosion; he died after twenty-eight days. At the time of his death, the number of leukocytes was 2 million, the number of erythrocytes 2 million, and the number of thrombocytes 10,400. Nothing is known about the fate of the victim shown on the right. Source: https://www.archives.gov/news/topics/hiroshima-nagasaki-75.

TABLE 4.4

Characteristic Symptoms and Health Consequences of Whole-Body Dose Radiation Administered over a Short Time

Equivalent Dose	Symptoms
Up to 0.25 Sv	No clinically detectable immediate effects. Late effects cannot be excluded.
0.25–1.0 Sv	Slight change in blood count. Headache and increased risk of infection. Late effects cannot be excluded.
1–2 Sv	Nausea and fatigue within a few hours after radiation, lasting a few hours up to one day – mild to moderate nausea (50% probability with 2 Sv) with occasional vomiting. Acute change in blood count due to decrease in leukocytes. Symptoms last about four weeks and consist of loss of appetite (50% likely at 1.5 Sv), malaise, and fatigue (50% likely at 2 Sv). Temporary infertility in men. Possible late effects: Cataract formation, anemia, development of malignant tumors, and leukemia. 10% deaths after 30 days (lethal dose (LD) 10/30).
2–3 Sv	Nausea occurs 100% at 3 Sv, occurrence of vomiting reaches 50% at 2.8 Sv. Other symptoms: Hair loss (50% likely at 3 Sv), malaise, and fatigue. Massive loss of white blood cells, risk of infection increases rapidly. In women, permanent sterility begins. 40% deaths after 30 days (LD 40/30).
3–6 Sv	Nausea and vomiting after a few hours, after a week: hair loss and fever, internal bleeding, and purpura. After two to three weeks, inflammation of the pharynx as well as stomach and intestinal ulcers. 50% deaths after 30 days (LD 50/30).
6–10 Sv	Nausea, vomiting, and diarrhea after a few hours. Within a few days diarrhea, internal bleeding, purpura, and inflammation in the gastrointestinal system. Stomach and intestinal tissue severely damaged, and bone marrow almost or completely destroyed. Initial symptoms within 15–30 minutes, lasting up to two days. Final phase with death by infection and internal bleeding. 100% deaths after 14 days (LD 100/14).
10–50 Sv	Immediate nausea and severe weakness. Rapid cell death in the gastrointestinal tract with massive diarrhea, intestinal bleeding, and water loss as well as electrolyte balance disorders. Death due to circulatory failure with fever, delirium, and coma. 100% deaths after 5 days (LD 100/5).
Above 50 Sv	Immediate disorientation and coma within seconds or minutes. Death after a few hours due to complete failure of the nervous system.

Source: https://www.cdc.gov/nceh/radiation/emergencies/arsphysicianfactsheet.htm (November 4, 2023).

This can have a direct effect on the entire nervous system, such that the brain's substance is attacked by the edema that forms and the cardiovascular system's control mechanisms are disturbed. At such high doses, death occurs after only a few hours.

These symptoms, caused by different radiation doses, all fall under the term *radiation disease*. While symptoms vary with different radiation exposure, at high doses all of these symptoms must be expected to occur at the same time. Typical effects are summarized in Table 4.4.

4.1.5.2 Stochastic Radiation Effects

Stochastic radiation effects are the possible late effects of radiation, which occur with a certain probability. These effects depend on the dose and on the physiological condition of the affected person. Stochastic damage is caused by radiation damage to the DNA of individual cells, leading to somatic mutation – the degeneration of organs or organ function – or, possibly, hereditary mutations in germ cells. Mutations are changes in the genetic code, which has been incorrectly encoded during cell division. In this way, respective genetic information changes, potentially leading to changes in the transmission of information for organ function or to the genome.

Mutations in the genetic code can occur for various reasons, including the random errors in transmission of genetic information that occur in all transmission processes. On the other hand, external influences can also lead to changes in DNA structure. There is a broad spectrum of possible mutation-triggering processes: chemical reactions between the DNA and externally supplied, foreign chemical substances (carcinogens); virus infections; non-ionizing ultraviolet radiation; the body's own ionizing, or also external radioactive radiation. The latter either contributes to the destruction of DNA sequences or leads to the formation of aggressive chemical radicals, which in turn react chemically with DNA molecules. The combination of random and internal processes leads to natural mutation rates in somatic and hereditary cells, with an incubation period or latency period between the first mutation and the onset of subsequent symptoms.

The natural mutation rate of higher organisms (mouse, human) is about $1.0 \cdot 10^{-8}$ to $2.5 \cdot 10^{-8}$ mutations per generation and base molecule.[46] On average, 99% of these mutations are repaired by the actions of enzymes or proteins, while 1% are maintained. The effective mutation rate is therefore $1.0 \cdot 10^{-10}$ to $2.5 \cdot 10^{-10}$ mutations per generation and base molecule. Assuming that the human genome contains approximately $3 \cdot 10^9$ bases,[47] an average of 0.3 to 1.0 miscopies occurs during the replication of the entire genome. The question remains as to how many of these mutations cause somatic or hereditary long-term consequences.

Under certain conditions, somatic mutations can lead to tumor formation in solid organs or to leukemia in blood. These types of cancer are usually caused by an accumulation of natural mutations within a cell, which can lead to the breakdown of its functionality. A single mutation is not enough. Progressive, uncontrolled growth of cell material occurs through various gene effects, for example, through damage to the repair enzymes which then multiply with each cell division. This change, from the initial genetic disorder to the development of a tumor, generally requires a latency period of several years. The latency period is determined by the time scale of the successive cell-division processes. In most cases, the probability of tumor formation increases with age.

In cases of genome mutation, transmission of altered information depends on the time scale of the generation change. These mutations are a natural process that determine evolution in biological life. Without DNA mutations there would be no further development in biological systems. Development is determined by the fact that negative genetic material – which is produced by mutation and damages body functionality – dies off, while positively mutated genetic material, which improves body functionality, has a greater chance of survival. These statistical selection processes in DNA mutations reflect Darwin's theory of evolution and are considered to explain multiplication and selection in biological species[48] (see Chapter 7).

Additional radiation may increase the stochastic mutation rate, which can result in possible germ cells for tumor formation or changes in the genome. This seems to be the case for the development of leukemia. However, this radiation can also act as an additional factor in which the probability of mutations increases multiplicatively – especially if several factors contribute to mutation-related damage such as the formation of tumors.

In general, the use of the LNT (linear no threshold) model, for determining the effect of radioactive radiation, assumes a linear relationship between dose and effect (see Chapters 4 and 8). Thus, even at low doses, there is the assumption of a certain probability of genetic damage. Other views assume improved cell-repair capacities at low doses and thus expect a hormesis effect (see Figure 4.2). Since the statistical basis for this view is not considered sufficient, the risk-minimizing approach is applied when formulating radiation protection laws. This is true despite the often misunderstood and complex mechanisms that are part of accurately comprehending radiation's affect, such as the bystander effect and the direct enzymatic-repair process. Therefore, the linear LNT model remains and is implemented without regard for a threshold value.

Among the most frequently discussed long-term effects of radiation are the development of tumors, or other cancerous phenomena in organs, and mutations in the genome leading to dramatic malformations in subsequent generations. Both possibilities are vividly discussed in the wake of reactor accidents such as Three Miles Island, Chernobyl, and Fukushima Daiichi. In principle, these consequences can also occur with long-term radiation exposure at relatively low dose rates.

It is frequently suggested that these effects can also be a consequence of tumor irradiation, as part of radiological leukemia therapy.[49] In addition, it is being suggested that such effects might cause increased cancer rates resulting from radiation-related environmental influences, such as suggested for cosmogenic radiation in mountain dwellers and radiogenic radiation in miners. The possible influence of residual radiation, on the local population, from nuclear power plants is also often cited. In addition to an increase in the cancer rate, an increase in the hereditary mutation rate may be possible, thus increasing the rate of miscarriages and malformations in the following generation and influencing evolutionary processes in the longer term.

Estimates, for long-term radiation damage and resulting mutations, are mathematically complicated and subject to considerable uncertainty. This ambiguity is partly explained by the lack of data as well as uncertainties in the available data itself. In addition, different statistical methods are used for data analysis and utilization. These methods evaluate different risk factors, factors which – in addition to the radiation effect – can affect tumor formation in slightly different ways. Mathematical models are then developed based on this data.[50]

Today's predictions are mainly based on the model proposals of the following institutions: Board on Radiation Effects Research (BEIR 2006) in the National Research Council of the US National Academies, International Commission on Radiological Protection (ICRP 1991), National Council on Radiation Protection and Measurements (NCRP 1993), Environmental Protection Agency (EPA 1994, 1999), and the United Nations Scientific Committee on the Effects of Atomic Radiation (UNSCEAR 2000, 2006, 2008, 2013). The following statements on the effects of somatic and hereditary mutations are based on the BEIR report of 2006 and the UNSCEAR report published in the same year.

4.1.5.3 Effects on Tumor Formation and Leukemia

In addition to random mutations, the chemical influence of carcinogenic substances and the natural environmental components of radioactive radiation determine the tumor rate in natural or non-event-related occurrences. The influence of carcinogenic substances in our environment dominates the cancer rate. However, an increase in radiation exposure such as by anthropogenic radiation sources will change the dose exposure rates. These changes may increase the radiation-induced processes leading to cell damage, causing it to become the dominant factor for leukemia and the formation of tumors. But at what dose is this threshold reached and how does it compare to the dose we are exposed to in daily life?

Cells affected by cancer are the so-called somatic cells. While they are not part of the genetic material, affected somatic cells can lead to organ mutation and change in function. It should be noted that the probability of tumor formation is smaller than the probability of an original change in the cell structure. This is because most cells can be successfully repaired, while more seriously damaged cells die so that, in either case, damaged cells are not available for tumor formation. At a lower dose, the risk of radiation-induced tumor development is relatively low, at least much lower than the risk of carcinogenic environmental influences such as air pollution, malnutrition, and smoking.

According to the LNT model, risk increases linearly with radiation dose. This means that even at low internal doses, the risk of induced mutation increases steadily (Figure 4.2).[51] The typical latency period for tumors can last from years to decades, depending on the age of the affected person. With a high radiation dose of more than 3 to 6 Sv, the chances of cancer developing later are relatively high, but at this dose, with increasing probability, deterministic radiation effects lead to relatively rapid radiation death before tumors can form. For these reasons, rapid tumor development due to high dose rates is extremely unlikely.

The low number of radiation accidents, in which many people were exposed to high doses, limits the possibility of reliable statistical analysis. The main source for our current information on the long-term effects of radioactive radiation at high-dose rates continues to be the victims of Hiroshima and Nagasaki[52] and, to a lesser extent, Chernobyl.[53] However, animal testing (mostly on mice and rats) largely complements the information obtained from such analyses. In addition, a growing number of Chernobyl victims are being examined, although access to data regarding the relatively speaking, highly contaminated liquidators is limited. At present, no conclusions about long-term stochastic effects can be drawn regarding the radiation victims of the reactor catastrophe at Fukushima Daiichi.

From 1950 to 1985, observation of Japanese bomb victims was first carried out by the American Atomic Bomb Casualty Commission and later with the greater involvement of Japanese institutions, by the Radiation Effects Research Foundation (RERF). Investigations included 91,000 people, 76,000 after 1986. Of these, 50,000 persons received a dose of less than 0.1 Sv, while there were 17,000 doses from 0.1 to 1.0 Sv. An additional 2,800 persons received potentially life-threatening doses from 1 to 4 Sv. Of those who were observed, 240 persons were exposed to a dose of more than 4 Sv.

These were, of course, people who were still alive in 1950 at the beginning of the observation period. Since most of the radiation victims who were exposed to high doses had died shortly after 1945, from the short-term effects of radiation, they obviously could not be part of these investigations. Therefore, the distribution of victims statistically recorded in this long-term study reflects, in a certain sense, the dose-dependence of the mortality rate. In these studies, a distinction was made between leukemia as a blood disease and tumor development in other organs.

Such investigations must also consider or correct for the natural human cancer rate. These cancers can be caused by other mutations such as stochastic errors in DNA division or chemical processes. Since 1950, 86,611 of the radiation victims from Hiroshima and Nagasaki with reliably estimated irradiation values have been constantly monitored for the formation of tumors. Fifty years later in 2000, 10,127 people had died of cancer; this is 12% of the group. In comparison groups not exposed to high doses, the natural mortality rate from cancer over this period was 9,647 cases. Accordingly, when computing the difference between these cancer rates, 479 people died from radiation-related cancer. This is 0.55% of the original study group, which is about four times the statistical uncertainty in the numbers.[54] Statistics are similar for leukemia. Of the 86,955 people whose initial radiation dose was sufficiently known, 296 died of leukemia in the following fifty years; this is 0.34%. When compared with the statistically expected leukemia rate of 203 cases, this is an additional rate of ninety-three persons or 0.11% of the persons who were exposed to a measurable radiation dose.

These first studies have since been supplemented by a series of other studies, in particular the Japanese Life Span Study (LSS) which attempts to cover a wider range of radiation damage. These results are regularly published in a series of reports called LSS Reports.[55] The LSS studies, recorded from 1950 to 2003, include 58,500 survivors from Hiroshima and 28,132 from Nagasaki. The reports contain important statistical information such as location, estimated radiation dose, time of death, and type of death. The excess relative risk (ERR) is then derived from the death rate and the organ affected by cancer.[56]

A mathematical-statistical model with different weighting factors was used. In the observation period from 1950 to 2003, a total of 58% of affected persons died. Of persons who were more than forty years old at the time of the bombing, 99.6% died within the observation period. In 2003, almost 80% of people under twenty years of age were still alive. It was found that 20% of the deaths were due to cancer and 1.4% to leukemia and blood disorders. The majority, 71% of those affected, died from other causes of illness, with 5% dying as a result of accidents.

Figure 4.20 shows the risk of tumor development due to additional radiation, as determined in the study. The linear increase of risk at higher doses is clearly visible. In the lower dose range, statistical uncertainties are too large for clear statements. Several theoretical models, for extrapolating the radiation risk in this low-dose range, lead to different risk assumptions. These uncertainties lie in the dose estimation, which can only be estimated by comparison with other known consequences of radiation such as hair loss.

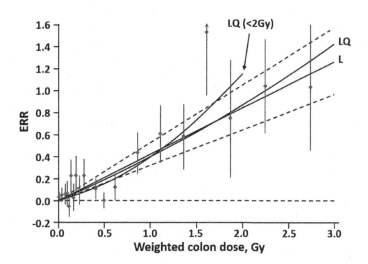

FIGURE 4.20 Figure from the LSS14 study, showing the excess relative risk (ERR) as a function of the whole-body dose. It was assumed that the more easily determined intestinal dose corresponded to the whole-body dose. Different lines correspond to a mathematical function for extrapolating the observed data in the low-dose range (L: linear; LQ: linear-square). Source: DOI: 10.1667/RR2629.1.

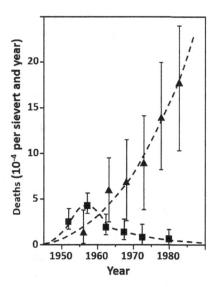

FIGURE 4.21 The risk of cancer formation, represented as deaths per sievert and year, for radiation victims from Hiroshima and Nagasaki in the years after the August 1945 bombing. The triangles show the deaths due to tumor formation while the squares indicate the deaths due to leukemia. While the death rate due to tumor formation increases steadily, the curve for leukemia-related deaths reaches a maximum after more than ten years and decreases steadily after about ten years. Redrawn based on : Krieger, (2007), 502.

In all of these studies, the majority of radiation victims had received only a very small dose. Therefore, this radiation-related additional mortality risk for different dose rates must be considered as a function of the date of death (Figure 4.21).

Today, the extreme difference in the risk curves is interpreted within the framework of two different risk models. These form the basis for model estimates regarding major radiation accidents. The model of leukemia behavior corresponds to the so-called additive model. In this model, the radiation-related leukemia rate is added to the natural leukemia rate. The curve demonstrates how radiation-induced leukemia occurs after a typical mutation-related latency period of two to five years, reaching a maximum mortality probability after ten years. The curve then falls back to the natural leukemia rate, which increases steadily with the age of the affected person. The probability of radiation-related leukemia is therefore directly proportional to the radiation dose. This proportion is added to the probability of age-related leukemia.

Different conditions are used in determining the risk of developing cancer tumors. The radiation-induced tumor rate does not show an additive maximum after the expected latency period of ten years but rather, increases steadily with dose rate and time. Thus, the radiation-induced tumor rate is proportional to the rate of spontaneous age-related tumor formation in individuals who have not been exposed to elevated radiation doses. This behavior is described by the so-called multiplicative risk model, which is now considered valid for all tumor types. Implementation of this model indicates that additional radiation exposure increases the number of tumor germs.[57] Age-related effects lead to an increased probability of tumor formation, which in this model follows the pattern of spontaneous tumor formation.

Studies of tumor rates following the reactor accidents at Chernobyl and Fukushima have not yet been completed. First analyses are contained in the UNESCAR reports compiled from the International Atomic Energy Agency, WHO, and UN. After Chernobyl, the average dose – for the population in Belarus, Russia, and Ukraine – was 52 mSv, of which 30 mSv was due to external radiation exposure and 22 mSv to internal radiation exposure; these are average figures with significant local variations. The average exposure of the liquidators was 100 mSv.

There are still major statistical uncertainties regarding the long-term health effects. This is primarily because a precise assessment of the health effects was only carried out for the liquidators. After two to three years, the leukemia rate appears to have risen to two to three times the normal rate. The US National Cancer Institute has summarized several studies[58] and currently cites 110 cases of leukemia from the 110,000 liquidators who were involved in the clean-up work.[59] The 2006 WHO report cites comparable figures for liquidators but also says that no increased leukemia rate was found in the total population of the affected areas.[60] Overall, results seem to confirm the expectations predicted by the Hiroshima and Nagasaki studies.

Fifteen years after Chernobyl, the UNSCEAR report from 2000 shows no observable increase in tumors and neither does the WHO study from 2006. This is probably due to the long incubation period necessary for tumor development. According to these summarizing analyses of the various available data and studies, longer-term studies seem necessary to provide statistically reliable information.

In contrast to these official reports, an unofficial summary was published in 2011 by a group of Ukrainian and Belarusian physicians.[61] According to the authors, their information is based on several local studies and data records, which were largely written in Russian and therefore are not easily accessible to the public. This work paints a dramatically different picture of the consequences of Chernobyl, particularly noting an increase in the incidence of cancer, infant mortality, and birth deformities in Belarus and Ukraine. Other consequences discussed are premature aging and higher mortality rates in adults. In addition, the death of more than 100,000 liquidators (by 2005) is predicted.

This information has been denied by official authorities. However, the authors claim this to be a part of the official misinformation and concealment tactics of the involved governments. The dramatic results do, however, contradict those of the Hiroshima and Nagasaki studies. In addition, the results suggested by the Yablokov report are disputed by many other studies.[62] The original editor, the Academy of Sciences of the State of New York, also dissociated himself from the study. Nonetheless, the study does form the basis for an expanded presentation against nuclear war, which is promoted by German physicians.[63] This work also reports a considerable increase in leukemia and cancer rates in Germany, especially in Bavaria. There are several other publications that evaluate the Yablokov statements as fact.

However, taking the quoted data at face value, one finds that it shows considerable statistical uncertainties. This is mainly due to a lack of reliable information, which is indispensable for a precise epidemiological study. Neither the exact radiation dose nor the exact location of the persons affected at the time of irradiation are documented with certainty. For example, an increase in thyroid-gland tumors was suggested, but the exact correlation between dose and effect was not demonstrated with sufficient accuracy. Along the same lines, if it is claimed that 1,000 to 3,000 additional children are deformed at birth, this is 40 to 80 children per year. These incubation rates are considerably higher than other quoted studies expected, which suggests that the Yablokov report has either not ignored epidemiological techniques and corrections or willfully overdramatized the data.

As with the Chernobyl data, a similar situation is occurring in the health analysis of the 25,000 radiation and clean-up workers of Fukushima Daiichi. Statistically speaking, only a few years have passed since the accident in 2011 and UNSCEAR reports from 2013 state that no radiation-related illnesses or deaths had been observed to date. However, a recent study suggests a gradual increase in tumor rates after 2008.[64] This development needs to be followed.

4.1.5.4 Long-Term Genetic Effects

In addition to the stochastic mutation rate of DNA molecules in male or female germ cells (sperm or oocytes), additional radiation-induced mutations can also occur in the germ cells themselves. In popular literature and the film industry, mutations have led to such dramatic changes as the Hulk and Spider-man. Godzilla has become synonymous with the consequences of the Hiroshima bombing and the subsequent bomb tests in the Pacific. Reality is, however, much different.

At the beginning of this chapter, it was estimated that an average of 0.3 to 1.0 miscopies occur each time a genome is replicated. However, the replication rate of the genome determines how many of these miscopies are passed on to the next generation. In order to take form, the female egg requires 30 replications, the male sperm about 400. This results in 9 to 18 mutations (0.3 to 1.0 x 30) per egg cell and 120 to 400 mutations (0.3 to 1.0 x 400) per sperm cell.[65] So, each zygote produced during fertilization should contain between 129 and 418 mutations. Assuming an average value of 300 mutations per zygote[66] and a current world population of about 7 billion people, each new generation has $300 \times 6 \cdot 10^9 = 1.8 \cdot 10^{12}$ or about 2 trillion mutations per generation. This means that the genetic structure of humanity is subject to constant mutation and every nucleotide in the human genome mutates every twenty years.

In view of the genome's considerable natural-mutation rate, it is difficult to detect additional radiation-induced mutations in germ cells. This is particularly true when considering the low doses involved in medical applications occurring over long periods of time. Even when considering the large number of descendants of the survivors of Hiroshima and Nagasaki, no statistical mutation rate could be determined with certainty.[67] The descendent generation was examined for many possible genetic defects such as pregnancy rate, miscarriages, infant mortality, growth effects, tumor rate, and chromosome defects. The number of children examined whose parents experienced an increased dose rate was 31,000, while 41,000 children of parents without additional radiation dose served as the control group. Yet, even with these well-defined parameters, the statistics are subject to error.

Today's information on genetic mutations that can lead to birth defects, chromosomal abnormalities, and other genetic abnormalities is based solely on animal experiments performed on mice and rats, which are chosen due to their high number of offspring. Considering the low average dose, of 0.14 Gy, received by long-surviving atomic bomb victims in Japan, the lack of observational data is understandable. Available data corresponds to predictions based on experiments with mice, and this data at least shows that the human genome does not have a higher sensitivity to radiation-related damage than that of smaller mammals.

The probability of radiation-induced genetic changes in the human genome is assumed to be 1% per Sv. Of these, 0.15% per Sv led to dominant changes in the first subsequent generation and a further 0.15% per Sv sequelae into the second generation and can thus be regarded as long-term mutations.

NOTES

1. Physical units for the dose are labeled as gray or sievert, which correspond to energy per mass or joule per kilogram as outlined in Section 4.1.2.
2. An interesting and depressing summary, on the introduction of the LNT method within the political environment of the 1950s, is given by Jack Devanney, *Why Nuclear Power Has Been a Flop*, 59, 120. See also Edward Calabrese, "Linear Non-Threshold (LNT) historical discovery milestones," *La Medicina del Lavoro* 113 (4), (2022): e2022033.
3. Annual dose, average over a five-year period.
4. https://www.ncbi.nlm.nih.gov/pmc/articles/PMC4519811/ (May 23, 2023).
5. http://www.nrc.gov/about-nrc/radiation/health-effects/info.html#dos (May 23, 2023).
6. A summary account of the first fifty years can be found in William J. Schull, *Effects of Atomic Radiation: A Half-Century of Studies from Hiroshima and Nagasaki* (New York: Wiley-Liss, 1995). Investigations still continue today, although most victims of the attacks are now deceased.
7. Report of the National Research Council, *A Review of the Dose Reconstruction Program of the Defense Threat Reduction Agency* (Washington, DC: National Academies Press, 2003).
8. During the first atomic bomb explosion in China, a cavalry attack was mounted – for the sake of a better impression.
9. However, only a small percentage of these soldiers are believed to have been close enough to ground zero to have been exposed to direct neutron- and γ-radiation.
10. Report of the UN Chernobyl Forum, *Health Effects of the Chernobyl Accident and Special Health Care Programs* (Geneva: World Health Organization, 2006), http://www.who.int/ionizing_radiation/chernobyl/whochernobylreport_2006.pdf (May 23, 2023).
11. T. Ohnishi, "The disaster at Japan's Fukushima-Daiichi nuclear power plant after the March 11, 2011 Earthquake and Tsunami, and the resulting spread of radioisotope contamination," *Radiation Research* 177 (2012): 1–14.
12. UNSCEAR 2013 report, *Sources, Effects and Risks of Ionizing Radiation* (New York: United Nations, 2014).
13. T. D. Luckey, "Radiation hormesis overview," *RSO Magazine* 8 (2003): 22–41.
14. The most recent BEIR report was published in 2006. National Academies, "Health risks from exposure to low levels of ionizing radiation," *BEIR VII Phase 2* (Washington, DC: National Research Council, 2006).
15. E. J. Calabrese and M. K. O'Connor, "Estimating risks of low radiation doses – a critical review of the BEIR VII report and its use of the linear no-threshold (LNT) hypothesis," *Radiation Research* 182 (2014): 463–474 and Kevin D. Crowley, Harry M. Cullings, Reid D. Landes, Roy E. Shore, and Robert L. Ullrich, "Comments on estimating risks of low radiation doses—A critical review of the BEIR VII report and its use of the Linear No-Threshold (LNT) hypothesis by Edward J. Calabrese and Michael K. O'Connor," *Radiation Research* 183, 4 (2015): 476–481.
16. Mohan Doss, "Linear no-threshold model vs. radiation hormesis," *Dose Response* 11, 4 (2013): 495–512.
17. Egon Pohl and Johanna Pohl-Rüling, "Die Strahlenbelastung der Bevölkerung in Bad Gastein, Österreich," *Berichte des Naturwissenschaftl. Vereins Innsbruck* 57 (1969): 95–110.
18. Dosimetry units gray (Gy) and sievert (Sv) will be introduced and discussed later in Section 4.1.2.
19. Klaus Becker, "Health effects of high radon environments in Central Europe: Another test for the LNT hypothesis?" *Nonlinearity in Biology, Toxicology, and Medicine* 1 (2003): 3–35 and Peter Deetjen, Albrecht Falkenbach, Dietrich Harder, Hans Jöckel, Alexander Kaul, and Henning von Philipsborn, *Radon als Heilmittel* (Hamburg: Verlag Dr. Kovač, 2005).
20. William F. Morgan and William J. Bair, "Issues in low dose radiation biology: The controversy continues. A perspective," *Radiation Research* 179, 5 (2013): 501–510.
21. Ionizing radiation carries enough energy to ionize materials and break up chemical bonds, releasing ionized chemical fragments into the biological system which could bond with large-scale body molecules.
22. There are also muons, a special type of massive elementary particles belonging to the lepton class, along with electrons and neutrinos.

23. The number depends on the intensity of the radiation flux.

24. In the United States, however, the dose is frequently given in another unit, the rad. This historical unit is no longer internationally valid, although it is still often used in the Anglo-Saxon linguistic area; 1 rad = 0.01Gy.

25. In the United States, however, the rem is still found as the standard unit for the equivalent dose, which results from the dose unit rad; 1 rem = 0.01 Sv.

26. The effective dose is approximately equal to the equivalent dose but has been obtained by a refined organ-specific method.

27. https://www.epa.gov/radiation/radiation-sources-and-doses (March 31, 2023).

28. According to the German Federal Ministry of Health, an X-ray check at the dentist corresponds to a dose of about 10 μSv. However, with about 0.6 X-rays per inhabitant and year, Germany occupies a top position in radiation exposure due to dental treatment.

29. https://en.wikipedia.org/wiki/Radiation_exposure (March 31, 2023).

30. This figure has probably declined in recent years due to the rapid change from the traditional tube television to the modern LCD flat screen. Tube screens work with high-energy cathode rays (electrons) that generate X-rays on common devices. Therefore, their dose had to be legally limited to 1μSv/h at a distance of 10 cm.

31. Absorption and the resulting increase in intensity with altitude can be calculated using the absorption formula introduced earlier, in Section 4.1.1.

32. http://doris.bfs.de/jspui/bitstream/urn:nbn:de:0221-201108016029/3/Bf_2011_BfS-SG-15-11Exposit ionFlugPersonal.pdf (May 25, 2023).

33. Guadie Degu Belete and Yetsedaw Alemu Anteneh, "General overview of radon studies in health hazard perspectives," *Oncology* 2021 (202): 6659795.

34. https://www.epa.gov/radiation/radiation-sources-and-doses (November 6, 2021).

35. Albert Einstein was awarded the Nobel Prize in 1921 for the quantum theoretical explanation of the photoelectric effect, as discovered by Philip Lenard, and not for the development of the theory of relativity, as is often assumed. This was because, according to the original conditions for the Nobel Prize, the work to be honored should have an application-oriented benefit.

36. American Cancer Society, *Global Cancer Facts & Figures*, 2nd Edition (2011), American Cancer Society, http://www.cancer.org/acs/groups/content/@epidemiologysurveilance/documents/document/acspc-027766.pdf (May 26, 2023).

37. https://www.epa.gov/radtown/radioactivity-tobacco (January 21, 2023).

38. https://www.nytimes.com/2011/05/01/opinion/01caldicott.html and https://www.helencaldicott.com/about/ (April 1, 2023).

39. Paracelsus, whose real name is Philippus Aureolus Theophrastus Bombastus von Hohenheim, was born as the son of a Swabian doctor in what is now the Swiss town of Einsiedeln. He is regarded as one of the most famous natural scientists and alchemists of the Renaissance and as the founder of toxicology as an independent science. The sentence comes from the book *Septem Defensiones* written in 1538 and reads in the original Latin formulation, "Omnia sunt venena, nihil est sine veneno. Sola dosis facit venenum."

40. This value is only an estimate based on the annual equivalent dose.

41. DNA (deoxyribonucleic acid) carries RNA (ribonucleic acid) and genetic information on the somatic cell function of the various organs of the body as well as the genetic information of the living organism.

42. A self-organizing system is an open system that, by exchanging energy and materials with its environment, changes its basic structure as a function of experience and environment.

43. Tumor cells are also subject to very rapid cell division and proliferation processes. For this reason, radioactive irradiation has a greater effect on tumor cells than on neighboring, slowly mutating, or long-lived cell material.

44. Ashley W. Oughterson and Shields Warren, *Medical Effects of the Atomic Bomb in Japan* (New York: McGraw-Hill, 1956) and William J. Schull, *Effects of Atomic Radiation: A Half-Century of Studies from Hiroshima and Nagasaki* (New York: Wiley-Liss, 1995).

45. Report of the National Research Council, *Assessment of the Scientific Information for the Radiation Exposure Screening and Education Program* (Washington, DC: National Academies Press, 2001).

46. The exact number of mutations is not yet known. While different methods in extrapolation produce slightly different numbers, all are within the given range. The rapid improvement in genetic engineering will probably soon lead to a reduction in this uncertainty range.

47. http://de.wikipedia.org/wiki/Genomics (May 27, 2023).

48. John H. Relethford, *Human Population Genetics* (Hoboken: Wiley-Blackwell, 2012), 77–100, 181–204.

49. Secondary tumor formation is less likely due to the significantly longer latency period.

50. Kenneth L. Mossman, *Radiation Risks in Perspective* (Boca Raton: Taylor & Francis, 2007).
51. It should be borne in mind that the chances of a purely statistically justified somatic mutation increase with life expectancy.
52. James V. Neel and William Schull, *The Children of Atomic Bomb Survivors* (Washington, DC: National Academy Press, 1991).
53. World Health Organization, *Health Effect of the Chernobyl Accident and Special Health Care Programs*, Report of the UN Chernobyl Forum Expert Group Health (Geneva, 2006).
54. In the beginning, Americans were used as comparison groups, but this left uncertainties that were difficult to capture statistically, since large differences in cultural lifestyles and living conditions do not necessarily lead one to expect comparable cancer rates among Japanese and American population groups. Follow-up studies, with Japanese comparison groups from the Hiroshima area with low levels of radioactivity, contain uncertainties relating to the possible consequences of low-dose exposure.
55. The latest of these reports is LSS14, K. Ozasa et al., "Studies of the mortality of atomic bomb survivors, report 14, 1950–2003; An overview of cancer and non-cancer diseases," *Radiation Research* 177 (2012): 229–243.
56. The excess relative risk factor (ERR) describes how much the relative risk factor (RR) is exceeded by an additional burden. This is calculated from the simple formula: ERR = (RR-1). In the case of radioactive irradiation, RR is the ratio of the overall risk – from additional radiation exposure – and the risk from natural background irradiation. The background risk depends strongly on local conditions (see Chapter 8). As a result, ERR data is often burdened with large statistical uncertainties (Figure 4.16).
57. Hanno Krieger, *Grundlagen der Strahlungsphysik und des Strahlenschutzes* (Wiesbaden: Vieweg+ Teubner Verlag, 2007), 501–507.
58. http://chernobyl.cancer.gov/leukemia_ukraine.html (May 27, 2023).
59. The total number of liquidators was considerably larger.
60. World Health Organization, *Health Effects of the Chernobyl Accident and Special Health Care Programs*, eds. B. Bennett, M. Repacholi, and Z. Carr (Geneva, 2006).
61. Alexey V. Yablokov, Vassily B. Nesterenko, and Alexey V. Nesterenko, "Chernobyl, consequences of the catastrophe for people and the environment," *Annals of the New York Academy of Sciences* 1181 (2011).
62. The Chernobyl report is very controversial in the specialist literature. The original publisher, the New York Academy of Sciences, now dissociates itself from the statements made there. Five scientific opinions, on the statements made, essentially came to the conclusion that the book has considerable deficiencies and contradictions. In a report commissioned by the New York Academy of Sciences, M. I. Balonov attributed the book with "very little scientific merit while being highly misleading to the lay reader." The estimated number of nearly one million deaths was described as "more in the realm of science fiction than science," http://www.nyas.org/publications/annals/Detail.aspx?cid=f3f3bd16-51ba-4d7b-a086 -753f44b3bfc1 (May 23, 2023).
63. Sebastian Pflugbeil, Henrik Paulitz, Angelika Claussen, and Inge Schmitz-Feuerhake, "Health effects of Chernobyl – 25 years after the reactor catastrophe," German Affiliate of International Physicians for the Prevention of Nuclear War (IPPNW), 2011.
64. Akiko Shibata, Shigehira Saji, Kenji Kamiya, and Seiji Yasumura, "Trend in cancer incidence and mortality in Fukushima from 2008 through 2015," *Journal of Epidemiology* 31 (12), (2021): 653–659.
65. F. Vogel and A. Motulsky, *Human Genetics: Problems and Approaches* (Berlin: Springer Verlag, 1997).
66. When egg cells and sperm cells unite, the zygote (Greek: *zygotos* = bound together) forms a cell with a complete set of chromosomes.
67. James V. Neel and William Schull, *The Children of Atomic Bomb Survivors* (Washington, DC: National Academy Press, 1991) and William J. Schull, *Effects of Atomic Radiation: A Half-Century of Studies from Hiroshima and Nagasaki* (New York: Wiley-Liss, 1995).

5 The Radioactive Universe

5.1 THE RADIOACTIVE UNIVERSE

Radioactivity is a natural and time-dependent phenomenon, as we have discussed. But what is its origin? What caused radioactivity to form? Some radioactive isotopes are older than our solar system. They are part of our environment, of nature and what we are as living beings. Like stable isotopes, radioactive isotopes were created in previous star generations prior to the formation of our solar system and are also forming now, through continuous nuclear decay and reaction processes on our planet. The number of presently known radioactive isotopes is greater than 3,000, and this number is constantly growing with new discoveries at particle accelerators and research reactors.[1] Of these, 289 isotopes have half-lives of more than 80 million years. These constitute nearly 10% of known radioactive isotopes and are more than the number of known stable nuclides, which is 254.

Radioactive elements decay, and this decay or lifetime is closely correlated with the internal structure of the particular radioactive atomic nuclei. Radioactive decay can occur within less than nanoseconds (10^{-9} s) or take billions of years (10^{+9} a). These isotopes are part of our natural environment. They have played a crucial role in the formation of the solar system and our planet, driving dynamical processes in the earth's interior and possibly also the formation and evolution of life, as is still being discussed today.[2] Isotope decay may change the information structure in our genetic material and may be a cause in our evolution as will be discussed in Chapters 7 and 8. In this chapter, we will present in more detail the detection, occurrence, and impact of natural radiation and its role and impact in our environment.

5.1.1 DETECTION OF COSMIC RADIOACTIVITY

The origin of radioactive isotopes is closely connected to the origin of all chemical elements in our universe. The chemical evolution of the universe began about 13.8 billion years ago within the context of the Big Bang theory. During the first second of the Big Bang, the radioactive neutron with a half-life of about 10.2 minutes decoupled from the stable proton through the freeze out of the weak interaction. During the third minute protons and neutrons underwent fusion, forming the heavy hydrogen isotope deuterium. Deuterium was rapidly converted by subsequent proton or neutron capture reactions to the helium isotopes ^{3}He and ^{4}He and small amounts of lithium isotopes ^{6}Li and ^{7}Li. During the first 200 million years of the universe, the synthesis of heavier elements had to await the formation of stars. Only stars provide the high density and temperature conditions necessary for forming heavier elements. After the formation of stars, the chemical evolution of our universe continued, gradually building up heavier and heavier elements and isotopes – both stable and radioactive, up to lead and uranium – through the coming and going of many star generations in the rapidly expanding universe.

Observational astronomy has revolutionized our ability to explore the depths of space and time back to the farthest points in the very early past. Part of this scientific development now includes high-powered optical telescopes, located on remote mountain elevations, and instruments stationed in Earth's orbit, such as Hubble which was named after astronomer Edwin Hubble (1889–1953). Hubble has now been successfully replaced by the James Webb Space Telescope (JWST), which probes the very early moments of star and galaxy formation. Satellites such as NASA's COBE (Cosmic Background Explorer), the WMAP (Wilkinson Microwave Anisotropy Probe), and the ESA's Planck

DOI: 10.1201/9781003435907-5

(named after Max Planck) measure the fading reverberations of the Big Bang in the electromagnetic spectrum, extracting information about conditions in the first second of the universe from structures and irregularities in the distribution of electromagnetic background radiation (Figure 5.1).

Cosmic background radiation was initially dominated by exceedingly energetic gamma radiation. This did, however, gradually dim with the expansion of the universe, shifting to longer and longer wavelengths. So, light expanded into the visible range, light that filled the universe with a bright shine before it progressed toward the dark age. Today, low-energy cosmic background radiation corresponds to a temperature of only 2 Kelvin, although observations still provide important information about the earliest period of our world's evolution.

The dark age of our universe corresponds to the period when the energy of cosmic background radiation had become longer, shifting to infrared and then toward still longer wavelengths beyond the visible range. Therefore, the universe was actually dark. Only with formation of the very first stars, and the onset of nuclear reactions in the center of these stars, was new light borne in a visible wavelength range that was not part of the cosmic background radiation. This new light lit the universe and ended the dark age as suggested in Figure 5.2.

FIGURE 5.1 Planck and Hubble satellites track the evolution of our universe from the Big Bang to the present day. Source: https://www.esa.int/ESA_Multimedia/Images/2013/10/Artist_s_impression_of_Planck and https://www.esa.int/Science_Exploration/Space_Science/Hubble_overview (March 13, 2024).

FIGURE 5.2 Artist conception of the first stars emerging at the dawn that followed the dark age, 200 million years after the Big Bang. Source: NASA/WMAP Science Team -https://www.nasa.gov/vision/universe/starsgalaxies/fuse_fossil_galaxies.html (May 29, 2023).

In this way, the first generation of stars began to shine. Their luminosity reflected the amount of energy produced by nuclear reaction in the stellar core, which was then transmitted to the surface. The spectrum of this light was determined by the excitation processes in the atomic shells of the first elements, the primordial elements formed in the Big Bang.

Systematically surveying all the stars in our galaxy with a network of smaller telescopes (Figure 5.3), scientists have found the earliest generations of stars in the halo of our galaxy. Thus far, these stars are the oldest to be observed and spectroscopically analyzed. The spectra from these first stars, and the many subsequent star generations, show the atomic excitation lines of the elements and their intensity provides information about the chemical history of our universe, for both stable and radioactive nuclei.

Hubble measures the element abundance in stars through spectroscopic analysis of their faint radiation, which reaches Earth from the past. Such measurements will be made in the future by the planned Giant Magellan Telescope (GMT) in Chile's Atacama Desert, as seen in Figure 5.3, as well as by the James Webb telescope as Hubble's successor instrument.

The JWST is the latest instrument and will expand our view by performing spectroscopic analysis of the earliest galaxies. This analysis will provide direct evidence on the status of the chemical evolution of elements that was occurring many billions of years back, when their light was emitted to eventually reach us today.

X-ray telescopes are stationed on satellites to measure the dust and gas remnants of stellar explosions, from novae to supernovae. These telescopes include the European instrument ROSAT (Roentgen SATellite)[3] and NASA's flagship X-ray telescope Chandra,[4] as seen in Figure 5.4. From emitted characteristic X-ray radiation, the telescopes extract the distribution of elements formed in these explosion processes. The latest generation of X-ray telescopes – such as NASA's NuSTAR (Nuclear Spectroscopic Telescope Array) and the ESA instrument eROSITA (extended ROentgen Survey with an Imaging Telescope Array) – not only observe exploding stars, they also measure the X-ray radiation emitted by matter from accreting and collapsing neutron stars and accreting black holes.

In addition to measuring the X-rays from these cataclysmic events, which are among the most powerful sources of energy in the universe, scientists are also detecting weak shock or gravitational waves that are reaching Earth after many light-years. These waves are being measured by high-precision interferometers set up in kilometer-long tunnels, such as LIGO in the United States and VIRGO in Italy (Figure 5.5).[5]

Detectors installed deep underground – such as SNO (Sudbury Neutrino Observatory) in Canada, Super-Kamiokande in Japan, and Borexino in Italy's Gran Sasso massif – count neutrinos reaching our Earth from the center of the Sun (Figure 5.6). These detectors, based on various principles,

FIGURE 5.3 On the left, the recently discovered oldest star in our galaxy; on the right, the planned GMT to be built in Chile's Atacama Desert, for measuring the elemental distribution in these early stars and the associated first phase of element synthesis. Credit: Timothy Beers, University of Notre Dame.

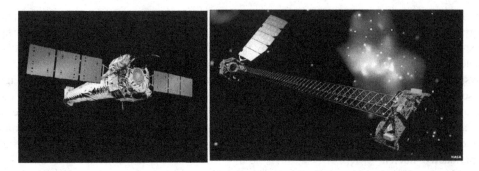

FIGURE 5.4   X-ray satellites Chandra and NuSTAR measure the X-rays emitted by explosive events in our Galaxy. Source: Caltech.

FIGURE 5.5   LIGO (left) and VIRGO (right) are two interferometry detectors that, together with other instruments in the United States and Europe, measure the gravitational waves reaching Earth from distant stellar explosions. The first successful observation was the collision of two neutron stars in 2017, which could also be observed as a stellar explosion in the optical spectrum. Source: LIGO Science Collaboration and The Virgo Collaboration/CCO 1.0.

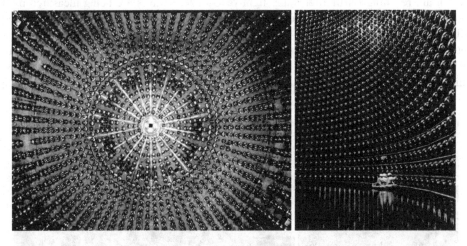

FIGURE 5.6   Photodiodes in the Borexino detector, which are filled with scintillator liquid (left), and in the Super-Kamiokande detector (right), which is filled with pure water. These photodiodes are designed to measure light flashes generated by neutrino events in their respective meter-high detector spheres. The detectors' dimensions can be seen on the right, where scientists are replacing one of the photodiodes. Source: The Borexino Collaboration, Kamioka Observatory, ICRR, University of Tokyo.

detect neutrino interaction with different kinds of detector material. For example, Borexino is a sphere filled with scintillator fluid; when a neutrino reacts with the liquid's molecules, a flash of light is triggered that can be detected and amplified by photodiodes. On the other hand, Super-Kamiokande is filled with 50 kilotons of water to detect neutrino-induced Cherenkov radiation[6]; while a neutrino is a neutral particle that won't be able to produce Cherenkov light, it can interact with an atom to produce a high-energy electron which is itself traveling faster than the speed of light in the water medium. High-velocity charged particles in a dense medium produce Cherenkov radiation – according to the Cherenkov principle, named after Russian Nobel laureate Pavel Cherenkov (1904–1990) – which in turn can be detected by photo-sensors. Other detectors, such as the SNO detector filled with 1,000 tons of heavy water, again detect Cherenkov radiation in a heavier D_2O medium.

In the future, similar detectors will measure neutrinos from distant supernova explosions or even study extremely cold neutrinos that were created in the first seconds of the Big Bang and have gradually cooled down with the expansion of the early universe. Even today, these neutrinos are filling the universe with radiation – another Big Bang echo that has continued for more than 13 billion years.[7] But while it was initially difficult to identify and map cosmic-ray background radiation, it is much more challenging to detect low-energy neutrinos due to their extremely low interaction probability with matter. Also due to their extremely low energy, these neutrinos cannot be measured by present means. New techniques need to be developed that are optimized to measure this energy and the interaction probability of these particles, information about which is presently only available through theoretical calculations. New detectors with extremely low internal-radiation background are required.

All these studies show that the early universe is filled with radiation of different kinds and different nature, photons and neutrinos. Today, both have lost their initial high-energy state with the expansion of the universe. Studying the nature and intensity of these cosmic background signals does, however, still provide us with a glance into the early stages of the universe and its gradual evolution. Developing and then implementing a rich portfolio of instrumentation, scientists are able to study this radiation in order to generate information about the evolution of our universe at the earliest stages.

While cosmic background radiation is the last remaining signal from the Big Bang, the radiation reaching us from the stars has been produced over the following billions of years, during our universe's evolution. This kind of radiation reflects the history of element formation, during which stable isotopes formed in slow processes, lasting billions of years inside stars, or radioactive isotopes formed in stellar explosions that sometimes lasted only seconds. The long-lived radioactive nuclei that developed in these stellar processes are still with us and continue to be present on our planet, where we are exposed to their decay energy.

These processes can be seen very directly from measurements made by ESA's INTEGRAL gamma-ray telescope, which systematically observes the distribution of radioactivity in our universe (Figure 5.7), while similar distributions can be documented for other radioactive isotopes.

The observed radiation is emitted by radioactive elements – such as ^{18}F, ^{26}Al, ^{44}Ti, and ^{60}Fe – which are particularly prominent at the galactic coordinates of supernovae. Since these isotopes have relatively short lifetimes when compared to cosmic timescales, they have in many cases been produced relatively recently. This is illustrated in particular by the example of Supernova Cassiopeia A (CAS A), which exploded about 360 years ago (Figure 5.8). The X-ray image taken by NuSTAR shows the dominant elements – magnesium, silicon, and iron – produced in this explosion. The shock generated by the supernova explosion ejected the elements outward, mixing them with interstellar dust. Also visible is the distribution of radioactive ^{44}Ti,[8] with a considerable amount remaining in the collapsing remnant star and thus becoming part of the resulting neutron star.

These observations are based on spectroscopic analysis of electromagnetic radiation, ranging from high-energy γ- and X-rays to low-energy radiation from ultraviolet, visible, and infrared light. Similarly, the chemo-physical study of rock samples from meteorites plays an extremely important

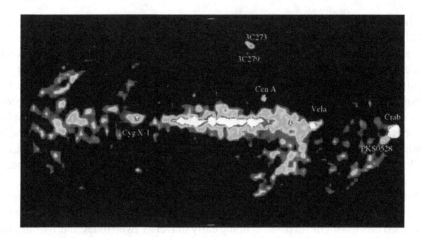

FIGURE 5.7   The radioactive universe. This figure shows the Milky Way and its distribution of gamma rays, from 1 MeV to 30 MeV, across our galaxy. Activity is concentrated at the center of the galaxy as well as at sites where supernovae and other stellar explosions have occurred. Courtesy of Roland Diehl, MPI for Extraterrestrial Physics, Garching, Germany.

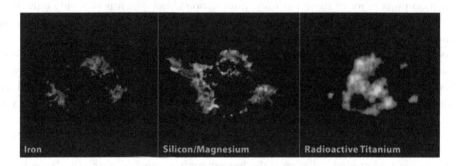

FIGURE 5.8   Supernova Cassiopeia A as seen from NuSTAR. The coloring of individual elements is artificial to highlight differences in distribution. Credit: NASA/JPL-Caltech/CXC/SAO.

role. Meteorites were formed by gradual condensation processes in the atmospheric winds of distant stars. These early condensates accumulated more dust particles during the formation of our solar system and rained down on earth after millions and billions of years. In addition, manned and unmanned space shuttles are bringing back samples of material from the moon, planets, and asteroids in our solar system. American chemist Harold C. Urey (1893–1981) – considered one of the founders of modern cosmochemistry during the 1940s and 1950s – reportedly motivated the Apollo Moon Program with his saying, "Give me a piece of the moon and I give you the history of the solar system."

Today's detection methods make it possible to study even the tiniest amounts of trace elements, from existing stable and long-lived radioactive elements, found in microscopic cosmic-dust samples. As shown in Figure 5.9, such methods identify and classify the origin of these dust samples as either stardust or products of stellar explosions.[9]

Observations yield information about the evolution of element synthesis over time, as well as the natural accumulation of galactic radioactivity. However, these observations give no indication of the underlying nuclear-physical mechanism that causes this buildup. Therefore, observations are complemented with direct measurements of those nuclear reactions that play a critical role in element synthesis. These measurements are made by simulating conditions that are as close as possible to stellar conditions. Simulations are carried out with particle accelerators, research reactors, intense

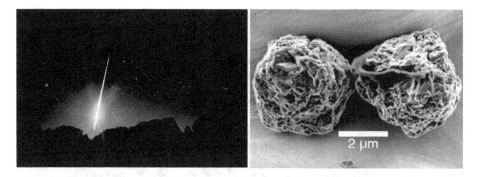

FIGURE 5.9   On the left, sky track of a falling meteorite. Credit: Wally Pacholka/Barcroft Media/Getty Images. On the right, a silicon-carbide dust grain from the Murchison meteorite that fell in Australia in 1969, as an example of the smallest trace elements studied today at the Institute of Cosmochemistry in the University of Chicago. Source: http://geosci.uchicago.edu/directory/andrew-m-davis (May 29, 2023).

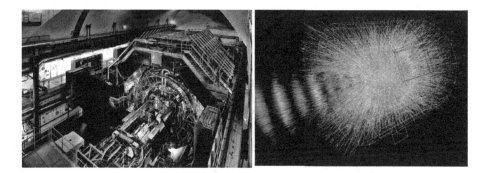

FIGURE 5.10   On the left, the ALICE detector at CERN for measuring quark-gluon plasma in the ideal-liquid state. This should provide clues to the state of matter in the first second of our universe. Right: Simulation of the collision of two lead atoms at relativistic velocity. The resulting particles are detected by the ALICE detector and their respective intensity and energy are used to infer conditions in the particle soup at the moment of collision. Source: https://home.cern/science/experiments/alice (May 29, 2023).

photon sources, and high-power laser instruments. As far as possible, the energy must correspond to the prevailing temperature in the interior of stars.

ALICE (A Large Ion Collector Experiment) at CERN and the RHIC accelerator at Brookhaven are where so-called collider experiments simulate the early quark-gluon phase of the universe. In these simulations, colliding particles fly at nearly the speed of light; these particles represent the particles from which all matter emerged that is observed in the universe today. Collapse of the short quark-gluon plasma phase manifests itself in an enormously high flux[10] of short-lived particle radiation. Measurement and analysis of this radiation determines the conditions and properties of the universe's early phase. Figure 5.10 shows a simulation of particle flux and the ALICE detector constructed to measure it. The detector consists of a large number of individual instruments specially designed for the detection of various specific types of particles.

Extremely slow nuclear reactions in stars guarantee their long lifetime. At smaller particle accelerators, reactions are measured at very low energies corresponding to temperatures in the stars' interior. The resulting count rates are extremely low, requiring an environment well shielded from natural radiation background so as to allow for sensitive detection of the reaction signals. Therefore, these measurements have to be taken deep underground in an environment that is nearly

FIGURE 5.11   On the left, an intense proton beam (from right to left) initiating nuclear reactions on nitrogen particles, using a vertical supersonic jet stream; from the author's collection. On the right, a fusion reaction at the University of Rochester's OMEGA laser plasma facility. The instrumentation around the centrally located plasma is clearly visible. Source: OMEGA facility, University of Rochester.

free from background radiation. Necessary facilities have been developed with the LUNA accelerator, in Italy's Gran Sasso mountain range, and in old mineral mines, such as the CASPAR at SURF (Sanford Underground Research Facility) in Black Hills, South Dakota. Here, a high-intensity beam of accelerated protons encounters a stream of nitrogen gas (in this case nitrogen gas ^{14}N) that exits a gas nozzle at supersonic speed (Figure 5.11). The interaction between protons and ^{14}N particles produces γ-radiation through nuclear reactions. Special detectors measure these reactions, in addition to the bluish-white light that is produced by atomic-excitation processes.

In parallel, experiments are also being conducted on hot plasma, compressed by peta-watt lasers, to achieve the nanosecond thermal conditions of the stellar interior. The goal is to observe fusion reactions between the positive ions that comprise the plasma (Figure 5.11). To measure the fusion conditions and products, the plasma must be captured and analyzed by collectors. For this purpose, sensors are placed close to the fusion site; these sensors measure the temperature, density, and temporal evolution of the plasma conditions, in addition to the reaction products.

Explosive processes in stars occur at a rate of seconds. Radioactive elements produced within this time appear stable and participate directly in subsequent nuclear reactions. Measuring such reactions requires that radioactive-particle beams be produced. For this, complex facilities must be constructed, first to produce the radioactive particles – up to several billion per second – and focus them, and to then accelerate or decelerate them to explosive energies. Several relatively smaller facilities, at the University of Louvain-la-Neuve in Belgium and at the University of Notre Dame in the United States, have pioneered the field, followed by TRIUMF in Canada and RIKEN in Japan. The two largest facilities currently under construction are FAIR (Facility for Antiproton and Ion Research) at GSI (Gesellschaft für Schwerionenforschung) near Darmstadt in Europe and FRIB (Facility for Rare Ion Beams) at Michigan State University in the United States (Figure 5.12).

Results of these measurements are all gathered on a computer network, in order to describe the entire element-synthesis process within the context of thermodynamic and hydrodynamic models of stars and stellar explosions. The goal is to then compare these results with the observed results obtained by the astronomy community.

For a more detailed understanding of the chemical evolution of our universe, production of the long-lived isotopes in exploding stars plays a particularly important role. These radioactive isotopes can be directly observed in radiation emissions, as shown in Figures 5.7 and 5.8. However, this isotope decay is also indirectly reflected in the time dependence of brightness observations.

FIGURE 5.12   On the left is a model of the FAIR facilities that are currently under construction. The synchrotron ring accelerator, for the generation of exotic particles, is located underground in the forest area seen in the upper right. In the foreground are experimentation halls, with storage rings and particle separators. Source: FAIR Center Darmstadt. On the right is a model of the FRIB at Michigan State University; in the foreground is the hall for the linear accelerator, for radioactive particle generation; in the background, the experimentation halls. Source: FRIB, Michigan State University.

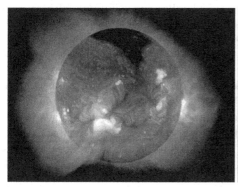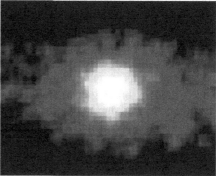

FIGURE 5.13   On the left, the Sun in an X-ray image from the Soft X-ray Telescope (SXT) stationed on the Yohkoh satellite. It can be clearly seen that, under the influence of magnetic fields, strongly convective plasma either erupts in prominences or appears as dark sunspots at cooler locations. At equatorial altitudes, the hot gas bound by magnetic fields shows up as a luminous corona of the Sun. At the poles, the gas cannot be magnetically bound. Source: wikimedia.org/wiki/File:Sun_in_X-Ray.png (May 29, 2023). On the right, the Sun is seen as a brightly glowing neutrino spot, which corresponds to the center of the Sun. This image was taken by the Japanese detector Super-Kamiokande, located at a depth of 1,000 m in the Kamioka mine in Hida City, Japan. The brightness of the pixels reflects the intensity of the neutrino flux directly produced by the nuclear reactions of the pp chains. Source: apod.nasa.gov/apod/ap980605.html (May 29, 2023).

Thus, the dimming light from the stellar explosion is correlated with the decline in energy emission that results from the decay of and therefore declining number of radioactive elements produced therein.

Other signatures are the neutrinos emitted in β-decay processes. The existence and decay of these radioactive elements – such as ^7Be, ^8B, and ^{15}O, which are being produced in the Sun's interior – can be measured by detecting the neutrinos generated as part of their decay process. Observations made at aforementioned neutrino detectors, such as SNO in Canada, Super-Kamiokande in Japan, and Borexino in Italy, directly supported the thesis that the Sun, like all other stars, produces its energy deep inside. This occurs through nuclear fusion, which balances the Sun's stability against gravitational forces. Figure 5.13 shows the neutrino flux being directly released by pp chains, a sequence of fusion reactions between hydrogen and helium nuclei at the Sun's center. This will be discussed in more detail later, in Section 5.1.2.

Spectral studies reveal the presence of heavy radioactive isotopes in the photosphere of the Sun. These isotopes are also found on Earth, as a part of the solar system. In the planetary system itself, the inner planets are mostly composed of solid material, while the larger planets such as Jupiter and Saturn consist primarily of gaseous material that has accumulated around a solid core. The material composition of Earth and other planets should thus be comparable to that of the Sun. Yet, chemical and geological fractionation processes have led to an accumulation of the heavy radioactive elements in the Earth's core and mantle. The decay of these elements determines the fate of the geological evolution of our planet as described in Chapter 6.

As part of our planet, we humans up-take new atoms and molecules into our bodies every day through breathing, eating, and drinking. While we excrete some of these, a large amount remains in our blood, organs, nerves, and bones. A small amount of this is radioactive and can influence biological development as will be discussed in Chapter 7.

5.1.2 ORIGIN OF THE ELEMENTS

As we have discussed, based on the results of many observations, measurements, and model calculations, a consistent picture of element synthesis has developed over the last several decades. In the first second after the Big Bang, two types of baryons – protons and neutrons – emerged from the quark-gluon soup of the early universe. Initially, they remained identical due to constant interaction with leptons, electrons, and neutrinos; with rapidly decreasing density, the weak interaction decoupled and protons and neutrons found their identity, with neutrons having a lower abundance due to their slightly higher mass.

Protons and neutrons sought to combine to deuterium but this heavy hydrogen isotope is very fragile, due to its low binding energy, and was immediately photo-dissociated through the high photon flux that occurred during the first couple of minutes of the universe. It was only in the third minute, after further expansion, that photons cooled to lower energies; it was then that neutrons and protons could undergo fusion to form deuterium ^2H. Through further nuclear reaction processes in the following few minutes, heavier primordial nuclei formed – such as helium isotopes ^3He, ^4He and lithium isotopes ^6Li and ^7Li.

The fact that there exist no stable nuclei with mass number $A = 5$ and $A = 8$ prevented the formation of the heavy elements beyond lithium in the Big Bang.[11] This fascinating fact demonstrates how closely the evolution of our universe is interconnected with and determined by the microscopic structure of atomic nuclei. The so-called primordial abundance distribution, resulting from the Big Bang, shows that 75% of elements remain as the hydrogen isotope ^1H and 25% as the helium isotope ^4He. Thus, the baryon mass in the universe consists mainly of protons and α-particles,[12] with lithium as a trace element that is difficult to detect.

Element synthesis came to a standstill after about ten minutes because the density of the universe had expanded to such an extent that no further reactions could take place due to the rapidly growing distance between particles. Only 200 million years later, the first massive stars, which were gravitationally formed from inhomogeneities in the mass distribution, generated light and the universe emerged from its dark epoch. As estimated from simulations, this first generation of stars had between ten and two-hundred solar masses or it is also possible that many thousand solar masses existed, which would eventually break up to form smaller constituents. Either way, these stars were so large that they could not stabilize themselves by internal energy production and so contracted under the influence of their own mass, whereby the first nucleosynthesis processes took place in the stars' interior. This involved a series of alpha fusion and capture reactions, which led to the buildup of carbon and oxygen.

This is especially remarkable because, with the conversion from primordial helium to carbon and oxygen in the first stellar generation, the islands of instability at mass $A = 5$ and $A = 8$ had been bridged! This opened new opportunities for nuclear reaction and fusion processes to take place, which led to the production of heavier elements up to mass $A = 40$. With the death of these stars

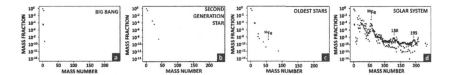

FIGURE 5.14   The abundance distribution of the elements, ordered by their mass number A. On the left are the primordial abundances as assumed from current Big Bang simulations. The two middle figures show the abundance distributions of the two oldest observed stars, with the left figure showing the probable abundance distribution of a distant supernova formed from one of the first stars. The right-middle distribution shows a second- or third-generation star, where the stellar spectrum already reflects elements up to strontium. On the right, the elemental distribution in our solar system with all known elements up to uranium and plutonium. Source: H. Schatz, "Rare Isotopes in the Cosmos," *Physics Today* 61 (2008), 40–45.

in supernova explosions, the newly formed elements were ejected into space, enriching primordial material with the first heavy elements (see Figure 5.14). This means that already in the first stars, the biologically critical elements carbon and oxygen had been formed.

The entirety of elemental diversity that we know today has been fed from the abundance distribution that formed in first stars, which slowly built up through complicated nuclear-reaction sequences in many subsequent stellar generations. Figure 5.14 shows snapshots of this element or isotope synthesis process starting with primordial distribution as it dominated the universe for 300 million years, followed by the observed isotope distribution in star systems that were more than 10 billion years old, and finally distribution in our solar system today.

In this process, of slow element buildup in multiple generations of stars, various production processes critically depended on specific temperature and density conditions as well as on the time duration of each phase in the evolution. These were the finely tuned components governing the life and death of stars during element formation. This buildup of elements is similar to the formation of molecules, which evolve from the chemical reactions of atoms. In these chemical processes, as in star development, the nature and rate of the chemical processes critically depend on the particular environmental conditions on Earth, on other planets, and in interstellar space.

All stars go through different phases of evolution. This depends on the initial abundances. Primordial stars are unique for their initial lack of carbon and oxygen, which plays an important role in later star generations. Further, while first stars have been somewhat distinctive due to their primordial seed material, all subsequent star generations go through similar nucleosynthesis patterns. The time scale of a star's evolution is determined by its mass, with massive stars reaching much higher density and temperature conditions and so burning through their fuel supply much faster. The various burning phases are determined by the available fuel as well as internal transport and mixing mechanisms, while the duration of these different phases depends on the mass of the star. The more massive the star, the faster the evolution proceeds. The first two evolutionary phases, hydrogen and helium burning, took between billions and millions of years, respectively – depending on the mass of the star. In the following phases, carbon, neon, oxygen, and silicon burned relatively quickly, making it difficult to observe stars in these later phases. Each phase of stellar evolution is determined by specific nuclear-reaction sequences; that is, stars are nothing more than giant gas clouds held together by gravity, with a fusion reactor in their core that acts as an engine to drive evolution.

As the presolar dust cloud contracted under the influence of gravity, the stellar interior heated and compressed to temperatures and densities where fusion reactions began between the light hydrogen atoms. These are the so-called pp chains, a series of fusion and capture reactions that convert ^{1}H hydrogen isotopes to ^{4}He, releasing bonding energy in the process (Figure 5.15). The released energy heats the stellar interior and increases internal pressure, which balances and

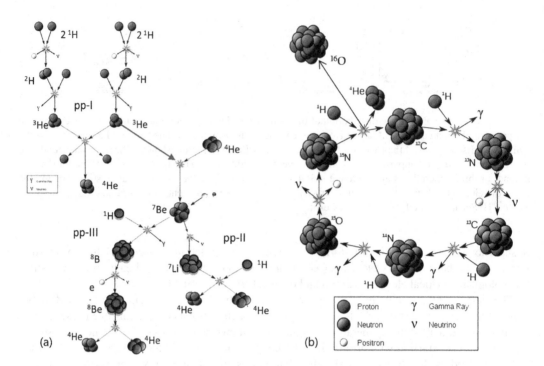

FIGURE 5.15 Stellar hydrogen-burning reactions. On the left, the three pp chains through which hydrogen is fused to helium. These reaction sequences determine the internal energy budget of our Sun and many other stars of comparable size. On the right, the CNO cycle or Bethe-Weizsäcker cycle, which is a catalytic reaction sequence in which four ^{1}H hydrogen isotopes are transformed to ^{4}He by trapping carbon and nitrogen isotopes. It should be pointed out that the CNO cycle does not exist in first stars due to the lack of carbon (C), nitrogen (N), and oxygen (O) in primordial matter. Source: Modified from https://commons.wikimedia.org/wiki/File :FusionintheSun_sv.svg (March 13, 2024) and https://commons.wikimedia.org/wiki/File:CNO_Cycle-tr.svg (March 13, 2024).

stabilizes the gravitational contraction of the protostar after sufficient fusion energy has been produced.

As part of the reaction chain, radioactive isotopes such as ^{7}Be and ^{8}B are produced. With this, ^{7}Be decays to ^{7}Li by electron capture from the star's plasma and ^{8}B transforms to ^{8}Be by β^+- decay. ^{8}Be then immediately decays to two α-particles – that is, two ^{4}He nuclei. This is the reaction mechanism that governs the energy production in our Sun, which offers a unique observational opportunity (Figure 5.16). Both decay processes also release neutrinos, which, thanks to their low interaction with matter, leave the Sun almost unhindered and can be observed on Earth by the aforementioned underground neutrino detectors. In this way, neutrinos offer a unique opportunity to look directly into the center of the Sun, while the human eye detects only the light of the Sun as produced in the photosphere.[13] Neutrinos take only a few minutes to reach the solar surface, while β^+-radiation – delayed by scattering processes – takes several hundred thousand years to travel the same distance.

Strong magnetic fields characterize magneto-hydrodynamic convections, protuberances, and cold sunspots on the Sun's surface, as seen in the X-ray image shown in Figure 5.13. We can again compare this with the image produced at the Japanese Super-Kamiokande, which shows neutrino radiation coming directly from the innermost part of the Sun and reaching our Earth unimpeded.

Our Sun has been in this stage of hydrogen burning for 4 to 5 billion years. It is estimated to be in approximately the middle of this development phase, so another 5 to 6 billion years will pass before major changes occur. These changes would result from the hydrogen supply being slowly depleted while the burning zone at the inner core would slowly expand radially outward, as the burned-out

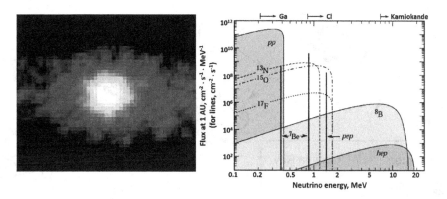

FIGURE 5.16 On the left, a brightly shining neutrino spot corresponding to the center of the Sun. On the right, the neutrino spectrum showing the contributions made by neutrinos, resulting from various weak-interaction processes in the pp chains. Source Credit: https://apod.nasa.gov/apod/ap980605.html (March 13, 2024), R. Svoboda and K. Gordan (LSU). Lower-energy neutrinos come mostly from hydrogen fusion $H + H \Rightarrow D$, where the conversion of di-protonium to deuterium releases a neutrino. Higher-energy neutrinos come mostly from the β +-decay of ^8B. Also shown are neutrinos from decay of the radioactive isotopes ^{13}N, ^{15}O, and ^{17}F, which originate from the weak CNO fraction in the solar burning processes and have only recently been directly detected by Borexino. Source: The Borexino Collaboration, "Experimental evidence of neutrinos produced in the CNO fusion cycle in the Sun," *Nature* 587 (2020): 577 (right).

core slowly contracts under its own gravity. Thus, the gas ball of the Sun will expand into a red giant star that will swallow the planets near the Sun – Mercury, Venus, and possibly Mars and Earth.

This type of hydrogen burning can only sufficiently balance gravitational contraction in smaller stars with masses up to 1.5 solar masses. This is because the first reaction of the pp chains $^1H + {}^1H \Rightarrow {}^2H + \beta + + \nu$ forms a deuteron, with conversion of a proton to a neutron. This weak interaction process is extremely slow, allowing for the Sun's long lifetime.

In more massive stars such as Sirius, which is the brightest star in the constellation of Canis Major, additional reactions must be initiated because pp chains are not sufficient to produce the energy required to stabilize the star against gravity. The necessary additional reaction sequence is the aforementioned CNO cycle or Bethe-Weizsäcker cycle, predicted (almost) independently in 1936 by Hans Bethe and Carl Friedrich von Weizsäcker.[14] Here, the carbon isotopes ^{12}C and ^{13}C and the nitrogen isotopes ^{14}N and ^{15}N serve as catalysts, allowing a reaction cycle – four capture reactions of a hydrogen nucleus with the emission of two β + particles, as shown in Figure 5.15.

A similar cycle develops from proton capture occurring at the oxygen isotope ^{16}O. The ^{16}O is either already present in the stellar interior or is formed by proton capture at ^{15}N from the first part of the cycle. As part of these cyclic reaction sequences, radioactive isotopes ^{13}N and ^{15}O (in the first cycle) and ^{17}F (in the second cycle) buildup and so are constantly present in the stellar interior, contributing to the stellar neutrino flux by β +-decay. Through this process, carbon isotopes present in the stellar interior are converted to the ^{14}N nitrogen isotope.

The hydrogen supply in the interior of a star, like the Sun or also in more massive stars, is eventually depleted and the burning zone expands outward, developing into a hydrogen-burning sphere that surrounds the burned-out core. In response to this relocation, interior pressure increases outward as a result of the radiative behavior; thus, the star inflates to find a new equilibrium. To the observer, the color changes from white-yellow to red as the star becomes a red giant.

On the other hand, the interior of the star can no longer be stabilized by the generation of fusion energy and so it contracts, again increasing its density and temperature. This contraction comes to a halt only with the onset of helium burning.

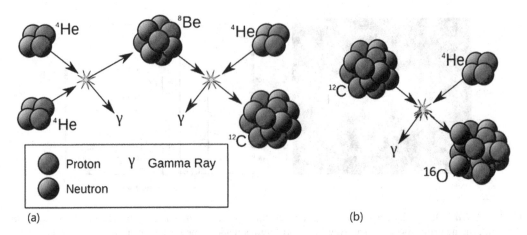

FIGURE 5.17 Nuclear reaction sequence of stellar helium, which burns via light ^{4}He nuclei that are converted to carbon ^{12}C by three-particle fusion (left). Through another alpha (^{4}He) capture, the oxygen nucleus ^{16}O is built up (right). Source: Wikipedia Commons.

FIGURE 5.18 On the left, the red giant star Betelgeuse, probably at the end of the helium-burning phase. On the right, Sirius in the hydrogen-burning phase, with its binary partner Sirius B or Mira, a white dwarf faintly visible in the lower left quadrant of the figure. Source: https://en.wikipedia.org/wiki/Sirius#/media/File:Sirius _A_and_B_Hubble_photo.jpg (March 13, 2024).

With helium burning, energy is produced by the fusion of three α-particles – the so-called triple-α-process – to form a ^{12}C carbon nucleus, and ^{12}C forms ^{16}O oxygen by another α-capture reaction (Figure 5.17). Thus, the helium burning in stars was and is the primary production site for the elements carbon and oxygen, which are the two most important elements for the emergence of complex organic molecules in our universe. Without these nuclear reactions deep inside the red giant stars, these two isotopes, on which all organic life is based, would not exist.

The best-known red giant is Betelgeuse (Figure 5.18), which is 437 light-years away from Earth and is the main star in the constellation of Orion. Betelgeuse is so huge that the orbit of Jupiter in our solar system would fit within it.

However, in addition to these reactions, which produce the energy in the red giant, there are a number of other nuclear reactions going on that ignite the ^{14}N ash from the previous hydrogen-burning phase. These reactions produce neutrons, which attach to other isotopes in the stellar-core material. In this way, heavier elements are slowly built through a sequence of neutron captures, which are followed by β-decay of the radioactive isotopes that formed in the process. This is the so-called s-process (slow neutron-capture process) through which a large fraction of the heavy elements above iron are formed by gradual neutron capture reactions, with neutron being produced by alpha-induced processes on light nuclei such as ^{13}C and ^{22}Ne (see Figure 5.23).

Only a few of the radioactive isotopes produced in this process survive this phase of the star's evolution, since their half-lives are relatively short compared to the duration of helium burning. A classic example is technetium (Tc), which was detected in red giant stars in 1952. Technetium has no stable isotopes. In the s-process of red giants, ^{97}Tc and ^{98}Tc can be formed,[15] which have half-lives of 2.6 and 4.2 million years, respectively.

Helium burning comes to an end when the helium supply has been converted into carbon and oxygen. Then, as a result of enormous gravitational forces, the star's interior contracts again under its own weight. For low-mass stars like our sun, this is the last phase. Since the gravitational energy being released is not enough to sufficiently heat the stellar interior, the next phase of nuclear reactions between carbon and oxygen will not be ignited.

Thus, the star's core continues to collapse until the pressure of the electrons, which are now crammed into the smallest space, stabilizes the contraction. This stabilization results from a phase transition to form so-called degenerate matter. With this process, the outer shell is ejected as a gas cloud and a so-called white dwarf remains, which essentially consists of only ^{12}C carbon and ^{16}O oxygen, as end products of the helium burning.

In the binary system[16] of Sirius, its second component is Mira – a white dwarf that shines only faintly thanks to its residual heat with a surface temperature of about 25,000 Kelvin. Mira's mass is comparable to the Sun, but its enormous internal density makes it the same size as Earth (Figure 5.18).

In massive stars, on the other hand, evolution continues with the carbon--burning phase, where two ^{12}C isotopes fuse with the emission of alpha particles, protons, and gamma radiation, producing a distribution of ^{20}Ne, ^{23}Na, and ^{24}Mg, respectively. However, this phase lasts only a few thousand years depending on the mass of the star. Since the carbon at the star's core has been consumed, the center of the star continues to contract under the influence of its own gravity. At this stage, the carbon-burning zone is confined to a slowly expanding shell around the core, while ^{20}Ne remains at the star's interior, constituting the bulk of the ash from the carbon-burning phase.

As the interior condenses, intense radiation causes the weakly bound ^{20}Ne isotope to dissociate into an oxygen ^{16}O nucleus and an α-particle. Due to released gravitational energy, the star's center heats up further, reaching a temperature at which nuclear fusion occurs between oxygen nuclei. This fusion process mainly produces silicon. These last two phases, of neon burning and oxygen burning, last only a few years. When the respective burning supplies are consumed, ^{28}Si is the main component of the remaining ash.

Silicon burning is considered the last phase of stellar burning. Similar to neon-burning processes, radiation-induced dissociation processes break up ^{28}Si. Here, light residual particles – such as protons, neutrons, and α-particles – simultaneously attach to the ^{28}Si, rapidly forming heavier nuclei. In this way, the elemental silicon distribution shifts to an iron-nickel distribution through rapid reactions, creating a new equilibrium. This new equilibrium is established within a few days, with long-lived radioactive isotopes – like ^{44}Ti ($T_{1/2} = 59$ years) and ^{56}Ni ($T_{1/2} = 5$ days) – also being produced. This stellar evolution can be represented schematically within the framework of a density-temperature phase diagram of the stellar interior. This as well as expansion of the outer envelope and resulting changes in observational parameters, as studied by astronomers, are presented in Figure 5.19.

In this final stage, the star's structure can be thought of as an onion. In the center is the silicon-burning zone. This center is surrounded by layers or shells: The interior is the oxygen-burning zone, followed by the neon-burning zone, which is in turn followed by shells characterized by carbon, helium, and hydrogen burning (Figure 5–20). This onion model is, however, somewhat simplified. In reality, strong convection takes place – due to the rapid drop in temperature – between the stellar interior and the outer shells. This can be observed in modern three-dimensional computer simulations.

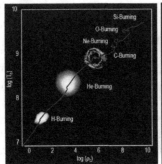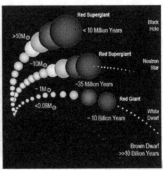

FIGURE 5.19 On the left, stellar evolution at the temperature-density phase. Temperature and density are shown in logarithmic scale. The curve plotted in red represents a simulation of conditions inside a star of 20 solar masses. The density increases from left to right – from less than 1 g/cm³ in the presolar cloud to more than 10^9 g/cm³ (gigagram) in the silicon burning phase. The temperature is shown on the vertical axis, increasing from 10^7 K or about 10 million degrees Kelvin in the protostar to nearly 10^{10} K or 10 billion degrees Kelvin. Also shown are the respective conditions for the various burning phases described in the text (created by the author). On the right, a schematic representation of the outer expansion of a star under the influence of the radiation pressure in the outer envelopes. As the surface enlarges, the color of the star shifts to red. The stars' end depends on the mass. Very low-mass stars do not ignite and are therefore called *brown dwarfs*. Jupiter, for example, could be considered a brown dwarf. Stars with masses in the solar range eventually end up as white dwarfs, after the helium or even carbon-burning phase. Massive stars evolve to the silicon--burning stage, which is followed by a supernova explosion where the main part of the mass is ejected. The remaining star remains as either a neutron star or a black hole. Illustration by Weinreich & Partners, Scientific Multimedia.

Since maximum binding energy occurs in the nickel-iron region, fusion processes no longer release energy in the star's center to stabilize against gravity. Instead, electron capture, in the extremely dense plasma, consumes the star's interior energy. This causes the stellar core to cool, destabilizing the star so that the interior suddenly collapses. Within seconds, the star's matter compacts under its own weight until it reaches the density of the atomic nuclei themselves, resembling something labeled as nuclear matter with a density of about $2 \cdot 10^{14}$ g/cm³, 200 trillion times denser than water at 1g/cm³. In this process, all of the existing atomic nuclei break apart and dissociate back to their individual parts – protons and neutrons. This creates an extremely high flux of free neutrinos, which carry energy to the outside of the star.

Nuclear matter follows different state conditions than normal matter. These conditions are currently the focus of various nuclear physics experiments, in order to determine the details. However, in simplified terms, we know that nuclear matter is like a rubber ball: It can only be compressed to a certain limit, then it recoils. The recoil creates an internal shock that radiates outward. And, although the outer shells have not yet noticed that the ground has broken away beneath them, when the shock hits them from the inside, they are compressed and heated in seconds.

Calculations have shown that in this process, the shock defaults because its initial energy is lost in multiple scattering events. However, radiation pressure, generated by the neutrinos that were produced in the collapse, supplies new energy and drives the shock wave farther in its outward expansion. The outer stellar envelope is thus explosively flung away, leaving behind only the compressed stellar interior – a neutron star with a dense atmosphere of highly radioactive titanium, iron, and nickel. Such an event, the violent death of a massive star, can be observed as a supernova explosion, which has occurred multiple times in our universe, each time seeding the interstellar matter with new heavy elements produced over the star's life cycle.

The Crab Nebula is a steadily expanding supernova remnant, which was first observed by Chinese astronomers in 1054. Signals in each of its wavebands emphasize the explosive expansion of hot, radioactive matter with a rapidly rotating, highly radioactive pulsar at its center. The energy

FIGURE 5.20   Three-dimensional simulation of a supernova explosion (Oak Ridge National Laboratory). Color coding represents density and temperature conditions. Lines represent the outward flow of material – thrown away from the star by the shock front – or the inward flow of material, which falls back onto the neutron star marked as a gray sphere. The heavy inner core thus remains part of the neutron star, while the original star's outer envelopes are ejected away, forming the seed material for subsequent star generations. Source: https://www.pinterest.com/pin/159103799309279634/ (March 13, 2024).

released in these nuclear processes is mostly high gamma-ray energy, which is converted – by Compton scattering, particle excitation, and other radiative transfer processes – into X-ray and other low energy electromagnetic radiation components, from ultraviolet to optical and infrared light, as shown in Figure 5.21.

In these extreme conditions of the rapidly expanding shock front, ejected protons and neutrons fuse to form highly radioactive ^{44}Ti and ^{56}Ni. The light curve of a supernova explosion vividly confirms this. During the explosion, energy is suddenly released and the light curve rises abruptly; this rise is usually not observed because astronomers only become aware of it after the explosion, when a new star appears in the sky. Then, all observations concentrate on this phenomenon to determine the progress of the light curve's intensity, duration, and decay. Decay is slow because a large part of the observed luminosity comes from energy released in the decay of freshly produced radioactive isotopes. So, as astronomers observe the decline of luminosity with time, they are able to accurately document the decay of various radioactive components that were produced in the supernova explosion, mainly radioactive nickel isotopes ^{56}Ni and ^{57}Ni. These decay within a short time to form radioactive cobalt isotopes ^{56}Co, ^{57}Co, and radioactive titanium ^{44}Ti (Figure 5.22).

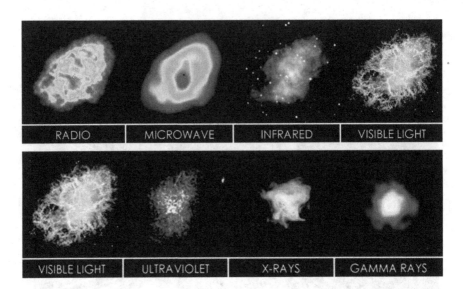

FIGURE 5.21 Crab Nebula as a classic example of a supernova, observed by Chinese astronomers in 1054. Shown are images of the expanding cloud at different wavebands of electromagnetic radiation, which characterize temperature and radiation distribution. While the ejected material is clearly recognized in the visible-light range, the X-ray range shows the fast-rotating neutron star as a pulsar whose atmosphere glows as a highly radioactive source in the gamma-radiation range. Credit: https://en.m.wikipedia.org/wiki/File:Crab_Nebula_in_Multiple_Wavelengths.png (March 13, 2024).

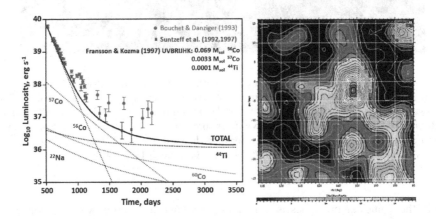

FIGURE 5.22 Left, the luminosity curve of supernova 1987A, observed in 1987 in the Magellanic Cloud, contributing decay curves of the various radioactive isotopes produced. Courtesy of Frank Timmes, ASU. On the right, distribution of ^{44}Ti as it appears in typical γ-radiation in the Cassiopeia A supernova, which exploded around 1635. Courtesy of Roland Diehl, MPI Garching.

In addition to the directly observable radioactive isotopes, many other long-lived radioactive elements are produced that do not contribute significantly to the light curve, due to their slow decay or lower abundance. As the shock wave expands, such elements owe their origin to the countless free neutrons that, within a second, attach to existing heavy nuclei in a so-called r-process (rapid neutron-capture process). A series of neutron captures moves the existing abundances toward the neutron-rich side of the line of stability, building an abundance distribution that is defined by the equilibrium between neutron capture and photodisintegration. This lasts for about a second because, with further expansion and cooling of the explosive front, these nuclei along the r-process path (Figure 5.23) decay back to the stability line by β$^-$- and β-delayed neutron decay while also

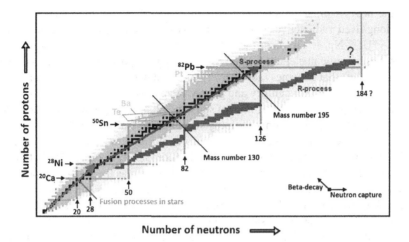

Number of neutrons ⟹

FIGURE 5.23 Schematic representation of an s-process reaction path, which occurs along the line of stability marked by the black boxes, and the r-process located far away in the neutron-rich region of the nuclide map and marked by red boxes. The s-process terminates in the lead-bismuth region, building s-process elements along the line of stability and also producing some long-lived radioactive isotopes along the reaction path. In the r-process, all of the abundance distribution that has developed along the reaction path decays back to stability, generating their own features in the abundance distribution; these include elements up to lead and bismuth but also beyond, with the production of many long-lived isotopes in the actinide region of uranium and thorium. These actinides are the origin of the natural decay series that decays to stable bismuth and lead isotopes through long chains of α- and β-decays. Credit: B. Sherrill, NSCL/MSU.

shifting the abundance distributions further, by parallel neutron capture of the neutrons released during the decay. Together with the s-process, this r-process is considered the origin of almost all elements heavier than iron. These include the precious metals so prized by mankind, from gold to platinum, as well as the actinides that are so demonized, from thorium and uranium to plutonium. It is a complex process and considerable research is still invested in the question of the actual r-process site as well on its contributions to our presently observed abundance distribution.[17]

However, the r-process could possibly occur in other places as well, besides supernovae. Today, the collision of two neutron stars (neutron-star mergers) is frequently discussed as a preferred alternative. According to theoretical calculations, this merger could provide even more favorable conditions for an r-process. Such models do, however, require the existence of numerous binary systems, of two neutron stars, in the early phase of the universe. A large quantity would have been necessary to account for the observed abundance of radioactive elements in our universe.[18]

Such an event, labeled GW170817/SSS17a, was observed for the first time in 2017 using gravitational-wave measurement. These measurements could then be confirmed by measuring the light curve in the optical domain. The light's observed color gradually changed from blue to red, corresponding to a decrease in emitted photon energy. According to the observers' interpretation, temporal behavior indicated the production of heavy elements during this event, the so-called lanthanides or rare earth metals. This was taken as direct confirmation that the heaviest elements observed, the lanthanides and actinides, were created by this type of explosive merger of two neutron stars.

Where the r-process occurred, either in supernovae or mergers, and whether there were possibly two r-processes is still the subject of intense debate. An exact answer can only be given after detailed analysis of the explosion conditions in each case. It can be clearly seen in Figure 5.23 that the s-process reaction path runs along the line marked by all the stable elements, while the r-process path runs along a path that includes many very neutron-rich unstable nuclei. It is primarily the decay of these neutron-rich species that produces the very long-lived radioactive actinide nuclei that dominate the natural radiogenic radioactivity in the earth today.

Decay of the heavy r-process elements feeds the range of the actinide nuclei including many of the heavy long-lived radioactive actinides such as uranium, neptunium, and thorium. These isotopes, ^{235}U, ^{238}U, ^{237}Np, ^{232}Th, form the starting point of the natural decay series discussed in Chapter 2. As these atomic nuclei are ejected into interstellar space with the star's explosion, they join the interstellar dust that becomes radioactively enriched with each supernova (or merger) explosion – on average, one or two per century and galaxy. As part of the interstellar dust, or also as part of the material that forms the next generation of stars and planetary systems, they slowly decay over billions of years in long chains of α- and β-decay sequences to stable bismuth and lead isotopes.

From the cataclysmic death of stars, in supernova explosions, we are left with interstellar dust that blows through space as a fine nebula. Finally, after completely cooling, gravitational attraction and the formation of molecules result in small granules that form as the condensation points for new star generations. Besides the ejected dust, neutron stars or black holes still remain, having formed from the collapsing stellar interior. Due to their enormous gravitational force, these extremely dense objects suck in interstellar dust, or possibly even whole neighboring stars, in a process called *accretion*. As matter falls onto these dense objects, high-energy X-rays are emitted that fill our universe (Figure 5.24).

This particular accretion of matter onto an object has an effect like pouring gasoline onto a fire. The accretion adds fuel to the hot and dense atmosphere of the neutron stars, which can lead to explosive nuclear processes on the star's surface. This freshly added nuclear fuel rapidly burns and is converted to higher masses in a matter of seconds via chains of light-nuclei fusion reactions. The fusion energy released in this process is directly detectable in space as extremely energetic flashes of X-ray radiation while the ash remains embedded in the neutron star. Such events are therefore called *X-ray bursts*, of which many hundreds have been studied since their first observation in the 1970s. Along with supernovae, they are among the most energetic events that can be observed in our galaxy.

The universe glows from the radiation produced by past generations of stars. The source of this radiation includes: neutrinos, which have filled the universe almost unhindered since its beginning and are slowly cooling as it expands; γ-radiation produced by the decay of long-lived radioisotopes; continuous low energy γ-radiation (511 keV) from the β $^+$-decay of short-lived reaction products; and X-rays, which come from gravitational accretion or from electromagnetic processes. In addition, the

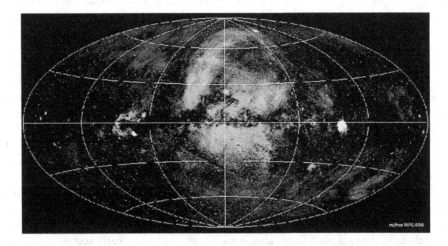

FIGURE 5.24 An X-ray image of our universe as seen by the European X-ray satellite ROSAT. The figure shows various intense X-ray sources along the Galactic equator, which are recognized as the remnants of supernovae accreting neutron stars or black holes. In addition, the figure also shows that the entire universe is filled with low-energy background radiation in the X-ray region. Source: ROSAT-All Sky Survey, DLR.

universe is filled with high-energy particle radiation, which is accelerated by the magnetic fields of our Sun or distant supernovae and eventually reaches Earth.

Interaction between these kinds of cosmic radiation and the Earth's atmosphere accounts for our exposure to cosmogenic radioactivity. There is a large range of possibilities by which highly energetic cosmogenic particles interact with the molecules, atoms, and nuclei in our atmosphere. This will be discussed in the following section.

5.1.3 ORIGIN OF COSMIC RAYS

Cosmic rays are a high-energy radiation field to which the Earth is constantly exposed. This radiation is being produced by nuclear reactions in cosmic objects such as the Sun, other stars, supernovae and other explosive stellar processes in our galaxy as well as other far-away galaxies. Since the universe is largely empty,[19] cosmic rays can pass unhindered through large areas of the universe without undergoing major absorption processes. Therefore, even radiation produced during the early phase of the universe can be seen, although it has cooled due to the constant expansion of the universe. For Earth and its inhabitants, the most important component of cosmic radiation is the so-called cosmic rays. While these cosmic rays are classified into many types, they are predominantly high-energy charged light particles but also high-energy photons that bombard the Earth's outer atmosphere.

Particle radiation makes up the largest fraction of cosmic rays. These ionized particles are the most influential component in Earth's atmospheric radiation, although they are mostly deflected in the magnetosphere by the Earth's magnetic field. Cosmic ray particles have different origins, which correlate with distinct energy ranges within the radiation spectrum. This observation confirms that the composition of particle radiation changes as a function of energy, since the abundance distribution of the particle flux is related to its origin. Most of the radiation in the low-energy range comes from the Sun. These particles are produced in solar storms or protuberances and are accelerated by the Sun's magnetic fields. The intensity of these protuberances and the associated particle flux are directly dependent on the Sun's activity, which is subject to an eleven-year cycle. This radiation has energies of up to 10^{10} eV (10 GeV) and intensities of up to 10^4 particles per square meter per second (Figure 5.25), which impinge on the outer layers of Earth's atmosphere.

Cosmic ray intensity drops rapidly, with increasing energy, over seven orders of magnitude – from 1 particle per square meter per second at 10 GeV to 1 particle per square meter per year at 10^{16} eV (10 TeV). Cosmic rays in this energy range are produced by countless supernovae explosions (one to two per century, on average) in our galaxy. This energy is accelerated by associated magnetohydrodynamic processes; cosmic jets from fast-rotating black holes or pulsars are also a possible source for this cosmic ray component. This high-energy radiation fills the interstellar space of the Milky Way, with a substantial increase in intensity at its center.

Cosmic rays also have higher-energy components, of up to 10^{20} eV, that originated in the extragalactic region of our universe, according to today's conception. These rays have traversed the enormous distances between galaxies[20] and their existence gives an indication of the dramatic early history of our universe, after its formation 13.8 billion years ago. The small fraction of this extremely high-energy radiation, in the range of less than 1 particle per square meter per century, is insignificant with regard to its effect on the radiation load of our atmosphere.

Cosmic rays consist of approximately 87% protons (hydrogen nuclei), 12% alpha particles (helium nuclei), and 1% heavier atomic nuclei up to iron. On average, about a thousand to ten thousand particles per square meter per second strike the Earth's outer atmosphere. However, this particle flux depends very much on latitude, since charged-particle radiation is largely mitigated by deflection in Earth's magnetic field. Since the Sun is considered the dominant source of cosmic rays, overall flux also depends on the Sun's activity, which varies according to conditions in the solar magnetic field.

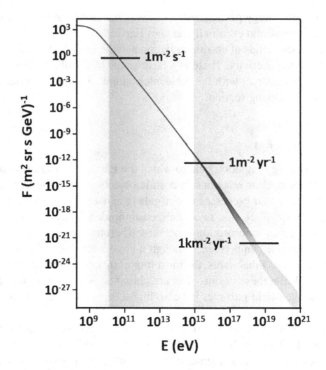

FIGURE 5.25 The flux of cosmic rays as a function of the kinetic energy of incident particles. Source: https://en.m.wikipedia.org/wiki/File:Cosmic_ray_flux_versus_particle_energy.svg (March 13, 2024).

The eleven-year solar activity cycle is clearly observed as a strong – often, almost ten- to twenty-fold – increase in dark sunspots,[21] while the cosmic-ray intensity varies by only 0.1%.

High-energy cosmic particles, so-called primary particles, interact with gas molecules in the atmosphere producing broadly scattered showers of secondary particles. These particle showers are initially, narrowly spread but if they reach the ground, they can extend horizontally for up to one square kilometer (Figure 5.26).

Secondary particles are mainly produced by the breakup or spallation of primary particles. Spallation products include protons, neutrons, muons, and pions, although heavy particles can also be produced in small numbers with the breakup of heavy primary particles such as iron.

The number of secondary particles depends on the energy of the incident primary particle; at solar energies, radiation showers can produce up to 10^{11} secondary particles per primary particle. Secondary radiation intensity gradually decreases due to collision, absorption, and decay processes, so that only a few percent of the initial radiation intensity reach the Earth's surface. However, particles from the radiation flux can directly undergo nuclear reactions with atmospheric atoms, resulting in the production of a significant amount of a number of radioactive isotopes such as ^{3}H (tritium), ^{10}Be, and ^{14}C in our atmosphere. These are deposited on the ground by rain and other atmospheric means and distributed by hydrological, geological, and biological processes, as will be discussed later in more detail.

5.1.4 Cosmic Neutrino Radiation

Next to the cosmic particle flux, neutrinos are probably the most interesting radiation field. The universe is literally filled with neutrinos. Neutrinos as well as their antiparticles, anti-neutrinos, are produced in weak interaction processes. These processes include β-decay and electron capture, as well as reactions and decay processes caused by other types of leptons such as muons. Neutrinos

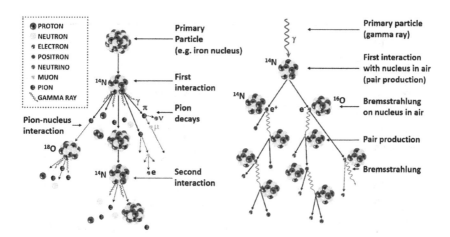

FIGURE 5.26 Development of cosmic showers – high-energy particles – due to spallation reactions in the atmosphere. After, Konrad Bernlöhr, https://www.mpi-hd.mpg.de/hfm/CosmicRay/Showers.html (June 5, 2023).

themselves are neutral particles. Therefore, collision processes and reactions only occur via the weak interaction, resulting in extremely low scattering or reaction probabilities.

The effective cross section for neutrino-induced scattering processes is in the range of 10^{-14} barn, which is considerably lower than reaction processes involving charged leptons such as electrons or uncharged hadrons such as neutrons. This means that neutrinos can traverse large masses of material almost unhindered, as mentioned earlier with the example of solar neutrinos. These solar neutrinos are produced by reaction and decay processes inside the Sun, during the fusion of hydrogen into helium, and then escape almost undisturbed from the Sun's interior.[22] Neutrinos can only be measured over a period of years using large-volume detectors deep within the Earth but even then, probability for their detection depends largely on their energy.

Our atmosphere's largest neutrino-flux incident comes from the Big Bang. As discussed before, these neutrinos originated during the first few seconds, by weak interaction processes, before separation of the baryons into protons and neutrons. Due to the expansion of the universe, neutrinos cooled down to extremely low energies of 0.1 meV. At this energy, they have filled the universe virtually undisturbed for 13.8 billion years. The flux of the cosmic neutrino incidence on Earth is – at about 10^{12} neutrinos per square meter and second – many orders of magnitude higher than the flux of cosmic-particle radiation.

Solar neutrinos are the second important component (Figure 5.27) of the neutrino flux. As discussed earlier, these neutrinos are produced during hydrogen burning in the interior of the Sun, primarily by pp chains, with a few percent also forming during decay reactions in the CNO cycle. Solar neutrinos have a much higher energy, of 1 MeV, when compared with cooled neutrinos from the Big Bang. The integrated flux of these solar neutrinos is comparable to that of Big Bang neutrinos – about 10^{12} neutrinos per square meter per second.

A much higher flux is expected from nearby supernova explosions. These neutrinos originate primarily from electron capture reactions involving iron and similarly heavy elements in the pre-supernova star. These reactions destabilize the star's center and initiate its collapse into a neutron star. Indeed, the explosion of supernova 1987A, in the neighboring dwarf galaxy of the Magellanic Cloud nearly thirty years ago, produced an enormous neutrino flux that was comparable to, or even larger than, the solar flux. However, only a few neutrinos from this flux have been detected on earth.

Anti-neutrinos, from the Earth's mantle and crust, are in this same energy range. These are the so-called geo-neutrinos, which are not really cosmogenic but rather radiogenic since they originated

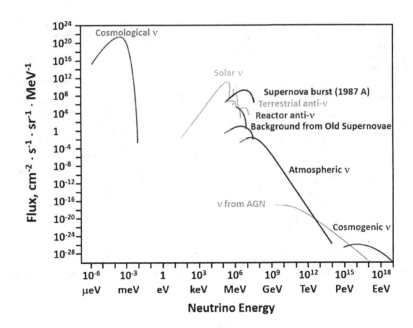

FIGURE 5.27 Flux of cosmic neutrinos as a function of neutrino energy. Marked are the different places neutrinos are known to originate. Image source: IceCube collaboration/NSF. C. Spiering, "Neutrino Detectors Under Water and Ice," (2020) in C. Fabjan, H. Schopper, (eds), *Particle Physics Reference Library* (Cham: Springer), https://doi.org/10.1007/978-3-030-35318-6_17 (March 13, 2024).

in the radioactive decay chains of long-lived actinides and from the decay of ^{40}K and other long-lived isotopes embedded in the geological material of our planet, as will be discussed in Chapter 6.

As enormous as these numbers are, the impact of neutrino radiation on the daily radiation exposure of humans is negligible. While neutrinos are the most intense cosmic-radiation component on Earth's surface, the probability of neutrinos being absorbed is spectacularly low. When considering the human radiation dose, they can be completely neglected. One can easily estimate that less than one neutrino per second is stopped in the body. With an energy of less than 1 MeV, this results in a dose rate of less than 10^{-15} Gy/s or 1 fmGy/s for a human body.

5.1.5 Cosmic Photons

A third component of cosmic radiation is photons. This electromagnetic radiation fills the universe and its reflection is seen as Cherenkov light in the night sky. In many cases, the photon flux is a uniform source of radiation. Most of this is absorbed by nuclear and atomic ionization, excitation, and scattering processes in the outer layers of the atmosphere. A higher flux of radiation that could penetrate the Earth's atmosphere would be expected if a supernova exploded near our solar system. Depending on the distance, this could significantly impact the high-energy photon flux at Earth's surface for a brief time, leading to the production of additional long-lived radioactivity.[23]

Cosmic photons cover a wide energy spectrum that ranges from thermal to optical radiation and, in the high-energy range, from ultraviolet to X-rays and finally γ-radiation. This energy radiates into the void of space from nearby and distant stars as well as supernovae. The energy distribution of this radiation is described by Planck's radiation law, which shows that the radiation spectrum emitted by each body is dependent on the body's temperature (Figure 5.28).

The largest flux of photons on Earth comes from the closest cosmic-ray source, the Sun. The Sun is the Earth's largest energy source, with a surface temperature of approximately 6,000 °C, a

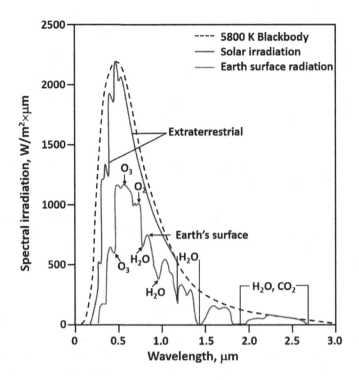

FIGURE 5.28 The Planck spectrum of solar radiation as a function of radiation wavelength. The dashed line represents the theoretical Planck curve, which corresponds to the radiation distribution of a body with a temperature of 5,800 K. The two solid lines correspond to the distribution modified by absorption effects at the outer surface of the atmosphere and at Earth's surface, respectively. Source: F. Incropera, et al., *Fundamentals of Heat and Mass Transfer* (New York: Wiley, 2007).

luminosity of $3.85 \cdot 10^{26}$ J/s and a thermal output of 73.5 MW/m². Looking at the radiation flux emitted from the solar surface, maximum radiation intensity is in the range of ultraviolet and visible light with a significant amount of infrared and thermal radiation (Figure 5.28).[24] It can be seen that at the low surface temperatures of the Sun, hardly any higher-energy radiation is emitted. Earth's atmosphere offers protective shielding since it absorbs most high-energy ultraviolet radiation with wavelengths below $\lambda = 0.4\mu m$. In addition, the atmosphere also absorbs a significant portion of photons with longer wavelengths via vibrational excitation of molecular-gas components, such as H_2O, O_2, O_3, and CO_2. These are the so-called climate gases that help balance earth's surface climate by preventing heat loss to outer space.

As in the case of neutrinos, the Big Bang reverberates the photon distribution that is filling the universe as the aforementioned cosmic background radiation. The enormous amount of radiant energy released by the Big Bang dominated the early universe as high-energy radiation for nearly 400 million years and still appears today as an echo of this tremendous event. The temperature at the beginning of the photon epoch was billions of Kelvin, which in the Planck spectrum corresponds to a radiation energy of up to giga-electronvolts. However, the universe cooled as it expanded, reaching a temperature in the 1,000 Kelvin range over the next 400 million years (Figure 5.29).

At temperatures around 1,000 Kelvin, the photon energy had reduced to only a few electronvolts. At this low-energy range, photons were no longer able to ionize the hydrogen and helium elements and prevent the recombination of free ions with the electrons so that atoms could form. In other

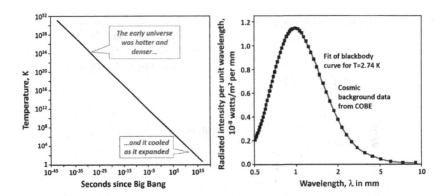

FIGURE 5.29   On the left, cooling of the universe from the early phase of the Big Bang to the present. This photon-dominated epoch occurred in the range between the first second and 400 million years. Shown on the right is the Planck spectrum of 3K background radiation; this is a radiation echo of the Big Bang, with a wavelength in millimeters. Source: Author's lecture notes.

words, space became too cold and since that time, atoms can only be ionized in the plasma of hot stars. Today, more than 13 billion years later, the universe has cooled to the point where the Planck distribution is at a radiation energy between 1.3 eV and 13 eV, corresponding to a wavelength distribution of 1 mm to 10 mm. Recent observations from the COBE and Planck satellites have accurately measured this distribution, showing (Figure 5.29) that background radiation corresponds to the energy distribution of an icy-cold body of 2.7 Kelvin, near the temperature of absolute zero.

In addition to low-energy radiation generated by the Big Bang and the Sun, higher-energy radiation is also present from the X-ray to γ-energy range. This has mostly resulted from explosive stellar events, such as supernovae to X-ray bursts that can be directly observed with satellite detectors outside the atmosphere. This radiation is not homogeneously distributed, but rather, it comes from the regions of greatest activity in stellar evolution and stellar death, such as the galactic center. Significant portions of this radiation, especially in the visible range, are absorbed by the dust and dark clouds located between the solar system and the center of the Milky Way. Without these dark clouds, the intensity hitting Earth would be many times higher.

Considering all of these together, electromagnetic radiation from the center of the galaxy and along the galactic plane reaches Earth in each of the wavebands, ranging from high-energy and thus short-wavelength γ- and X-radiation into the low-energy or long-wavelength range of radio waves. Figure 5.30 shows this distribution of radiation sources for all length and frequency wavebands, from longwave radio radiation to shortwave gamma radiation. Observable results are based on satellite measurements of the different wavelength ranges, meaning radiation hitting Earth's outer atmosphere. For most of the radiation, the atmosphere is opaque and the corresponding part of the spectrum is absorbed in the higher layers of the atmosphere - especially the ionosphere and the mesosphere. This again shows the importance of Earth's atmosphere as a radiation shield for the planet.

In addition to the Galactic Center, other point sources of high radiation intensity can also be seen (see Figures 5.7 and 5.21). These can mostly be attributed to gas and dust clouds, the remnants of a supernova explosion in our galaxy. The radiation comes from both the pulsar – a rapidly rotating neutron star, as the central remnant of the explosion – and the rapidly expanding highly radioactive cloud of matter that was ejected in the supernova. Although such explosions mostly produce high-energy electromagnetic radiation, the photon flux is mostly very small and completely absorbed in the atmosphere. If the supernova explosion had taken place in the vicinity of a solar system closer

FIGURE 5.30   Distribution of electromagnetic radiation sources, ordered by frequency or wavelength ranges, over the galactic plane, with radiation reaching its maximum intensity in the Galactic Center. This is true for all frequency ranges except X-rays and visible light, which are absorbed by interstellar dust accumulations between the solar system and the Galactic Center. Credit: Astrophysics Science Division, NASA.

to Earth, there would be, depending on the distance, a considerably larger radiation flux on Earth's surface than we observe at present.

NOTES

1. Michael Thoennessen, *The Discovery of Isotopes: A Complete Compilation* (Heidelberg: Springer, 2016).
2. Henrik Svensmark, "A persistent influence of supernovae on biodiversity over the Phanerozoic," *Ecology and Evolution* 13/3 (2023), doi:10.1002/ece3.9898.
3. ROSAT operated as an international observatory from 1990 to 1999.
4. Chandra is operated jointly by NASA and ESA.
5. Interferometers rely on splitting a beam of light into two beams, using mirror reflection, and subsequently merging them to produce an interference pattern between the two electromagnetic waves. Large-scale laser interferometers such as LIGO and VIRGO are designed to detect gravitational waves that have originated in far distant stellar explosions such as the collisions of neutron stars or black holes. When the gravitational waves pass through the interferometer, it creates discrepancies in distance between the laser beams, which have to travel through the extended tunnel system. This affects the timing and interference patterns, thereby providing information about the event.
6. In the cooling water of open swimming-pool reactors and in nuclear power-plant decay pools, Cherenkov radiation is seen as the greenish-blue shimmering light produced by the release of ionizing radiation. This radiation is mostly comprised of free electrons that can move at a speed greater than the speed of light in this medium.
7. There are so many Big Bang neutrinos in the universe at such a density that, at any moment, 10 million would fill a volume comparable to a human body.
8. B. Grefenstette, F. Harrison, S. Boggs, et al., "Asymmetries in core-collapse supernovae from maps of radioactive ^{44}Ti in Cassiopeia A.," *Nature* 506 (2014): 339–342.

9. Benoit Côté, Maria Lugaro, Rene Reifarth, Marco Pignatari, Blanka Világos, Andrés Yagüe, and Brad K. Gibson, "Galactic chemical evolution of radioactive isotopes," *The Astrophysical Journal* 878 (2019): 156.

10. Radiation flux is typically defined as the number of radiation particles per intersected area and time.

11. Pioneers in this field are, for example, Russian-American physicist George Gamow (1904–1968) along with the Belgian priest Georges Lemaître (1894–1966), who are considered the intellectual fathers of the Big Bang theory while also providing essential work on radioactivity. Gamow long held the view that all elements were created in one swoop with the Big Bang. This was stated in his famous work named after the three authors: "R.A. Alpher, H. Bethe, H.G. Gamow," *Physical Review* 73, 803 (1948). Only in the 1950s did Gamow change his views.

12. However, the baryon mass – i.e., our atoms – forms only a small part (4.6%) of the total mass of the universe. The biggest part is formed by so-called dark matter (Dark Matter) with 24% and dark energy (Dark Energy) with 71.4%. Presumably, dark matter is made up of other types of particles that are difficult to detect, such as leptons. Our understanding of dark energy is still largely speculative; it is assumed that dark energy corresponds to masses that repel each other and thus expand the universe.

13. Likewise, other parts of the Sun's electromagnetic spectrum – X-rays, UV, and infrared radiation – are produced in the photosphere.

14. Michael Wiescher, "The history and impact of the CNO cycles in nuclear astrophysics," *Physics in Perspective* 20 (2018): 124–158.

15. Paul W. Merrill, "Technetium in the stars," *Science* 115 (1952): 479–489.

16. More than 75% of all stars are binary or multiple star systems, as can be concluded from rotational variations in the luminosity of the observed systems. Only a few stars are single star systems like our Sun. In the case of Jupiter, our largest planet, the mass has not been sufficient to ignite nuclear reactions in the interior.

17. John J. Cowan, Christopher Sneden, James E. Lawler, Ani Aprahamian, Michael Wiescher, Karlheinz Langanke, Gabriel Martínez-Pinedo, and Friedrich-Karl Thielemann, "Origin of the heaviest elements: The rapid neutron-capture process," *Reviews of Modern Physics* 93 (2021): 015002.

18. Indeed, such an event was observed in 2017. At LIGO, gravitational waves were observed at a distance of between 85 and 160 light-years. These observations were also confirmed in the optical-wavelength range. Analysis of the data is consistent with theoretical-model predictions for the r-process. Therefore, astrophysicists tend to assign a larger role to neutron-star merger in the production of heavy elements than was previously thought. M. R. Drout, et al., "Light curves of the neutron star merger GW170817/ SSS17a: Implications for r-process nucleosynthesis," *Science* 358 (2017): 6370.

19. That is, apart from those types of matter that have not yet been fully investigated. These include dark matter (Dark Matter) – presumably a type of elementary particle – and dark energy (Dark Energy) whose nature remains largely a matter of speculation to this day.

20. According to today's cosmological estimates, which are based on measurements from the COBE satellite, our current visible universe has a diameter of approximately 93 billion light years and contains, apart from our Milky Way, up to 200 million other galaxies.

21. These are cooler surface zones that correspond to a condensation of the solar dynamo's magnetic field lines. Due to their lower temperature, the sunspots appear darker to the observer.

22. For purposes here, the so-called neutrino oscillation has been ignored. There are three types of neutrinos, which are assigned to respective charged-lepton particles: the electron neutrino, the muon neutrino, and the tau neutrino. Neutrino oscillation means that one type of neutrino can change into another after a certain time. This is the case with solar neutrinos, which originate as electron neutrinos but partially transform into tau neutrinos on the long way from the center of the Sun to the Earth.

23. This is discussed as a possible explanation for the increased abundance of the long-lived radioactive iron isotope ^{60}Fe in the manganese-rich deposits of the Pacific Ocean's layers. The depth at which these deposits were detected would correlate with a near-Earth supernova about 3 million years ago. Such an event would result in a significantly higher radiative flux in the Earth's atmosphere, the effects of which on the genetic evolution of biological life are still unknown. K. Knie, "Astrophysics, traces of a stellar explosion," *Physics in Our Time* 36, 1 (2005): 8.

24. The wavelength of a photon λ is inversely proportional to the photon's energy with c as the speed of light and h as Planck's quantum of action.

6 Our Radioactive Planet

6.1 OUR RADIOACTIVE PLANET

Neither the depths of the universe nor those of our planet are directly accessible to us. Yet, we interpret Earth's interior as well as cosmic phenomena through the observation of proxies. These are secondary sources of data from which one can conclude, on the basis of theoretical interpretation, desired information by means of generally valid physical or chemical laws. On the basis of this information and interpretation, various structural models of the Earth's interior and the geological evolution of our planet are built. The process, like all scientific processes, is fraught with statistical uncertainties. These include uncertainties that arise in the interpretation of data and the complexities of the models.

In 1912, German polar explorer Alfred Wegener (1880–1930) challenged a dominant notion of the 19th century, that the Earth's crust was firmly connected to a largely solidified interior. Wegener posed a new theory, the dynamic model of continental drift. According to this model, the continents float as a crust on Earth's liquid inner core. Only after long resistance in the 1950s and 1960s, his idea eventually prevailed due to extensive geological and geographical studies and measurement. Over the following decades Wegener's theory extended so that today we have a generally accepted model of plate tectonics. The following observations are based on these ideas about our dynamic Earth.

External geodetic observation of Earth's orbit and gravitational pull gives a radius of 6,370 km and a total density averaging $\rho = 5.5$ g/cm^3. The average density at the surface is, however, only 2.5 g/cm^3. From this, we can calculate that the density in the Earth's interior is about 12 to 14 g/cm^3. This is more than four times the density of granite and almost twice that of pure iron.

A more detailed understanding of Earth's structure requires geological and chemical information on the distribution and homogeneity of the material inside our planet. Other questions concern the origin and geological evolution of Earth as part of the evolution of our solar system. Our understanding of the stars and the universe is mostly based on electromagnetic radiation, which we have obtained spectroscopically as discussed in Chapter 5. This spectroscopic information is supplemented by chemical and physical analysis of isotopic abundances in meteoric rocks (cf. Chapter 5.1.1).

If we want to take an analytical look into the interior of the earth, spectroscopic methods have only limited applicability. For such purposes, seismographic measurements can be used to study the density distribution in Earth's interior, analyzing the structures of different zones at the Earth's core, mantle, and inner crust. Direct measurement of the chemical structure of Earth's material is, however, limited to the uppermost crust since deep boreholes can only penetrate a few kilometers at best and do not even reach the inner crust.[1] On the other hand, when lava erupts from greater depths, direct chemical analysis of the ejected material is possible.

In addition, important information can be gained by determining the age of crystal structures from Earth's various early layers. Radiochemical methods can measure the abundances of both radioisotopes and their daughter isotopes, which are embedded in these rock crystals. From the abundance ratio of parent and daughter isotopes, one can directly determine the age of these rocks and obtain information about changes that have occurred over Earth's history. To do this, one must have basic information about the radioactive components of the elements found in the rock, particularly the half-life and the ratio of their stable daughter isotopes.

DOI: 10.1201/9781003435907-6

Since Earth's inner region is not directly accessible, chemical measurements are limited to the crust. However, within the context of our ideas about Earth's geological evolution, chemical measurements can be extended by including information on the chemical composition of meteoritic rocks, the chondrites. These reflect the chemical structure of presolar dust, from which at least the inner planetary system formed.[2] Meteoritic rocks are primarily comprised of an iron-nickel alloy (Fe-Ni) and silicates, with a light trace of iron and magnesium. This provides a first basis for the assumption that material within the Earth also consists of two components.

The most important information on the current structure of Earth's interior comes from seismic measurements. Various measuring stations are used to determine the distribution, direction, and speed of seismic waves moving along the Earth's crust and through the Earth's interior. These waves are triggered by local earthquakes or, in the past, by nuclear bomb tests. Travel time, attenuation, and reflection of the various seismic waves[3] indicate density and phase changes within the Earth material.

Results of these seismic-wave studies indicate a shell structure at the Earth's interior. The seismic profile of the Earth's body shows two discontinuities; these are interpreted as boundary lines between, on one side, the Earth's crust and mantle and, on the other side, the Earth's mantle and core. Thus, the Earth's structure can be roughly divided into three sections: crust, mantle, and core. With more refined methods, further discontinuities have been identified, which are either chemical in nature or indicate changes in the plasticity of Earth's material. These discontinuities reflect phase transformations. Such transformations are indicative of change in either the chemical composition or the Earth material itself, which changes into a liquid or solid phase under the influence of pressure or temperature conditions. Seismic methods thus provide the most important information on the present state and structure of Earth's interior.

Seismic studies are broadly complemented by measurements of the Earth's geomagnetic field and its variations. The existence of such a field requires that an electric current be generated in the Earth's interior, which in turn necessitates a liquid metallic Fe-Ni structure. Evidence of periodic variations in the field comes from the differential orientation of iron fragments as they are observed in magnetic rocks. Since the magnetic field arises from geodynamic processes in the Earth's interior, more detailed information on the fluid conditions at Earth's center can be extracted from this analysis. If the age of the ferruginous fragments can be determined, it may even be possible to draw conclusions about temporal changes in the Earth's magnetic field.

Geothermal heat is also an important information source. According to current ideas, heat production in the Earth's interior is 42 to $47 \cdot 10^{12}$ W, or 42 to 47 terawatts. This figure is primarily derived from an estimate of the heat flow between Earth's interior and surface. This flux is determined by the thermal conductivity, or conduction of the rock, and by the geothermal gradient $\Delta T/\Delta x$, which signifies the temperature change ΔT over some depth range Δx.

The geothermal gradient is determined from temperature measurements at various depths in a borehole. Laboratory measurements can determine the average thermal conductivity k of the rock layer that has been drilled through, using the heat-flow equation: $\Phi = k\Delta\dfrac{\Delta T}{\Delta x}$, which allows to estimate heat transport from Earth's interior to the outer shell and therefore also allows to calculate the cooling rate of our planet. Thermal conductivity ranges from $k = 2$ to 5 W/m• °C depending on rock type and characteristics.

This heat flux depends on the tectonic conditions at the drilling site and varies between 0.03 and 0.13 W/m^2, with an average value of 0.08 W/m^2. These values suggest an interior heat source but say nothing about the origin and distribution of this heat. The question is whether this heat flux is the residual heat of a slowly cooling body, whose original heat came from the planet's early formation, or whether other heat sources still exist that are contributing to the heating of the Earth's body. Such heat sources could be frictional processes, between the different Earth layers, or also chemical as well as radioactive-decay processes in Earth's interior.[4]

In 2005 an important scientific breakthrough considerably expanded the range of available information from which we are able to make observations about our Earth: exoplanets of distant solar systems were observed in various stages of development. From the physical-chemical properties of these exoplanets at different stages in their development, direct conclusions can be drawn about the evolution of our planetary system.

Of greater importance for our question was a second breakthrough from which scientists are now also able to directly determine the internal radioactivity of our Earth. This is done by measuring anti-neutrinos originating from the natural uranium-thorium decay series. These are geoneutrinos, which provide a direct look into the radioactive composition of our planet's interior.

These measurements were successfully carried out with the neutrino detectors Borexino in Italy and Super-Kamiokande in Japan[5] (cf. Chapter 5.1.1). Results directly demonstrate that 50% or more of Earth's heat comes from radioactive decay of heavy actinides. Further measurements are planned at the SNO detector in Canada to determine neutrinos present from decay of long-lived potassium radioisotope ^{40}K. Its occurrence is predicted in large quantities for Earth's mantle and especially for Earth's crust. Such statements are based on a series of systematic studies regarding the ^{40}K to ^{238}U ratio of various rock types in the crust; this ratio established the K/U ratio at 10^4.

The potassium isotope ^{40}K, with a half-life of $1.25 \bullet 10^9$ years, decays predominantly via β^--decay to ^{40}Ca. Only 10% of ^{40}K decays via electron capture to ^{40}Ar, which mostly occurs in its excited state at 1.46 MeV before decaying to its ground state via γ-emission. In total, 1.32 MeV are released during the β^--decay of ^{40}K to ^{40}Ca with an additional 1.5 MeV released during the decay to ^{40}Ar. In this process, 1.46 MeV are released as γ-energy. The long-lived actinide isotopes ^{232}Th and ^{238}U convert to lead isotopes via chains of α- and β-decay. This releases an average of 4 to 5 MeV of energy per α-decay, which converts directly to geothermal heat via scattering processes.

These predictions make it clear that radioactivity is Earth's dominant internal energy source. This has a determining influence on our planet's development, structure, and living conditions. The ratio between radiogenic heat production at the Earth's interior and heat loss through Earth's crust is the so-called the Urey ratio (after American chemist Harold Urey). According to current ideas, the Urey ratio is 0.57, as an average over Earth's history. Accordingly, freshly generated radiant heat replaces only 57% of heat lost. In Earth's early history, this replacement value was much higher. Thus, since formation, the interior has been slowly cooling – a process that continues to accelerate as internal radioactivity subsides.

6.1.1 Evolution of the Planetary System

Our Earth is part of the solar system located in our galaxy, the Milky Way. As such, our planet is far from the galaxy's radiation-filled center, where a high-energy glow of radioactivity is largely shielded by dark dust clouds located between our Sun and this galactic center. Our solar system was formed nearly 5 billion years ago by the contraction of a protostellar molecular cloud comprised of gas and dust. According to current ideas, this cloud looked similar to the Orion Nebula, which is considered the birthplace of young stars.

The gas and dust material in the cloud contained the ashes of many previous star generations in the then 9-billion-year-old universe.[6] The abundance distribution of the elements in this early solar – or presolar – system had also been formed by previous star generations through many nuclear-burning processes (see Chapter 5, Figure 5.14). Over the next relatively short 10,000 years, this cloud collapsed to become a faint star, the protosun.

According to our present models, the early cloud was rotating slowly. Yet, with time, this cloud developed by contraction into a rapidly rotating star, around which a flattened disk of gas and dust matter formed. The young star contained about 99.8% of the total mass, while the disk contained only 0.2%. From this disk, our planetary system formed. The details of the subsequent evolution of this dust disk into an ordered planetary system are not yet fully understood. Presumably, evolution involved a balance between the central star's gravitational attraction and the centrifugal forces of rotation. In addition, there was the influence of the protosun's magnetic field, which caused the flow of mostly electrically charged particles near the protosun to split in two. A part of these particles was accreted into the central star, while a part was deflected and emitted as a gas jet. This emission resulted in a mass and dust-free zone around the central star - thus separating the protosun and planetary disk (Figure 6.1).

Within the dense dust cloud, there were countless collisions between individual dust particles. Many of these dust sprinkles were covered by a layer of ice, due to the high hydrogen, oxygen, and even water content of the early cloud. This in turn led to the so-called snowball effect, where particles stuck to each other and formed the condensation nuclei for planet formation. These so-called planet embryos could already reach sizes comparable to those of the Moon or the smaller inner planets such as Mercury, Venus, Earth, and Mars. Computer simulations of these processes show that the formation of a planetary body the size of the Earth would only take a few million years.

Bodies of this size already had considerable gravity and could attract gases and dust particles from the cloud surrounding them. These accretion processes swept through the planetary environment. Within tens of millions of years, the giant outer planets of our solar system formed. These included today's Jupiter, Saturn, Uranus, and Neptune, consisting mostly of gas.

The changing mass in these accretion processes often led to changes in the orbit of the planets. This resulted in collisions with smaller planetary embryos or asteroids. In this way, either larger massive asteroids were captured or parts of planets were knocked out, as for example during the formation of our Moon, according to common conception. Such processes formed inner planets from the massive dust and rock materials of leftover planetary embryos. This mechanism took several 10 to 100 million years and explains the difference in the chemical composition of the inner, small rocky planets as compared with the outer, larger gas planets.

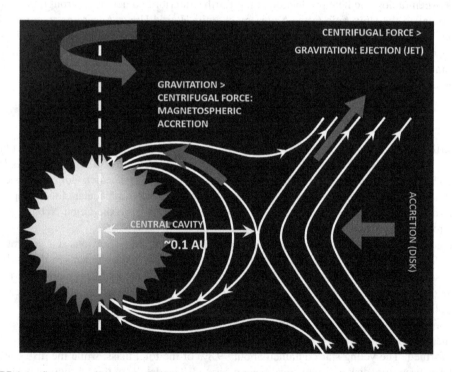

FIGURE 6.1 Structure of magnetic-field conditions in the early solar system, causing the separation of field lines. As the field split, a part remained connected to the central star and a part remained connected to the planetary disk. This also caused a further separation of the charged dust particles, which followed the course of the magnetic field lines. According to Muriel Gargaud, et al., *Young Sun, Early Earth and the Origins of Life* (Berlin: Springer, 2012).

It is assumed that planetary embryos had an average mass of 1 million (10^6) kg. With an accretion time of 10 million years, formation of the Earth body with its present mass of $6 \cdot 10^{24}$ kg required an enormous number of impact processes per day – an estimated 280 million! This released huge amounts of heat: at the typical speed of 30 km/s for asteroids, with a per impact energy of about $4.5 \cdot 10^{14}$ J, this is equivalent to ten times the heat of the Hiroshima bomb ($1.26 \cdot 10^{23}$ J) per day. This corresponds to a thermal power of $1.5 \cdot 10^{18}$ W or also 1,500 PW.

As the outer, giant planets steadily increased in mass, the associated gravity increased between them. These planets continued to change orbit and drew a new furrow of destruction through the planetary disk, which was filled with dust, boulders, and smaller planetary embryos. Many of these components were either caught or thrown out of their orbits, plunging toward the Sun. In this way, the inner planetary system was exposed to a hail of asteroid projectiles. This period in our planetary system's formation is the so-called Hadean after Hades, the realm of the dead in Greek mythology. Countless craters on the surface of the moon still serve as evidence of this time and while these craters also existed on Earth, later geological changes have largely covered them.

Earth's slow formation took place 4.57 to 4 billion years ago. This time frame has been derived from analysis of the oldest known rocks, such as acasta gneiss and zircon crystals, which crystallized from early magma. In addition, there is analysis of the carbon-rich chondrites, small trace elements in meteoric rocks found on Earth or collected by space probes. A portion of the chondrites' composition is, for example, CaAl crystals. These trace elements of long-lived radioactive isotopes form an internal clock driven by the radioactive decay mechanism. They include: ^{26}Al ($T_{1/2}$ = 717,000 a), ^{41}Ca ($T_{1/2}$ = 100,000 a), ^{60}Fe ($T_{1/2}$ = 1.5 Ma), ^{87}Rb ($T_{1/2}$ = 45 Ga), ^{147}Sm ($T_{1/2}$ = 100 Ga), and ^{182}Hf ($T_{1/2}$ = 9 Ma), as well as radioactive actinides that were embedded in the condensing or crystallizing material due to their chemical properties. However, all these radioactive isotopes decayed to other isotopes with different chemical properties such as ^{26}Mg, ^{41}K, ^{60}Co, and ^{60}Ni, respectively. In addition, ^{147}Sm, ^{182}W, and the actinides slowly converted to the various stable lead isotopes ^{206}Pb, ^{207}Pb, and ^{208}Pb through decay chains that are still observable today.

The abundance comparison of the different lead isotopes, as end products of the natural actinide-decay series, as well as the isotope ratios of the daughter elements of other radioactive components allow scientists to determine the age of the chondrites and their CaAl inclusions, which are the oldest known material in our solar system. A time frame for the various condensation and crystallization processes that occurred during the formation of the planet can also be determined from these materials. This means that an internal clock exists as an enormously important tool for tracking the geological evolution of our planet and the entire solar system.

How did geological developments proceed, and what does this mean for the inner radioactivity of our planet? Already in the early Hadean phase – when the surface, which was heated by perpetual meteorite bombardment, still consisted of liquid magma – a first separation of the different rock components began. The denser materials, which included liquid Fe and Ni as well as heavy metallic elements, sank to the center under the influence of Earth's gravity. Lighter rock types, such as silicates, had a higher melting point and so remained as solid condensates on the magma surface, where they slowly cooled to form Earth's mantle.

The time scale for this separation into a metallic interior and mineral-dominated mantle[7] can be established by analyzing the radioactive decay of nonmetallic hafnium ^{182}Hf into the metal tungsten ^{182}W. The half-life of ^{183}Hf is $T_{1/2}$ = 9 Ma and, in general, it can be said that after five half-lives, the original radioactive isotope has practically disappeared since after that period only about 3% remains according to the decay law. Therefore, hafnium would have remained as a mineralogical element in the silicate mantle during separation and, after decay, tungsten would have accumulated as a metal in the Earth's interior. From the tungsten abundance observed in today's mantle, it can be inferred that this separation of materials, between Earth's core and mantle, must have occurred about 30 million years after the formation of the planet.

This result is supported by comparing the abundance of neodymium isotopes in meteorite material with lava material from the inner-mantle zone. Samarium is a component of the silicate-rich mantle material. Two long-lived samarium isotopes decay by α-emission to neodymium: ^{146}Sm to ^{142}Nd with a half-life of 103 million years (Ma) and ^{147}Sm to ^{143}Nd with a half-life of 106 billion years (Ga). Chondrite inclusions in meteorite material show a lower ^{142}Nd/^{143}Nd ratio than in terrestrial mantle material. From this it is also calculated that the separation between mantle and core must have taken place early, 30 million years after formation of the Earth body.

6.1.2 RADIOACTIVITY IN THE EARLY SOLAR SYSTEM

As we have seen, the decay of long-lived radioactive elements plays a special role in understanding the early evolution of the solar system. While these decay processes provide scientists with an internal clock for the age determination of different materials, they also drove the many physical and chemical processes that determined our solar system's evolution. In addition, these same decay processes contribute to the habitability and evolution of our planet even today.

In the beginning, 4.5 billion years ago, there were the long-lived radioactive nuclei ^{40}K, ^{232}Th, and ^{238}U, together with other isotopes with half-lives of 10 million to 100 million years. As we have also discussed, there were still a whole series of short-lived radioactive isotopes with half-lives between 0.1 and 10 million years, which have long since decayed due to their relatively short lifetimes. The identification and abundance of these early radioactive elements is mostly based on analysis of the isotopic ratios of stable daughter nuclei found in meteorite material from the early solar system. These results show that there were a significant number of radioactive elements and that the intensity of their decay radiation had a major influence on the first phase of our planetary system's formation.

The initially most important radionuclides were short-lived elements such as ^{10}Be (beryllium), ^{26}Al (aluminum), ^{36}Cl (chlorine), ^{41}Ca (calcium), ^{60}Fe (iron), ^{92}Nb (niobium), ^{107}Pd (palladium), ^{126}Sn (tin), ^{129}I (iodine), ^{135}Cs (cesium), ^{142}Nd (neodymium), and ^{182}Hf (hafnium), which, because of their relatively short lifetimes, provided through the release of decay heat an enormous energy source for the early phase of planet formation. These isotopes have different nucleosynthesis histories and over extended periods of up to 50 million years they contaminated the original protosolar cloud with relatively short-lived radio activities, the decay products of which are now found in fixed proportion to the original stable isotopes of the respective elements. Some of these products are expected as a result of various burning phases in massive stars; some are predicted for lower-mass stars, as products of the s-process; and some are attributed to high-neutron-flux nucleosynthesis processes such as the r-process.

One of the original theses for the birth of our solar system is the Big Bang theory, whereby a supernova exploded in the region of the protosolar cloud. Also, more recently, rare neutron-star merger events are being discussed as possible triggers for the formation of the solar system.[8] This explosion is said to have enriched the cloud with radioactive elements and also, stimulated/accelerated its contraction through compression by the external shock wave. Other theories suggest that a stellar wind of massive (larger than twenty-five solar masses) Wolf-Rayet stars was the source that enriched the protosolar cloud environment with radioactive elements such as ^{26}Al, which in turn enriched the dust particles with radioactive depositions.[9] According to this theory, the massive star continued to evolve into a supernova whose shock front triggered the collapse of the protosolar cloud.

A detailed analysis of these scenarios, which account for the injection of radioactive elements, is extraordinarily complex and therefore not described in this framework. More extensive background and details about this emerging development can be found in the technical literature.[10]

Of the numerous short-lived isotopes that were present in the protoplanetary solar system (Figures 6.2 and 6.3), ^{26}Al ($T_{1/2} = 717,000\ a$) particularly stands out. ^{26}Al had a special role in the thermodynamic and chemical evolution of planetesimals, which were the first building blocks of the planetary system. The abundance of stable ^{26}Mg daughter nuclei found in meteoritic samples allows us to conclude that the initial ^{26}Al abundance present in the protosolar cloud was seventeen times

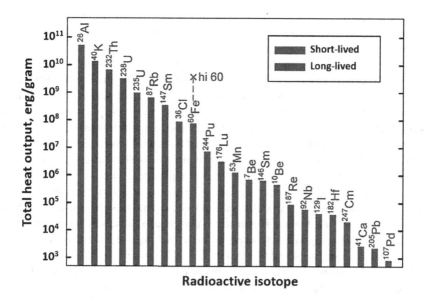

FIGURE 6.2 Heat energy per gram released with the complete decay of radioisotopes located in the early solar system. Radioisotopes are ordered according to their respective contribution to total decay-energy production. The original abundance of these isotopes is derived from meteorite measurements. Uncertainties exist for ^{60}Fe, which is why two proposed values are shown. After, Lugaro, et al., 2018.

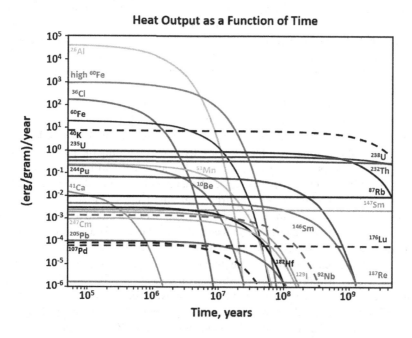

FIGURE 6.3 Annual decay energy per gram released by the different radioactive isotopes in the early solar system as a function of time, starting 50,000 years after time zero. It can be seen that most of the short-lived isotopes had already decayed after 10 million years, so that only long-lived isotopes like ^{40}K, ^{232}Th, and the two uranium isotopes ^{235}U and ^{238}U should still be present in today's planetary system and therefore also on Earth. The latter then still contribute to the internal heat balance of Earth and other planets. Originally released ^{26}Al decay heat is shown to be about 50,000 times higher than the heat released today by decay of ^{40}K and the long-lived actinides. After Lugaro, et al., 2018.

greater than the average ^{26}Al abundance in our Milky Way. This large abundance of ^{26}Al, with its relatively short half-life, released an enormous amount of decay heat, an amount that was considerably greater than the decay heat released by other radioactive isotopes.

The decay of ^{26}Al heated the inner core of planetesimals – small planets with diameters greater than 10 km – to several thousand degrees. In this way, the melting point of rock was reached,[11] causing a change in the original consistency. Mixing and the formation of new minerals occurred as well as evaporation of the ice chunks that had been incorporated into these planetesimals. Geochemical processes inside the planetesimals were triggered such that silicates formed and possibly also other organic molecules such as amino acids, which built up from the carbon present in the melt.[12] There is still considerable discussion around possible consequences for the further geochemical evolution of the planets as they perhaps slowly built up from these planetesimals.

It is, however, already clear that radioactivity determines, in a very complex way, the various aspects of planetary evolution. In particular, radioactivity determines the evolution of Earth's inorganic and possibly also organic matter. This process is essentially influenced by the heat evolution that occurs during radioactive decay. While this heat led to the melting of the original planetesimals, decay heat generated by remaining long-lived isotopes in the Earth's material still serves as an internal heat source. This source has far-reaching consequences for the evolution of our planet and for its present state. Decay heat is responsible for the partially liquid state of Earth's mantle (cf. Chapter 6.1.4), which as a circulating magma mass provides the engine for plate tectonics (cf. Figure 6.8). Thus, radioactivity is, albeit indirectly, the driving force for our planet's volcanic activity. In addition, the Earth's rotation, together with its iron core which is partially melted by decay heat, provide the geomagnetic field that protects our developing organic life from the deadly radiation of the outer space.

6.1.3 STRUCTURE OF THE EARTH

The structure of today's Earth can be described on the basis of results obtained with seismological techniques. Within the concept of a layer model, Earth's core has a radius of about 3,450 km and consists of an iron-nickel alloy, which is solid in the interior but liquid at the exterior. The core comprises about 15% of Earth's total volume. Above the core lies the so-called mantle, a not quite solid silicate layer more than 2,900 km thick. This mantle is plastically deformable and thus has fluid properties.

The mantle comprises 84% of Earth's total volume. The upper region of the mantle is called the *asthenosphere*. On top of it rests the solid crust, the lithosphere, which comprises only 1% of Earth's total volume. Friction and exchange, between the viscously mobile mantle layers and the lithosphere, are particularly important for our considerations regarding the planet's radioactivity.

Figure 6.4 shows a cross-section of Earth from the outer crust to the metallic center, which consists primarily of iron, nickel, and some trace elements of heavy metals. The figure also depicts temperature and density conditions resulting from measurements and various physical-model calculations. Temperatures in the Earth's core are seen to be between 5,000 and 7,000 °C and, while the inner core is solid due to the enormous pressure, the outer layer of Earth's core is largely liquid.

Density in the solid inner core is above 12 g/cm^3 and drops to 10 g/cm^3 in the outer liquid region of the core's interior. These conditions can be understood within the context of the Earth's interior, which is predominantly metallic in composition. Iron content dominates, with more than 80%, while the nickel content is 5–20% at its maximum. On average, the Ni/Fe ratio is about 0.057 to 0.082. Below 5%, other chemical components are also present. These are particularly found in the outer liquid portion of Earth's core, where the density is about 10% lower than would be expected for a pure iron-nickel melt. The outer core must therefore contain some percentage of light elements, which include oxygen and silicon but also carbon, sulfur, and hydrogen as possible components.

This proportion is suggested by the chemical affinity to early iron-nickel meteoritic material. The difference in composition, between the Earth's outer and inner core, is probably due to chemical fractionation that occurred during the slow crystallization of the inner core. During this time, light

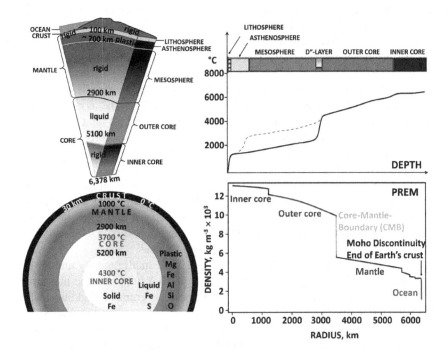

FIGURE 6.4 Earth's internal structure divided into different layer thicknesses, which are characterized by chemical composition and physical conditions such as temperature and density of the Earth material. Source: Wikipedia.

elements rose to the top, due to their lower specific gravity, and thus remained in the liquid residual melt.

In addition to the stable elements, long-lived radioactive isotopes play a role in the chemical history of the core. Short-lived radioactive isotopes such as ^{26}Al and ^{60}Fe are also of great geological interest because their decay may have contributed to the initial heating and melting of the Earth's core. The fraction of long-lived radioactive components of the uranium and thorium decay series is probably small because chemically they behave differently than iron and nickel. Nevertheless, it cannot be excluded that larger amounts of these heavy materials sank into the outer core under the influence of gravity, where they then mixed with the Fe-Ni melt found there.

It is estimated that the amount of radioactive ^{40}K is significantly higher than that of the other long-lived components since potassium has a special chemical affinity for sulfur. At the Earth's core, estimates of heat conduction and cooling behavior suggest a maximum potassium-abundance value of 29 ppm. This corresponds to an average activity of 1 Bq/kg. With about $1.9 \cdot 10^{24}$ kg for the total mass at the Earth's core, this gives an activity of $1.9 \cdot 10^{24}$ Bq.[13] Since the released decay energy of ^{40}K is about 1.4 MeV, this would correspond to a radiogenic heat production of $2 \cdot 10^{11}$ W or 0.2 TW (tera-watts) in the Earth's core.

A quasi-solid inner mantle encloses Earth's core. This inner mantle is a very hot 2,900 km thick silicate layer. It also contains a high proportion of metal oxides (magnesium and iron) and uranium-thorium actinides, which have a high chemical affinity for the silicate and oxide-rich mantle material. These actinides contribute significantly to the observed anti-neutrino flux. According to the number of measured geoneutrinos, about 50% of Earth's total heat (~20 TW) comes from the radioactive decay of actinides.

Since radioactive actinides have primarily accumulated in the mantle, the total accrued activity can be estimated. With an average decay energy of 5 MeV in the natural decay chains, this currently corresponds to a total activity of more than $2 \cdot 10^{25}$ Bq – an enormous value that, nonetheless, only represents a quarter of Earth's original radioactivity. This activity corresponds to $4 \cdot 10^{42}$ particles and translates to about 10^{15} tons of uranium with an average half-life of 4.4 billion years. However, this amount was originally twice as large since activities decay exponentially with time according to the decay law.

Besides the actinides, large amounts of potassium still exist in the mantle, comprising about a 0.1% share of mantle material. Potassium exists as K in atomic form and as oxide K_2O,[14] of which 0.01% is radioactive potassium ^{40}K with a half-life of $1.25 \cdot 10^9$ years. So far, the neutrinos originating from ^{40}K decay have not been measured, but several designs are currently being studied for the next generation of neutrino detectors since direct measurement of ^{40}K geoneutrinos would be a breakthrough in understanding Earth's origin and geochemical development.[15]

Information can be obtained from elemental analysis of lava material that has reached Earth's surface from various layers of the mantle due to volcanic activity.[16] On average, the mantle is expected to have a K/U ratio of 13,800, with a potassium value of 280 g per one gram of mantle material. This corresponds to an activity of 8.9 Bq/kg or an approximate total activity of $2.6 \cdot 10^{25}$ Bq or 26 YBq, at a mass of about $3 \cdot 10^{24}$ kg for the mantle region.[17]

The ^{238}U abundance is found to be 20 ng/g. Because of its small quantity and longer half-life, this gives an activity of 3.5 Bq/kg with a total activity of 10 YBq. From the measurements, one determines similar quantities of thorium, with 80 ng per one gram of cladding material. Since thorium has a long half-life of 14 billion years, this corresponds to an activity of 33 Bq/kg and a thorium-related activity of about 10 YBq in Earth's mantle. Thus, the total activity due to decay of the uranium and thorium chains is 20 YBq, which agrees well with anti-neutrino measurements. When adding to this the total ^{40}K activity of 26 YBq, the total radioactivity of Earth's mantle is nearly 50 YBq.

These radioactive components contribute to the continuous heating of the mantle. Currently, heat output from actinide decay is equivalent to about 11 TW, to which the decay of ^{40}K contributes another 6 TW. For comparison, a nuclear power plant has a typical net output of one gigawatt (GW). A thermal output comparable to the radioactive decay in Earth's mantle would require nearly 20,000 nuclear power plants. Currently, there are 439 reactors worldwide with a total output of 377 GW net power.[18]

It is important to note that the relative contribution of these elements, to the heat production in Earth's mantle, has changed over time because of their different half-lives. Also, total heat output has decreased over the last 4.6 billion years (Figures 6.3 and 6.5) due to the decay of these radiogenic isotopes. Radioactivity is not, however, the mantel's only heat source. Initially, most of the heat came from the release of gravitational energy with contraction of the planet. Internal radioactivity began to gradually dominate heat output after the first billion years.

Because of its long half-life of 14 billion years, ^{232}Th activity has so far remained relatively constant throughout Earth's 4.6 billion years. On the other hand, about 50% of original ^{238}U activity has decayed. For ^{40}K, with a much shorter half-life, only 10% of the original activity remains. The majority of remaining ^{40}K material has decayed to calcium and 10% to the noble gas argon. So, although it is estimated that thorium produced only 10% of the heat generated in the beginning, the thorium chain now dominates today's internal heat production. Conversely, ^{40}K was initially the dominant isotope for decay-heat production, but today, it provides only about or less than 50% of the total heat.

Figure 6.5 shows, in addition to the decay curves of the dominant radiogenic isotopes, the spectrum of anti-neutrinos predicted on the basis of expected radioactivity in Earth's mantle. The spectra are normalized to the number of measured anti-neutrinos from the ^{238}U and ^{232}Th chains.

Considering the high atomic weight of actinides and their decay products – which are among the heaviest known atoms – a gradual, gravitational chemical fractionation is expected due to their high specific gravity. The associated radioactivity accumulates in the lower part of the mantle, where the Earth's gravitational field reaches its maximum.[19] This is not, however, true for

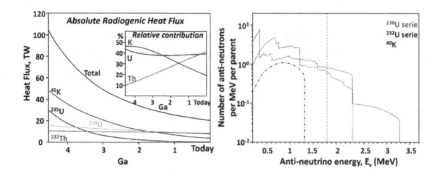

FIGURE 6.5 The left figure demonstrates the decline of Earth's heat from radioactive decay. Measurements are gathered from neutrino and lava measurements, which determined K/U and U/Th abundance ratios. On the right, the simulated spectrum of anti-neutrinos expected from Earth's interior, against which the measured number of anti-neutrinos from ^{238}U and ^{232}Th decay can be compared. Source: http://kamland.stanford.edu/GeoNeutrinos/geoNeutrinos.html (March 13, 2024). Note: Ricardo Arevalo, Jr., William F. McDonough, Mario Luong, "The K/U ratio of the silicate Earth: Insights into mantle composition, structure and thermal evolution," https://www.sciencedirect.com/journal/earth-and-planetary-science-letters" \o (Go to Earth and Planetary Science Letters on ScienceDirect), (2009): 361–369.

the distribution of the long-lived ^{40}K isotope, for which a more uniform distribution is expected. However, convective processes do affect the overall distribution of the various radioactive components.

The mantle makes up about two-thirds of Earth's mass; average density of the mantle layers is 3 to 6 g/cm³, much less than that of the Earth's core which is compressed by internal pressure. Mantel temperature is 3,000 °C, which is lower than the core but still very high because of the high decay heat of the radioactive elements. The inner mantle is solid due to the high pressure and increased melting point, as compared to the solid Ni-Fe core. This solid state only changes when pressure gradually drops toward the Earth's surface. The thick outer-mantel layer (the asthenosphere) measures 700 km and is therefore largely viscous.

In this layer, there is a K/U ratio of approximately 19,300. This ratio results from a potassium mass fraction of 100 g per gram of Earth material, with 5.3 ng for ^{238}U and for ^{232}Th about 20 ng per gram of Earth material. Together, this corresponds to an activity of 3.2 Bq/kg for ^{40}K, 0.9 Bq/kg for ^{238}U, and 0.8 Bq/kg for ^{232}Th. This ratio also contributes to heat production in the upper-mantle region, with 0.5 TW coming from the radioactive decay of ^{40}K potassium and 1 TW each from the decay chains of uranium and thorium.

Radiogenic heat production in Earth's interior is determined by activity from the estimated occurrence of each long-lived isotope, as found in the various layers (Table 6.1). The upper layer of the mantle, the asthenosphere, is covered by a relatively thin, solid crust, the lithosphere. This is not a closed crust but rather, is divided into seven major and several minor lithospheric plates, also called *tectonic plates* or *continental plates*. Plate zones include the individual continents and also vast areas covered by oceans. Plate boundaries mostly extend along the oceans and are characterized by the so-called reef zones or deep-sea channels.

The thickness of the lithosphere varies, with the thickness of the continental crust ranging from 30 to 70 km, while ocean crust is considerably thinner, averaging from 7 to 30 km. Crust material consists of silicates and oxides (quartz and feldspar) with only relatively small amounts of magnesium and iron. At the surface of Earth's crust, these rock types are subjected to constant chemical weathering and transformation processes through exchange with the atmosphere. This is the so-called cycle of rocks and today, there are no rocks that remain unchanged since Earth's first crust formations.

TABLE 6.1

Radiogenic-Radiation Activity in the Earth's Core and Mantle, and the Corresponding Heat Generation Calculated for Different Zones of the Interior of Our Planet

	Activity [10^{24} Bq]			Heat Generation [10^{12} W]		
Isotope	^{40}K	^{232}Th	^{238}U	^{40}K	^{232}Th	^{238}U
Core	2	-	-	0.5	-	-
Mantle	27	10	10	6	5	6
Asthenosphere	0.5	1	1	0.5	1	1
Total	30	11	11	7	6	7

The structure of Earth's core, mantle, and crust appears to be static in the short term but in actuality, it is constantly in motion. This motion results from the fluid mantle components and the associated hydrodynamic conditions. With this, physiochemical mixing processes occur between the mantle and crust. These processes continuously enrich crust material with long-lived radioactive material from the mantle. In addition, the dynamics of Earth's interior have a number of other consequences that determine the geological transformations that are a part of Earth's history and are decisive for the development of life on our planet.

The driving force or engine behind these developments is the ceaseless release of long-lived radioactive decay energy in Earth's mantle. Without this internal energy source, Earth would have largely cooled to a dead, rigid body.

6.1.4 RADIOACTIVITY AS GEOLOGICAL ENGINE

In the early Earth phase, release of gravitational energy caused heating in the Earth's interior. In addition, the decay of long-lived radioactive isotopes (actinides and ^{40}K) also generated heat, providing most of the energy that led to high temperatures in the interior. With Earth's gravitational contraction came the compression and solidification of its hot inner core, leaving an outer liquid-hot metallic zone. This outer core was heated further by actinide enrichment and the associated decay energy, as described above.

At temperatures above 5,000 °C, atoms are largely ionized so that they become electric-charge carriers. Due to the Earth's rotation and the temperature gradient between the mantle and inner core, convection currents and internal circulation develop in the liquid Fe-Ni shell. These processes generate electric currents that induce the geomagnetic field, the so-called geodynamo (Figure 6.6).

Models of convection-induced geodynamic behavior mostly reflect the observed behavior of the geomagnetic field (Figure 6.7). In particular, this includes the geomagnetic field's spontaneous polarity reversals, which have frequently occurred during Earth's geological history. The outer core's fluid state is a critical prerequisite for this. The geomagnetic field has existed for about 3.8 billion years, which is the age of Earth's solid inner core. The persistence of this field requires a permanent source of heat, which today is composed of the radiant heat generated by the inner core, the frictional energy at the mantle surfaces, and the decay energy of radioactive isotopes as supported by direct neutrino-flux measurement from Earth's interior.

The field's present flow conditions can easily cause cyclonic and anticyclonic eddy currents to be altered and these, in turn, determine the field's main direction. Such changes would cause polarity reversal of the geomagnetic field. Ferruginous rocks from Earth's crust have been studied and according to paleomagnetic measurements of residual magnetization, polarity reversal must have occurred at more or less regular intervals over the last several 100,000 years, with the last occurrence 780,000 years ago according to the latest findings. The geomagnetic field has decayed by more than 10% since

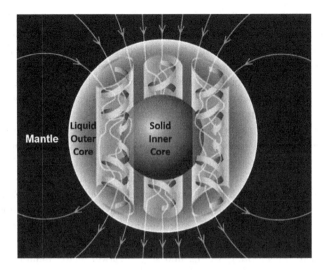

FIGURE 6.6   According to dynamo theory, the basic requirements for a magnetic field are convection currents driven by strong temperature gradients, planet rotation, and an electrically conductive core. All of these conditions are present for Earth but not for some of the other planets. For Venus, rotation is too slow, and for Mars, the current thesis is that the formerly liquid core has solidified. Source: https://en.wikipedia.org/wiki/Dynamo_theory#/media/File:Dynamo_Theory_-_Outer_core_convection_and_magnetic_field_generation.svg (March 13, 2024).

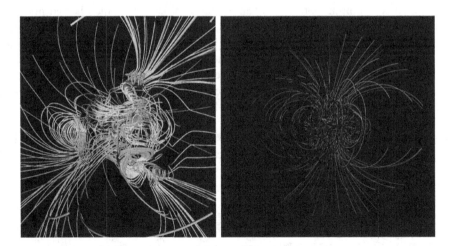

FIGURE 6.7   Simple model of the geodynamo: yellow and blue areas represent cyclonic and anticyclonic vortices in the core's flow. On the left, the chaotic flow of field lines inside Earth's core; on the right, the total geomagnetic field, which is a dipole field outside the Earth. Source: https://en.wikipedia.org/wiki/Dynamo_theory#/media/File:Geodynamo_In_Reversal.gif (March 13, 2024) and http://www.astropage.eu/wp-content/uploads/2016/09/erdmagnetfeld2.jpg (March 13, 2024).

its first determination around 1830 by German natural scientist Alexander von Humboldt (1769–1859). This is interpreted as unusually fast decay, to the effect that we are presently again in a reversion phase.

The geomagnetic field has an enormous influence on Earth's external radiation conditions. Without the Earth's magnetic field there would be no life-friendly environmental conditions on the surface, since the magnetic field largely deflects the sun's high-energy cosmic radiation and thus largely reduces the biosphere's external radiation exposure (cf. Chapter 6.1.8).

It is intriguing that Mars, a much smaller planet, lost its magnetic field a billion years after the planetary system was formed, according to recent research and analysis of Martian samples. Mars has a smaller iron core that cooled faster than that of the larger Earth. As the outer iron core changed from liquid mass to a solid block, the induction that maintained the magnetic field broke down. Measurements of Martian rocks younger than 3.5 billion years show that the previously existing magnetic field had largely disappeared. Without the magnetic field's protection, cosmic rays could penetrate unhindered into the atmosphere and to the surface. This led to the breakup and destruction of molecules and the loss of light atmospheric gases, at a rate of 1 kg/sec – a process that continues to this day. And so, Mars cooled and lost its atmospheric and surface water content, which possibly led to the collapse of initially positive life conditions that may have existed 3.5 billion years ago.

Also during the early phase, Earth's mantle was in a largely liquid state as it was being heated to over 10,000 °C by accretion and contraction processes, as well as the radioactive decay of ^{26}Al, ^{60}Fe, and other short-lived radioactive isotopes. But the decay times of these radioactive isotopes are relatively short and so internal radiogenic-heat production decreased rapidly. Consequently, internal heat was only maintained by the decay energy of longer-lived ^{40}K isotopes and the actinide decay chains. This resulted in the gradual cooling and solidification of the mantle zones, as described above.

Despite solidification, mantle material has a certain plasticity that leads to an inert fluid behavior with extremely high viscosity.[20] This causes complex thermodynamic behavior because the mantle layer is being heated from below, by heat exchange with the hot Earth core as well as its released radioactive-decay energy, while also being cooled from above by interaction with the lithosphere. As discussed, the entire inner-Earth region is also gradually cooling.

Under these thermodynamic conditions, materials in the mantle region demonstrate convective behavior. Similar to a boiling pot, there is an exchange process by which heated material rises upward from the deeper mantle region into the asthenosphere, where it sinks back down after cooling via heat exchange with the lower lithosphere. This convective behavior is extremely slow, with one circulation taking about 100 million years. The convection velocity was initially much faster because decay processes released considerably larger amounts of heat (Figure 6.4). This interpretation does, however, still leave many unanswered questions particularly related to the cooling rate of the mantle layers, which probably requires a somewhat more complex and multi-layered convection behavior (Figure 6.8) than can be discussed here.

Earth's convection motion subdivides into individual convection zones, which exist within the respective mantle components. This will be the case if the radioactive isotopes are distributed

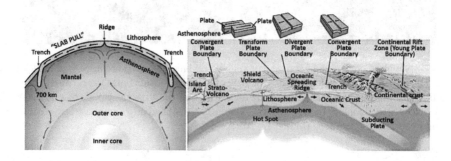

FIGURE 6.8 Convection in Earth's mantle: This involves not only heat transfer between Earth's core and crust but also an exchange of material between the mantle and crust, through friction between the upper asthenosphere of the mantle and the lithosphere. Through this mechanism, the lithosphere is expected to accumulate radioactive materials from the inner region of the mantle. Source: https://en.wikipedia.org/wiki/Marine_geology#/media/File:Oceanic_spreading.svg and https://www.wikiwand.com/en/List_of_tectonic_plate_interactions#Media/File:Tectonic_plate_boundaries.png (March 13, 2024).

according to their composition by geochemical separation processes, as assumed here. Accordingly, heavy uranium-thorium components will accumulate in the inner region of the mantle. In the case of the potassium isotope ^{40}K, there is a more pronounced uniform distribution. If, however, the elements of the uranium-thorium chains are also uniformly distributed, convection is likely to occur over several mantle layers. Distribution is difficult to verify and based on element-distribution measurements from lava material and the theoretical model assumptions extracted from these studies. The distribution question remains a subject of scientific debate.

Convection energy is transferred to the lithosphere through friction with the rigid crust, thus generating conditions for the tectonic motion of the continents, as seen in Figure 6.8. At the so-called divergent plate boundaries, mantle convection drives the continents apart via the outflowing of magma at trenches or so-called rifts along the middle-ocean zones. The outflowing magma mass then cools, forming a new crust. This process is called *sea-floor spreading*. As a result, the Atlantic Ocean widens by a few centimeters every year.[21] Global plate distribution, the course of the various rift zones, and the new cooled-magma layers cause continental drift.

Convergent plate boundaries exist along continental coasts where two plates collide. In this process, the heavier oceanic plate sinks upon contact with the firmer continental-plate zone and is forced down into the asthenosphere, the lower mantle layer. This process is called *subduction*. The plate folds, forming a deep-sea trench off the continental coast; in turn, the affected coastal region folds upward into a mountain range. The process can be observed especially on the west coast of the American continent, where strong earthquakes and volcanic activity – caused by friction processes between the different crustal zones – characterize both the rift zone and the subduction zone.

Subduction involves extensive material exchange as cold crustal material sinks deep into the mantle zone. According to some assumptions, this material even sinks to the Earth's outer zone, while hot magma from the interior rises to the surface as new crustal material. This occurs predominantly along the oceans' rift zones as depicted in Figure 6.9.

The average rate for generation of new crustal material at the surface is about 3.4 km^2 per year. In today's oceans, no lithosphere material is found that is older than 200 million years, which is roughly equivalent to the geological age of the Atlantic and Pacific oceans. The lithosphere material being pressed into the mantle is heated by friction, compression, and heat exchange with the outer

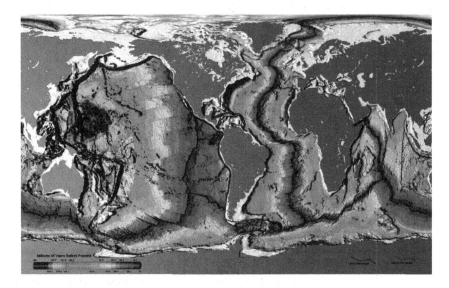

FIGURE 6.9 Evolution of the oceanic rift zones over time, causing continental plate movement. Color coding depicts the age of the oceans' bottom layers, with red zones indicating the most recent evolution. Source: Wiki Commons.

mantle's deeper material. Through these processes, the material reaches temperatures of 1,200 °C at a depth of 100 km. At the bottom of the asthenosphere, at a depth of 500 km and a temperature of 2,000 °C, most of the material begins to melt, mixing with material in the deeper mantle layers.

In addition to the lithosphere's slow convection process, volcanic processes contribute to material exchange between the mantle and crust. Deep-seated pockets of magma, called *plumes*, are often connected to the surface by channels, and it is through these channels that volcanic eruption occurs. Volcanoes thus provide a view of the mantle material's composition. Chemical analysis of this magma shows the expected mantle composition, of silicate and oxide minerals, with admixtures of iron, calcium, magnesium, and potassium.

Mantle convection and the formation of new crustal material, as well as direct transport of magma from the depths of Earth's mantle, carry radioactive material upward and into the lithosphere. Thus, the crust gradually accumulates radioactivity. Of course, during subduction, some of this material sinks back into the mantle, providing balance in the transport mechanism. However, especially in early times, crust formation was so much more prevalent than subduction that today the concentration of radiogenic material in the crust is more than 200 times greater than in the mantle.

The slow growth of radioactivity in the crust does not proceed by chemical fractionation of individual elements, as in the mixing processes between the lower mantle sphere and the outer liquid region of the Earth's core. Rather, an exchange of rocks and minerals occurs (Figure 6.10). These rocks and minerals are brought up from the liquid region of the asthenosphere and crystallize by cooling. This means that the radioactive isotopes are chemical components of these various minerals. In this way, radioactive isotopes make up the crustal rock and occur in the upper layers of Earth's crust, depending on rock type and local abundance variances.

Many of these radioisotopes are only present in small amounts and therefore do not have a large effect on mantle or crustal thermals. However, their importance lies in their various geophysical applications. The relative abundance distribution of different parent and daughter components, found in the varying rock and mineral types, can be used to form detailed determinations about the timing of these mixing and crystallization processes. This data can then be compared with respective model predictions.

In particular, the actinides ^{232}Th, ^{235}U, and ^{238}U constantly feed the decay chains, due to their long lifetime, and thus contribute to the permanent heating of Earth's crust. These decay products are mainly α- and β-particles that have only a short range in rock. Their kinetic energy is therefore transformed into heat by the deceleration process. However, in many cases secondary nuclear reactions occur between the relatively high-energy α-particles with an average energy of 5 MeV and atomic nuclei in the Earth material.

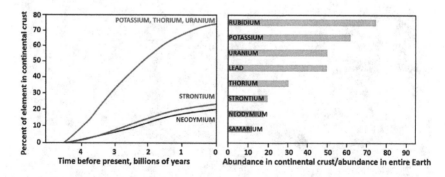

FIGURE 6.10   Model predictions for the gradual accumulation (left) and present percent abundance (right) of radioactive materials (compared to total abundance) in the Earth's crust, due to mantle convection. In these calculations, it is assumed that only 50% of the mantle is involved in the convection process. Source: Robert and Barbara Decker, "Volcanoes and the Earth's Interior," *Scientific American*, 1982.

A small percentage of the α-particles cause secondary nuclear reactions. These (α,n) reactions produce a steady flux of neutrons within the Earth's crust and at the surface, averaging 10 to 30 neutrons per minute. Although this is a very small flux, it is directly detectable and can lead to neutron-induced tertiary reactions. In addition, higher neutron fluxes may also occur depending on the intensity of the α-sources, which in turn depends on the abundance of uranium and thorium.

Apart from the decay series, decay of radioactive ^{40}K is a major contributor to the heating of the crust, due to its high abundance in rock material. Here decay energy is mainly released as short-range β-radiation. However, a ten-percent fraction also occurs as 1.45 MeV γ-rays, which is stopped by Compton scattering processes in the deeper zones of the crust. While most of the γ-rays are absorbed in the rock material, a fraction of this energy does escape and will directly contribute to the general radiation load at the surface.

All of these decay processes produce noble gases in the Earth's interior: α-decay leads to accumulation of helium gas (^4He); radioactive ^{220}Rn ($T_{1/2} = 55\ s$) from ^{232}Th, and ^{222}Rn ($T_{1/2} = 3.82\ d$) from the ^{238}U decay chains are continuously being formed and can diffuse from the surface into the atmosphere; in addition, the decay of ^{40}K produces the noble gas ^{40}Ar, which is chemically inert and also gradually diffuses out of the rock and into the atmosphere. The latter process is the main source for the accumulation of argon, the third most abundant component in our atmosphere.

Together, these observations and results demonstrate that geothermal heat – extracted from near the surface by local heat pumps with depths of 400 m or from crustal depth layers of several thousand meters by geothermal power plants – is a product resulting from long-lived radioactive element decay (cf. Chapter 11.4, Volume 2). The heat output estimated today corresponds to about 40 to 50 kW/km^2. Without these radioactive decay processes, crustal cooling would have occurred long ago – and Earth would be a much colder and less friendly habitat.

Volcanic eruptions are a special case for the rapid transport of long-lived radioactive materials from the depth of Earth's mantle to the surface. The amount and activity of the ejected materials depend on the type of volcanic eruption, since ejected material is a mixture of material from deeper mantle layers and volcanic surface rock. Accurate data are understandably difficult to obtain: Collection and analysis of ejected lava rock require relatively rapid access so as to prevent secondary chemical fractionation and mixing processes with surface material. Nonetheless, since as early as 1964, it has been observed that radioactive lead from volcanic eruptions has been enriching the atmosphere.[22]

This radioactive lead is produced from the decay of radioactive radon gas ^{222}Rn, with a half-life of 3.8 days, from the ^{238}U decay chain. The noble gas radon is released in large quantities during volcanic eruption. The ^{222}Rn decays mainly via the α-decay chain to ^{210}Pb with a half-life of 20 years and via two subsequent β-decays to ^{210}Polonium (Po). This ^{210}Po decays via α-emission to the stable lead isotope ^{206}Pb with a half-life of 138 days. Volcanic activity releases an average of more than 1.85•10^{15} Bq of volcanic ^{210}Po into the troposphere and 3.7•10^{15} Bq into the stratosphere each year.[23] This seems small compared to Chernobyl's emission of 5.2•10^{18}Bq and Fukushima's 8•10^{17}Bq, but the amount only represents radioactivity associated with the emission of radon gases.

A balance between volcanic production and natural decay establishes an equilibrium abundance of these radioactive substances in the atmosphere. Measurements of aerosols in the atmosphere over the Antarctic Ocean have found an equilibrium activity of 0.044 Bq/m^3 for ^{222}Rn and 0.00001 Bq/m^3 for ^{210}Pb. These volcanic measurements from over the ocean are particularly useful because there, the atmosphere is least affected by radon emission from the land mass. However, less data is available from the study of radioactive solids, which are not only found in the liquid lava mass but also in dust and ash fallout.

A more detailed investigation took place during the eruption of the Mount St. Helens volcano in March 1980.[24] The volcano is located near the Pacific Ocean in southern Washington State, where samples were taken for more than a year. These were analyzed for actinide content from the radioactive uranium and thorium decay chains. The samples contained an average concentration of 1 ppm of ^{238}U and 2.4 ppm of ^{232}Th, corresponding to an activity of 13 Bq/kg and 10 Bq/kg, respectively.

These are average rock levels (Table 6.2) and do not reflect excessive increases in local radioactivity. However, from the data thus obtained, the ^{210}Po emission released by the Mount St. Helens eruption ranges from $1.1 \cdot 10^{13}$ to $1.85 \cdot 10^{14}$ Bq. While this is below the overall annual average for volcanic production, the value is comparable to the natural ^{210}Po content of the atmosphere. These values are also comparable to the annual production of ^{210}Po from coal burning, fertilizer production, and metal smelting, which all provide anthropogenic contributions to the overall radioactive budget (see Chapter 10).

6.1.5 RADIOGENIC RADIOACTIVITY AT EARTH'S SURFACE

Radioactive elements do not, however, only determine the heating of Earth's deeper layers. Their activity also occurs at the Earth's surface. As in the deeper layers, radioactivity at Earth's surface mainly results from decay of the long-lived radiogenic isotopes ^{238}U, ^{232}Th, and ^{40}K and their radioactive daughter isotopes.

Due to their chemico-mineralogical behavior, these elements often occur in very different abundances depending on the rock and soil material. Table 6.2 lists the most common rock types and the typical concentration ranges in ppm, providing information about the most important radioisotopes in the respective minerals. Also given are the decay activity and dose rates of the various radiogenic components. These dose rates are considered as part of the natural radiation exposure for humans and animals.

From the various isotope abundances, one can calculate their activity using the known decay constants. This gives the following conversion values: 1 ppm $U^{nat} \approx 25.4$ Bq/kg; 1ppm ^{238}U ≈ 12.4 Bq/kg; 1ppm ^{232}Th ≈ 4.1 Bq/kg; on average 1% $K^{nat} \approx 311.7$ Bq/kg. This does not take into account radioactivity of the daughter elements. Each of the daughter elements must be present, in equal activity with the slowly decaying parent isotope, in order for radioactive equilibrium to be established. To calculate the total activity of soil material, the activity of the parent isotope, as given in the table, must be multiplied by the number of radioactive isotopes in its decay chain.

This calculation is straightforward for ^{40}K, since there is only a single radioisotope. However, for the ^{238}U decay series, there are a total of 14 radioisotopes and for the ^{232}Th decay chain, there are 10 radioisotopes with one branch in the decay chain. In addition to activity, Table 6.2 also gives the typical hourly dose rate in gray. Gray is calculated from the activity and energy released, by the respective decays, into the surrounding medium. Conversion is done by means of standardized empirically obtained factors, which are recorded in UNSCEAR reports[25] on the basis of recent measurements. According to these, an activity of 1 Bq at ^{238}U corresponds to a dose rate of 0.46 nGy/h, and 1 Bq at ^{232}Th corresponds to a dose rate of 0.623 nGy/h. The ^{40}K activity of 1 Bq corresponds to a dose rate of 0.0414 nGy/h, while 1% of K^{nat} natural potassium in soil material corresponds to a dose rate of 12.9 nGy/h.

For the biological effect of radiation on humans, one has to convert the absorbed dose of radiation into the equivalent dose. This calculation is done via the radiation-related weighting factor (cf. Chapter 3). For α-particles, this conversion factor is $Q = 20$. However, in the case of radiation from the ground, a reduced factor applies since only the portion of γ-radiation emitted into the atmosphere should be considered, the so-called local dose. The effective conversion factor for this local dose is 1.15 Sv/Gy so that a dose of 1 Gy corresponds to a local dose of 1.15 Sv.

The average occurrence of uranium in Earth's crust is relatively high at 2–4 ppm and so, uranium is much more abundant than antimony, cadmium, mercury, silver, or tin. Earth's crust thus contains about 10^{14} tons of uranium. Of this, about 10^{10} tons are present in ocean water as dissolved uranium salts; this is considered the largest potentially accessible uranium source. Soil itself, such as forest and farmland soil, contains up to 11 ppm uranium, a significant portion of which comes from fertilizer use (see Chapter 12, Volume 2). Rock and stone contain uranium in hundreds of different minerals. Of these, uranium is, of course, particularly strong in crystallized

TABLE 6.2

The Abundance Range of Natural Radioactivity Components in Mineral and Stone Material as Discussed in the Text and the Corresponding Activity and Dose Rate for Estimating Biological Impact

Minerals	Concentration			Activity			Dose Rate		
	^{238}U [ppm]	^{232}Th [ppm]	K_{nat} [ppm]	^{238}U [Bq/kg]	^{232}Th [Bq/kg]	^{40}K [Bq/kg]	^{238}U [nGy/h]	^{232}Th [nGy/h]	^{40}K [nGy/h]
Feldspar	0.2–3	0.5–10		2.5–37	2–41		1–17	1.3–26	
Quartz	0.1–5	0.5–10		1.26–62	2–41		0.5–25	1.3–26	
Amphibole	1–30	5–50		12–372	21–205		6–170	13–130	
Magmatite	0.3–20	4–90	0.6–5.4	4–252	2–369	180–1,683	2–116	2–234	7.5–70
Sandstone	0.5–3	1–9	0.01–6.7	6–38	4–37	3–2,081	3–17	6–23	0.1–86
Claystone	1.5–8	10–17	0.5–4.3	25–101	41–70	153–1,352	8–46	26–44	6.3–56
Carbonate	0.3–2.3	0.3–1	0.02–1.5	4–29	0.5–12	6–480	1–13	0.3–7.8	0.3–20
Bauxite	2–27	49		25–340	201		12–157	127	
Slate	3–1,200	100		38–15,120	410		17–6,960	260	
Phosphate	50–900	1–5		630–11,340	4–21		290–5,220	6–13	
Soil	1–20	4–23	0.05–9.5	13–252	17–94	16–2,960	6–116	11–60	0.6–123

uranium oxide UO_2 and uraninite U_3O_8, which we have already encountered under the name pitchblende (see Chapter 3).

Other such minerals are uranophane, a calcium-uranium silicate, and carnotite, a potassium and uranium-bearing mineral which, in addition to ^{238}U and ^{235}U activity, also harbors ^{40}K radioactivity. Of the general rock types, particularly granite is very uriniferous – hence the high natural radioactivity in granite mountain ranges. In addition, phosphate-bearing rocks such as autunite, selenite, and torbernite, and also the minerals present in sandstone, such as monazite and lignite (lignite), are highly uranium-bearing and therefore of great importance as secondary sources for uranium mining.

Long-lived radioactive thorium occurs in various forms, often together with uranium as daughter elements ^{227}Th and also ^{231}Th of the ^{235}U actinide chain, as well as ^{230}Th and also ^{234}Th of the ^{238}U uranium chain. Conversely, radioactive ^{232}Th is most commonly found as the long-lived parent isotope of the thorium chain. For chemical-mineralogical reasons, thorium frequently occurs in small amounts together with uranium in minerals. However, thorium is in general a relatively abundant element, comparable to the occurrence of lead and molybdenum and much more abundant than the occurrence of uranium. Thorium occurs at 2.5% abundance in monazite, a phosphate sand, and at up to 2% abundance in allanite, a silicate mineral, as well as in thorite, a thorium silicate.

The total activity of uranium, thorium, and their decay products vary between about 10 and 40 Bq/kg, depending on the geologic setting. For the distribution of uranium and thorium in the United States, see Figure 6.11. It is clear that zones of elevated radioactivity correlate with granite-bearing mountainous areas such as the Appalachian Mountains in North and South Carolina, Tennessee, and Georgia, and also the Rocky Mountain ranges of Idaho, Montana, and Colorado. Sedimentary radioactivity is predominantly present in the southwestern United States, in Arizona, southern California, Nevada, and Utah as a former ocean area.

In contrast, the occurrence of uranium and thorium is very low in volcano-rich areas, such as along the Cascade volcanic chain in Washington, Oregon, and northern California. This is also true in areas around the Great Lakes where remnants and sediments of the glacial lake Agassiz still cover the entire area. Similarly, marsh and coastal areas show low uranium-thorium abundance, as in Florida.

Radioactive decay of ^{40}K in the mantle and crust provides the third largest contribution to Earth's heat production, the seventh most abundant element in Earth's crust, and the sixth most abundant salt dissolved in ocean water. Potassium is found in a wide variety of minerals such as feldspar, an important component in granite, as well as in carnotite and the ocean sediment carnallite, which is widely and readily used as a fertilizer additive. Likewise, it exists as the natural potassium-chloride

FIGURE 6.11 Distribution of the radioactive-actinide elements uranium and thorium in the United States. Elevated enrichment is particularly found in the mineral-rich areas of the Southwest and Rocky Mountain states. Of note is the low occurrence in the volcanically active zone along the west coast in Washington, Oregon, and northern California, which is part of the Pacific Ring of Fire volcanic zone. Source: US Geological Survey DDS-9 (1993).

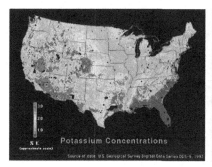
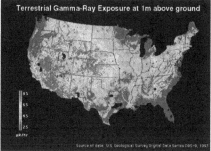

FIGURE 6.12   Accumulation of ^{40}K compared to general γ-radiation exposure in the United States. From the distribution similarity, it can be seen that the major portion of the γ-radiation load is associated with the occurrence of ^{40}K in soil. A high proportion of ^{40}K can again be found in the American Southwest's ocean-sediment areas. However, the prairie areas of Nevada and Colorado are also characterized by elevated ^{40}K occurrence. Of note is the amount of ^{40}K activity in the Mississippi River basin; this is thought to be derived from alluvial salts, which more heavily impacted Midwest areas. Source: US Geological Survey DDS-9 (1993).

mineral salt KCl, also known as sylvite, and is offered in the food trade as a health-promoting salt alternative to normal table salt (NaCl).

Potassium is an important component of soil with an average contribution of about 20,000 ppm. In addition, potassium often occurs as a mineralized sand component that is leached by moisture. Potassium's efficiency depends on weather conditions so that heavy, moist, clayey soil contains large amounts of potassium, while dry sandy soil has only a relatively low amount of free potassium. These soil conditions also determine the distribution of the radioactive ^{40}K component. Figure 6.12 shows the distribution of potassium, and hence its radioactive ^{40}K component, over the United States.

In comparison, the figure also shows the γ-activity measured one meter above the Earth's surface. The comparison clearly demonstrates that much of local γ-radiation exposure comes from the decay of natural ^{40}K. The greatest radiation intensity and thus radiation exposure occurs in the desert areas of the American Southwest, from Nevada (Death Valley), Arizona, and New Mexico to western Texas with the Llano Estacado. Similarly, an elevated potassium occurrence is recognized in the central region of the United States, in western Colorado and Montana. Volcanic areas in the American Northwest as well as the lowlands in the American Southwest show a very low potassium occurrence. On average, ^{40}K activity values between 100 and 1,000 Bq/kg were measured, depending on the region.

Similar conditions for the occurrence of soil radioactivity are also present for Europe.[26] However, studies show that variations are much less dramatic (Figures 6.13 and 6.14).

The decay of these long-lived radioactive elements produces a significant fraction of the radiation burden in the soil while also producing geothermal heat in the crustal region. The contribution of each nuclide can be calculated separately from the abundance, decay energy, and decay rate. Respective abundances are difficult to measure locally and are, therefore, determined from the rock structure of particular geological formations.

Radioactivity in the oceans is closely related to the natural radiogenic activity of the Earth's surface. Oceans cover most of our Earth's surface, $3.6 \cdot 10^8$ km^2, and have an average depth of 3,800 m. This corresponds to a total volume of $1.37 \cdot 10^9$ km^3. The coastal zone has an average depth of 150 m and extends 150 km into the sea. The surface zone of the oceans, with depths ranging from 10 to 200 m, is highly turbulent due to interaction with atmospheric winds. This interaction results in a uniform mixing of surface water and the salts dissolved in it and, at this depth range, temperature, density, and salinity (the amount of salt dissolved in a body of water) are constant.

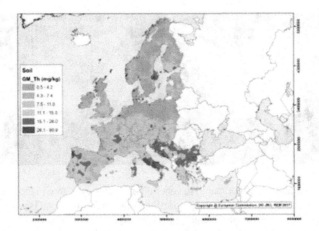

FIGURE 6.13   Distribution of uranium and thorium in European soil material in units: mg per kg of soil material. It is noteworthy that the corresponding abundance occurs more clearly in Scandinavia and northern Germany. Source: https://remon.jrc.ec.europa.eu (June 18, 2023).

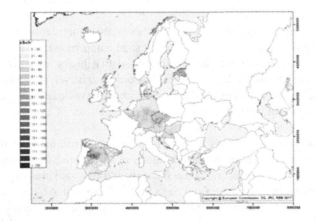

FIGURE 6.14   Map displaying the gamma dose per hour that a person may receive from terrestrial radiation. Source: https://remon.jrc.ec.europa.eu (June 18, 2023).

Density and salinity are higher at greater depths, as lighter water tends toward the surface. Salinity correlates directly with radioactivity, which mostly comes from dissolved salts and minerals in seawater. Distribution of the radioactivity is taken care of by ocean currents, which are driven by atmospheric winds. This can actually be monitored by tracing radioactive water spills and measuring their distribution and dilution in ocean water (see Chapter 15, Volume 2).

The average salinity of ocean water is 3.5%. A large proportion of the radiogenic components come from dissolved minerals, which contain uranium, thorium, and their decay products. This corresponds to an average activity of 1.6 Bq/l. However, the largest fraction comes from decay of the ^{40}K contained in numerous seawater salts, with an average concentration of about 400 mg/l. This translates to a radioactive ^{40}K content of 0.045 mg/l, which in turn corresponds to a natural radioactivity of 12 Bq/l. The average total load is thus an activity of 13.6 Bq/l, of which 88% comes from the decay of ^{40}K.[27]

In low-salinity waters such as the Baltic Sea, with an average salinity of about 1%, activity levels are much lower at 3.4 Bq/l. In waters with very high saline, such as the Dead Sea, potassium content is 28%, which is eight times higher than normal ocean water. Economically, this allows Israel and Jordan to produce potassium salts for use in the food industry as well as in skin and respiratory therapies.[28] At these levels, the Dead Sea has an average activity of nearly 96 Bq/l, more than seven times higher than average ocean water content. In addition to these radiogenic radioactivities in marine waters, there is a small amount of certain cosmogenic radioisotopes, such as tritium ^{3}H, beryllium ^{10}Be, and carbon ^{14}C (cf. Chapter 6.8), which also dissolve in seawater.

In this context, it should be mentioned that the radioactivity of one liter of seawater, converted to the total ocean volume, results in an activity value of $1.4 \cdot 10^{13}$ Bq or 14 tera-becquerel. Except in the narrow surface region, emitted particle radiation is absorbed in the water and passes into the atmosphere as helium gas. Gamma radiation has a slightly larger range, with a surface activity of about 1,200 Bq/m^3 and an assumed range of 10 cm in water.

6.1.6 GEOLOGICAL FORMATIONS WITH HIGH RADIOACTIVITY CONCENTRATION

Geological rock formations vary in important ways. High levels of radiation can be present when there is a significant accumulation of uranium, thorium, and/or potassium-bearing minerals, either in solid form or in dissolved form as in the case of hot springs. Several areas worldwide have been identified as having extremely high concentrations of natural radioactivity. More remote, as yet unknown locations with similar concentrations may still be discovered.

A particularly striking example is the Oklo area in Gabon, West Africa. There, in 1956, uranium-rich rocks were discovered and mined, primarily by French mining companies. Francevillette is a rock interspersed with a uranium-vanadate mineral compound, named after the nearby provincial French capital Villette (Figure 6.15). In 1972, during the mining of this natural uranium deposit, an unusual uranium-isotope composition was discovered at various locations along the uranium-rich rock vein. This natural abundance of the ^{235}U isotope, at 0.717%, is slightly lower than in other uranium deposits (0.720%). Scientists from the French Atomic Energy Commission CEA (Commissariat à l'énergie atomique et aux énergies alternatives) saw this as the first evidence for the existence of natural fission reactors, which break down larger accumulations of ^{235}U by natural nuclear fission.[29]

As early as the 1950s, natural reactors were discussed as a possibility. It was clear that their existence would require several geological conditions. The dimensions of the deposit must be at least 70 cm so that neutrons, released during the fission process, can be thermalized by scattering so as

FIGURE 6.15 Left, an example of Francevillette, an orange-yellow tinted mineral containing uranium vanadate and other uranium-oxide chemical compounds. Source: https://en.m.wikipedia.org/wiki/File:Francevillite-Chervetite-Curienite-64420.jpg (March 13, 2024). On the right, the uranium vein whose isotopic composition indicates the existence of a natural fission reactor in the rock. The person is indicating the so-called natural reactor, Oklo reactor 15; yellow precipitates in the rock are uranium oxides. Source: https://apod.nasa.gov/apod/ap021016.html (March 13, 2024).

to cause fission in additional ^{235}U nuclei. In addition, the ^{235}U deposit must be enriched to at least 3%, which was quite possible in Earth's early history when ^{235}U had not yet decayed to the extent it has today. Furthermore, the rock should contain light elements that serve as moderators, but not as absorbers, for the released neutrons. These were the conditions for the uranium deposits at Oklo where hydrous sandy rock with more extensive uranium accumulations were present. Water served as a moderator, and the lack of lithium and boron deposits reduced neutron absorption.

Along the uranium-rich rock vein, mining identified a total of seventeen natural reactors that were formed two billion years ago by geological evolution. The chemical origin of these reactors is an interesting chapter in Earth's history, underscoring the complex relationships between living and dead matter.

Formation of the Oklo reactors relates to the rapidly increasing oxygen content in the Earth's atmosphere two billion years ago. This increase was mainly due to the development of plant life, which, through photosynthesis, transformed the previously carbon-dioxide-rich (CO_2) atmosphere into an oxygen-rich (O_2) atmosphere. Prior to this development, uranium was only present in chemically stable reduced forms of water-insoluble compounds. However, the increase in oxygen led to oxidation processes in surface rocks, which among other things made uranium-rich minerals water-soluble. It was only when these rocks were saturated, thus releasing their oxygen content, that the oxygen content in the atmosphere increased.

As soluble mineral salts, uranium compounds – such as monazite, thorite, and uranite – became part of the hydrothermal cycle through rainwater and so could be washed to greater depths. Through this alluvial process, uranium solutions accumulated locally and reached a critical density, providing the prerequisite for natural fission processes.[30] The Oklo reactors were active for more than 100,000 years until the natural ^{235}U deposit was reduced, by fission and also decay processes, to such an extent that the self-controlled reactor operation could not continue due to lack of fuel. This time period has been calculated from the occurrence of fission products in the rock deposits.

Being dependent on the water supply, there were periods when the fission process was interrupted because the neutrons were not sufficiently slowed. Fission was, however, restored during periods of high rainfall. Furthermore, it is assumed that due to the released radiation energy, Oklo must have been a pulsed reactor – in modern terminology – generating a thermal power of about 100 kW. This is comparable to the power of modern research reactors. After half an hour of operation, heat would have caused the water accumulated in the sandstone to evaporate; the chain reaction then stopped because the neutrons could no longer be thermalized. After the sandstone was allowed to cool for about 2.5 hours, water collected again, thermalizing the decay neutrons so that the chain reaction started again.

This sequence took place over several 10,000 years, producing a neutron flux of 10^9 neutrons per square centimeter per second, with a total flux of 10^{21} neutrons per square centimeter and 15 GW total energy release. In the process, 5 to 6 tons of ^{235}U were converted into lighter decay products. It is noteworthy that these decay products remained at the reactor site for over two billion years, despite relatively unstable geological conditions, and were not dispersed over a larger area by leaching or diffusion processes.

The occurrence of natural reactors is a fascinating phenomenon that has led to numerous debates about their overall occurrence and their impact on the early development of biological life on Earth (see Chapter 7.1.3). Oklo reactors have fallen victim to uranium mining over the last forty years and are now gone. So far, no other natural reactors have been found. However, their existence may well be possible in uranium-rich areas of the world not yet developed by mining conglomerates, such as central Australia.

Three examples show areas where natural radiation exposure is well above the general average of 2.4 mSv/yr or 275 nSv/hr: the Ramsar area in northern Iran, Guarapari sand beaches on the Atlantic coast of Brazil, and Paralana hot springs in southern Australia. These and other regions serve as examples demonstrating that high natural-radiation levels, well above today's legal limits, do not seem to pose a danger for the local population, at least as can be measured to date (Figure 6.16).

Ramsar is located on the southern coast of the Caspian Sea in northern Iran and holds the world's record for natural radiation intensity. Local radiation exposure in the city's center is 250 mSv/year, which is almost eighty times higher than the average radiogenic-radiation exposure at Earth's surface. This is almost one third of annual radiation exposure in interplanetary space. The radiation's origin is traced to natural hot springs with high mineral-salt content. These radium leachates result from uranium-bearing rocks brought to the surface by hot water. The hot water quickly evaporates and this typically results in rapid limestone precipitation, referred to as travertine. Due to the high radium and uranium content of the water, the limestone is also very radioactive and yet, this stone was used as an easily worked building material for Ramsar's older buildings. In addition, the radioactive water is used in local spas and bathing facilities, and the remaining water percolates to enrich local groundwater.

The Guarapari beaches in northeastern Rio de Janeiro provide another example of significantly high amounts of naturally occurring radioactivity, with annual radiation levels ranging from 30 mSv to 175 mSv. This means that for beachgoers the annual average for local exposure is almost sixty times higher than a typical human being's average annual exposure. In fact, this dose is much higher than the legal limit of 20 mSv for professional radiation workers.

Hourly exposure of 20 Sv/h at the Copacabana Beach in Rio de Janeiro is almost ten times higher than the global average value. Here, the phenomenon of high radioactivity affects the Brazilian Atlantic coast between Rio and the state of Bahia. Radioactivity comes mainly from ^{232}Th-rich monazite outcroppings in the mountainous zone near the coast. In addition to thorium, radioactive daughter isotopes along the decay series are also present. However, significant differences have been found between sand-deposit measurements taken along the coast. This is presumably caused by seawater dilution such that ^{232}Th activities range from 57 to 41,200 Bq/kg, while ^{226}Ra activities range from 11 to 4,070 Bq/kg.

The Paralana hot springs are located in the arid, northern range of the Flinders mountains in southern Australia. These underground springs are fed by Mount Painter where water circulates through rock layers that contain high levels of uranium, thorium, and potassium and so are radioactively heated.[31] These radioactive rocks heat the water to almost 60 °C, with a radioactive heating

FIGURE 6.16   On the left, the Ramsar hot springs in Iran with limestone precipitates at the water's edge. Source: https://itto.org/iran/attraction/ramsar-thermal-springs/ (March 13, 2024). On the right, the white beach at Guarapari on the Brazilian Atlantic coast. In both cases, high levels of natural radioactivity are present from uranium-radium and uranium-thorium precipitates. Source: https://serranegrapousada.tur.br/blog/guarapari-saiba-tudo-sobre-essa-cidade-turistica-do-espirito-santo/ (March 15, 2024).

power of about 0.035 mW/kg. While the hot water carries dissolved uranium salts, abundances indicate local fluctuations with uranium concentrations detected from up to 660 ppm, for activity up to 11,000 Bq/m^3. The average dose rate of the soil material is 2.5 mSv/year, which is comparable to the world average. However, long stays near source ponds in the vicinity of the hot water can lead to considerably higher radiation exposure.

Other areas with unusually high radiation levels include the coastal belt of Karunagappally, Kerala, India where radiation also comes from thorium-containing monazite depositions. Here, median outdoor-radiation levels are more than 4 mSv/a but in certain locations along the coast, levels as high as 70 mSv/a have been measured. There are other locations with high radiation levels, namely in China. In the United States, radiation levels remain within the ranges shown in Figure 6.11 and 6.12. At high altitude conditions, a second natural-radiation component must, however, still be considered, namely cosmogenic radiation.

6.1.7 DEVELOPMENT OF THE ATMOSPHERE

The Earth's first atmosphere formed early. It developed through natural outgassing processes from water vapor H_2O, carbon dioxide CO_2, and nitrogen N_2. Over a longer period of time, other gases dissolved from the still hot liquid surface, especially noble gases like argon, helium, and radon, which are formed by slow radioactive decay. Composition of the early atmosphere can be estimated from the composition of volcanic gases, which consist of about 26% water vapor H_2O, 60% N_2, 6% CO_2, 6% CH_4, and 2% SO_2, with trace amounts of 0.1% H_2S and 0.005% HCL (Figure 6.17). Thus, the early atmosphere was a rather toxic and foul-smelling affair due to its sulfur content.

The outgassing process accelerated during the Hadean phase due to constant meteorite bombardment. Impact from bombardment broke up the crystal structures of solid materials, thereby releasing bound, residual gas components. In addition, enormous amounts of dust, gas, and water were ejected into the atmosphere. Meteorites also brought new material with different elemental and isotopic abundance distributions to Earth's surface and into the atmosphere.[32]

At the beginning of these processes, the surface had temperatures of about 2,000 °C and only cooled slowly over the following 100 million years. During this cooling process, condensation of the atmospheric water fraction also occurred. Together with high CO^2 content, this caused a greenhouse effect[33] that kept the atmosphere at high temperatures, up to 1,000 °C. In this phase, atmospheric pressure was much higher than today.[34] The water-gas fraction made up 270 bar and the CO_2 fraction was between 40 and 140 bar, while the fraction of nitrogen N_2 with other residual gases was about 1 bar, which corresponds to today's value. For comparison, the CO_2 fraction of today's atmospheric pressure is $3.5 \cdot 10^{-4}$ bar, which is more than 10,000 times lower.

Over the following 100 million years, dense cloud cover rained down and the first oceans formed. This process was completed 4.4 billion years ago. Surface temperature had cooled to 200 °C and high atmospheric pressure prevented water from evaporating. A mineralogical carbon cycle kept the atmospheric CO_2 component in equilibrium, whereby the chemical combination of magnesium silicates, CO_2, and water led to the formation of silicon dioxide SiO_2 and bicarbonates $2(HCO_3^-)$. In a second step, bicarbonates reacted with calcium to form calcium carbonates (gypsum and marble) but also released a third of the CO_2 that had been absorbed in the first step. Through this chemical process, rocks containing silicates are still converted to limestone today.

This cyclic process slowly freed the atmosphere from its high CO_2 content and so, the greenhouse effect gradually reduced and Earth's surface cooled. However, considerable volcanic activity in this early Earth age compensated for decreased CO_2 content by releasing large amounts of CO_2 and SO_2. This led to temperature fluctuations at the Earth's surface. When CO_2 levels were low, the Earth cooled completely and iced over. Volcanic activity and glaciation caused the mineralogical carbon cycle to freeze out. This again led to an increase in atmospheric CO_2 content with subsequent warming. In this way, over millions of years, new equilibrium conditions were established.

After about two billion years of Earth's history, the slow mineralogical carbon cycle was replaced by a much faster biological carbon cycle, which involved the onset of biological life and the spread of plant life. The biological cycle is based on the sunlight-induced breakdown of CO_2 into glucose, as a plant nutrient, and oxygen. This process is called *photosynthesis*, $CO_2 + H_2O + h\nu \Rightarrow C_6H_{12}O_6 + O_2$ and occurs primarily in summer when foliage has full sunlight. In winter, when leaves fall and decompose, the reverse is true. Organic material, such as glucose or methane, is converted back to water, carbon dioxide, and energy $h\nu$: $CH_4 + O_2 \Rightarrow CO_2 + H_2O + h\nu$. This process symbolizes all chemical CO_2-releasing decay by organic material.

Photosynthesis also influences the present composition of Earth's atmosphere. Rapid breakdown of CO_2, by chlorophyll and sunlight, releases very large amounts of oxygen so that today, the oxygen gas O_2 is the most important component in our atmosphere, along with nitrogen gas N_2 (Figure 6.17).

Inhaled oxygen is also converted to CO_2 gas and released into the atmosphere in other respiratory processes, such as breathing in humans and other living organisms. These various processes establish an equilibrium for the CO_2 content in the atmosphere. CO_2 content is based on the amount of plant chlorophyll and the intensity of incident sunlight, therefore it depends on the spread of plants over the Earth's surface as well as time of year. This CO_2 balance in the atmosphere determines the climate at Earth's surface. Other factors influencing climate, besides solar-radiation intensity at Earth's surface, are the reflectivity of Earth's surface, the so-called albedo, and the atmosphere's water-vapor content. Major disturbance of this equilibrium will result in climate changes that will eventually produce a new equilibrium. This is evidenced by the analysis of paleoclimatic conditions in Earth's history. Although Earth's climate evolution is not our topic, the carbon cycle represents an important component within the context of radiological conditions in Earth's atmosphere.

The Earth's atmosphere today is the result of this chemical development. As described, this involves the exchange of gaseous elements or molecules, by chemical reactions or diffusion, from the Earth's surface. Former processes included carbon cycles, which determined the carbon dioxide

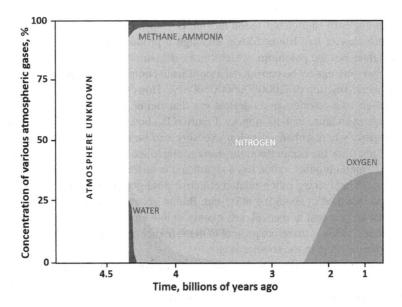

FIGURE 6.17   Gradual depletion of CO_2 and CH_4 in the atmosphere, by mineralogical-chemical processes in Earth's early phase, followed by the buildup of oxygen O_2 from the biological-chemical process of photosynthesis in plants during the following two billion years. Source: http://www.fas.org/irp/imint/docs/rst/Sect19/Sect19_2a.html (June 20, 2023).

and oxygen content in the atmosphere. Chemically inactive noble gases such as ^{4}He and ^{40}Ar originated from the decay processes of actinide chains as well as ^{40}K.

These gases can diffuse into the atmosphere from the lower depths of Earth's crust. In the deeper zones, these elements are physically frozen into the various rock structures and are only released during volcanic eruption. Noble gas elements also include short-lived ^{222}Rn, which is formed as part of the ^{238}U decay chain and can penetrate to Earth's surface as a noble gas, where it continues to rapidly decay. ^{222}Rn is, however, only found in trace amounts as a local equilibrium abundance between diffusion and decay.

Of these noble gases, argon is the most important component, accounting for nearly 1% of atmospheric gas; 99.6% of atmospheric argon is ^{40}Ar, a decay product of ^{40}K in Earth's crust and mantle. This observed large abundance of ^{40}Ar in the atmosphere is due to diffusion of ^{40}Ar gas from Earth's outer crust, as determined by the diffusion probability of argon from the crust. However, this gas also correlates with mantle convection, the tectonic exchange of crustal material and mantle material discussed in section 6.1.3. In this process, fresh ^{40}Ar can reach the upper crustal region. The ^{40}Ar also enters the atmosphere during volcanic eruption, together with decay gases stored in mantle material. This process was particularly effective during the high volcanic activity of the early Earth phase. The average outgassing rate of ^{40}Ar into the atmosphere is assumed to be $2.2 \cdot 10^9$ g/a, or 2.2 million kg per year.

The noble gas helium He, on the other hand, shows a very low value of only 5 ppm, although it is produced in large quantities in the Earth's crust by radioactive decay within the uranium and thorium chains. About 7 ppm of helium remain as part of natural gas deposits in Earth's crust.[35] The low helium value in the atmosphere is mainly due to the high velocity of these light gases, which also include hydrogen H_2 and helium ^{4}He since they readily escape the Earth's gravitational field and diffuse into space.

Neither ^{40}Ar nor ^{4}He are radioactive isotopes. They are stable noble gases, produced in the first case as an end product of radioactive decay and, in the second case, as a by-product of an entire decay chain. Radon isotope ^{222}Rn ($T_{1/2} = 3.8\ d$) is a radioactive partial product of the uranium decay series, which diffuses as a noble gas that moves chemically unhindered from the Earth's crust into the atmosphere where it decays by α-emission to ^{218}Po.

Another radiologically significant radon isotope is ^{220}Rn, a product of the thorium series. The ^{222}Rn has a much shorter half-life of 55.6 s, although it produces much higher follow-up activity through its daughter isotope polonium, which can be deposited as dust. Radon is mainly found in soil layers and therefore cannot be considered a significant component of atmospheric gases, with an average atmospheric fraction of 0.000000000000001%. However, it accumulates in closed caves, tunnels, and cellars as a colorless and odorless gas that can be easily inhaled there. In the Earth's crust, the average abundance is $4 \cdot 10^{13}$ mg/kg. Figure 6.18 shows a radon map of the United States, highlighting regions where enhanced radon exposure can be expected. In ocean water, where it occurs by emission from the ocean floor, the average abundance is estimated to be $6 \cdot 10^{16}$ mg/l.

Despite these small amounts, radon has a significant contribution to the natural radiation output at Earth's surface. The average radon-related effective dose per person is about 1.1 mSv/year, followed by cosmic radiation at about 0.4 mSv/year. Radon occurs predominantly in areas with high uranium and thorium content in the soil, i.e., mainly in low mountain ranges made of granite rock. In the United States, radon is therefore present in much higher concentrations in northern mountainous states and prairies than in the southern states.

Atmospheric composition dilutes and changes with distance from the ground.[36] Therefore, for our discussion, the composition of the atmosphere in the lower region – up to 10 km altitude (troposphere) and up to 50 km altitude (stratosphere) – is of direct interest. Figure 6.19 shows how the atmosphere is composed of the major gases, with the respective abundance as a percentage of the total volume. This distribution is valid up to an altitude of 25 km and only changes in the upper stratosphere and the following mesosphere. When considering the gases coming from Earth's crust that have accumulated in the atmosphere, the radioactive load appears to be minimal, with the

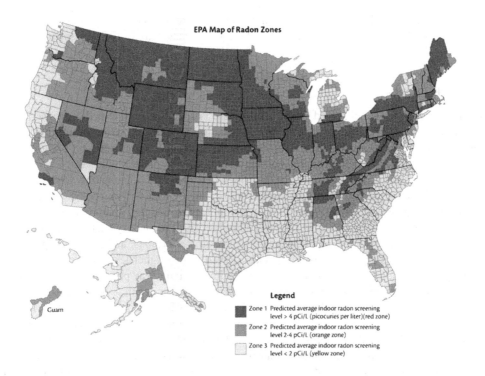

FIGURE 6.18 Radon map of the United States. The regions marked in red show an average activity of more than 0.3 Bq/kg, the orange zones between 0.15 and 0.3 Bq/kg, and the yellow zones less than 0.15 Bq/kg of material. Source: https://www.epa.gov.

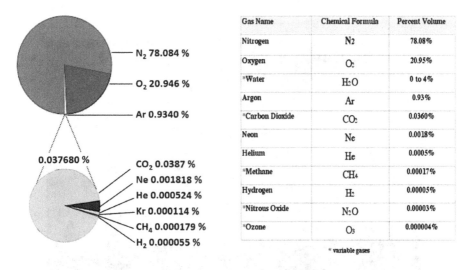

Gas Name	Chemical Formula	Percent Volume
Nitrogen	N_2	78.08%
Oxygen	O_2	20.95%
*Water	H_2O	0 to 4%
Argon	Ar	0.93%
*Carbon Dioxide	CO_2	0.0360%
Neon	Ne	0.0018%
Helium	He	0.0005%
*Methane	CH_4	0.00017%
Hydrogen	H_2	0.00005%
*Nitrous Oxide	N_2O	0.00003%
*Ozone	O_3	0.000004%

* variable gases

FIGURE 6.19 Chemical composition of Earth's atmosphere, which consists of 99% nitrogen and oxygen gases. Water vapor is neglected as atmospheric moisture. All other gas components show an extremely low proportion but can still play an important role for the Earth's climate, including CO_2, CH_4, and also O_3. Source: https://en.m.wikipedia.org/wiki/File:Atmosphere_gas_proportions.svg (March 13, 2024).

exception of radon. Its abundance is balanced by the equilibrium between emission and decay rates and is generally low. However, the sun contributes a significant share of the atmosphere's radiation load.

6.1.8 THE RADIOACTIVE ATMOSPHERE

The natural radioactivity observed in the atmosphere is specifically due to the intense radiation spectrum emitted from the sun as well as cosmic radiation from distant stellar events. For eons this intense radiation has transversed the universe unimpeded and is therefore called *cosmogenic radioactivity*, in contrast to the radiogenic radioactivity of crustal and mantle material.

The origin of cosmic radiation has been described in Chapter 5.1.3. Here, only long-term effects of radiation bombardment on the atmosphere's radioactivity budget will be presented. In the upper layers of the atmosphere, primary particles of gas molecules undergo spallation or breakup processes, producing particle showers with up to 100 billion (10^{11}) secondary particles (Figure 6.20).

The bulk of this secondary radiation consists of neutrons, protons, and pions. Despite initially high energies, protons and neutrons are quickly slowed in the higher levels of the atmosphere by scattering and reaction processes and neutralized by impact and decay processes. Protons can form chemical compounds with oxygen or ozone, producing water molecules. As light particles, these can also backscatter and escape the gravitational potential of Earth to be reflected back into interplanetary space. On the other hand, as long as neutrons do not interact (cf. Chapters 3.1.5 and 5.1.2), they will eventually decay to protons, electrons, and anti-neutrinos with a half-life of about 9 minutes. Finally, pions are a special class of hadrons, which consist of only two quarks – in contrast to the baryon particles, proton and neutron, with three quarks each. Pions have a relatively short lifetime and decay by different processes into muons, γ-photons, electrons, and neutrinos.

The muon and electron flux of a radiation shower decreases by secondary interaction so that only about 100 muons per square meter and second can be measured at Earth's surface. These muons are very energetic and have a high radiation potential at the surface. There they cause a high neutron and

FIGURE 6.20 High-altitude radiation showers over London in a simulation of the particle showers – of protons, neutrons, pions, and muons – triggered by primary cosmic rays. Source: Rebecca Pitt, http://www.ep .ph.bham.ac.uk/DiscoveringParticles/detection/cosmic-rays/ (June 26, 2023). These showers frequently occur and mostly correlate with solar activity, which can cause interference in radio communications. In reality, these showers remain invisible because the individual particles are only a few femtometers (10–15 m) in diameter.

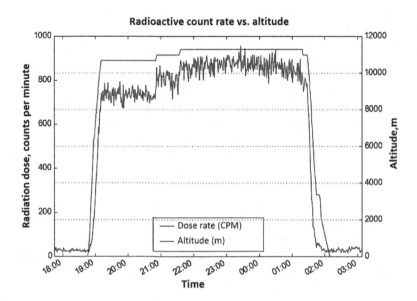

FIGURE 6.21 Radiation dose (count rate of incident radiation) as a function of time, from entering an airport one hour before departure to leaving an airport one hour after arrival. The average dose rate at flight altitude is approximately 5 to 6 mSv/h. Courtesy of Nick Touran; https://www.hatisnuclear.com/radiation-on-flights.html

γ-radiation background, in the energy range up to 20 MeV, by reaction processes with materials. This background impairs the measurement of weak radioactivity components, such as neutrinos. Therefore, measuring stations are installed up to 2,500 m deep in the Earth's subsurface in order to successfully absorb the background radiation and thereby allow for neutrino detection (cf. Chapter 5.1.1).

The intensity of cosmic radiation decreases rapidly with decreasing height due to these absorption processes, according to the absorption law described in Chapter 4 with an average absorption constant μ. Thus, radiation intensity decreases exponentially the closer one is to sea level. As a result, the dose estimated at sea level is approximately 0.05 μSv/h. This does, however, vary with latitude, since cosmic radiation is focused by the geomagnetic field and so is particularly strong in the polar regions.

Conversely, it can be estimated that the intensity of cosmic radiation doubles with every 1,500 meters of altitude. Thus, mountain dwellers at 3,000 m altitude receive four times the radiation dose of those who live near the coast. Altitude radiation also has a significant impact on air travelers. Normal flight altitude is 10 km on average. This means an approximate 100-fold higher dose for passengers and aircrew as compared to sea-level dose (Figure 6.21). An eight-hour transatlantic flight from Frankfurt to Chicago corresponds to an effective dose of 40 μSv at a rate of about 5 μSv/h. Thus, five transatlantic round trips correspond to an effective dose of 400 μSv, which more than doubles the average annual cosmic dose of the traveler.

Flights over polar regions increase the dose. This is because cosmic radiation is particularly intense at polar zones due to the funneling effect of the geomagnetic field as discussed in section 6.1.10. A twelve-hour flight from Frankfurt to Tokyo thus corresponds to an effective dose of 100 μSv.

The enhanced cosmic ray flux in polar regions is also reflected in the well-known aurora borealis celestial phenomenon, which is enhanced during increased solar activity (cf. section 6.1.10).

Cosmic rays and their secondary decay products contribute to the production of long-lived radioactive isotopes in the Earth's atmosphere, which can be deposited in the Earth's crust by physical and chemical exchange processes. These cosmogenic radionuclides include the carbon isotope ^{14}C ($T_{1/2} = 5,730$ a), radioactive tritium ^{3}H ($T_{1/2} = 12.3$ a), the rarer beryllium ^{10}Be ($T_{1/2} = 1.39 \cdot 10^{6}$ a), and, in smaller amounts, the heavier long-lived aluminum and chlorine isotopes ^{26}Al ($T_{1/2} = 7.17 \cdot 10^5$ a) and ^{36}Cl ($T_{1/2} = 3.01 \cdot 10^5$ a).

The most important component of cosmogenic radioactivity is undoubtedly the neutron-rich carbon isotope ^{14}C, which decays by β^--decay to the nitrogen isotope ^{14}N. The decay energy is relatively low at 152 keV, which means that ^{14}C has only a low dose rate. However, the half-life of ^{14}C is relatively long at 5,730 years. For this reason, ^{14}C is used for the age determination of archaeological and historical finds containing organic components (cf. Chapter 16.4, Volume 2).

The carbon isotope ^{14}C is partially produced by spallation reactions, between the high-energy primary particles of cosmic rays and gases, in the upper atmospheric layers. A larger fraction of ^{14}C isotopes is, however, produced by neutron capture processes since neutrons appear in large abundance as secondary particles in the cosmic-ray shower. The neutrons thermalize by collision on the surrounding gas particles and eventually undergo nuclear reactions with them. The most important reaction is $1n + {}^{14}N \Rightarrow {}^{14}C + {}^{1}H$ or, in nuclear physics terminology, ^{14}N(n,p)^{14}C. However, other reactions on carbon and oxygen isotopes with lower abundance – such as ^{13}C(n,γ)^{14}C and ^{17}O(n,α)^{14}C – also contribute to the ^{14}C production. The production rate for ^{14}C reaches its maximum at altitudes between 9 and 15 km in the upper troposphere and lower stratosphere. This rate increases toward high latitudes (i.e., toward the poles) because there the high-energy cosmically charged particles are focused by the geomagnetic field and so preferentially enter the atmosphere.

From the thermal cross section of the ^{14}N(n,p)^{14}C reaction, as well as the abundance of nitrogen atoms and the average neutron flux at these altitudes, one can directly estimate the production rate following the equations given in Chapter 3. Accordingly, the rate is 22,000 ^{14}C isotopes per second per square meter. Taking the decay rate of $\lambda = ln2/5,730\ y = 1.21 \cdot 10^4$ per year, the decay rate of ^{14}C radioactivity is about 2.66 Bq/y m^2. Averaged over the total volume of Earth's lower atmosphere, this gives an annual value of 1.4 PBq/y, or 1,400 trillion becquerel per year.

The ^{14}C behaves chemically like a stable ^{12}C atom, combining with oxygen in the atmosphere to form ^{14}CO$_2$. Radioactive ^{14}C is thus subject to normal biological-chemical exchange processes between the atmosphere, the ocean, and the Earth's crust. And so, due to the various production and absorption processes, an equilibrium develops between the stable ^{12}C abundance and the ^{14}C abundance in the atmosphere.

Of particular importance in the exchange processes is the carbon cycle. This cycle transfers ^{14}C into plant material and thus into other biological, especially animal and human, body material through the normal feeding cycle. Since all living things constantly exchange carbon with the atmosphere, through respiration and other metabolic mechanisms, the same distribution ratio for carbon isotopes is established in both the atmosphere and living organisms. And so, a balance between production and absorption exists, which fixes atmospheric ^{14}C activity at a total activity level of $1.4 \cdot 10^{17}$ Bq (140 PBq).[37]

Accordingly, the atmosphere contains one trillion more stable ^{12}C atoms than the radioactive ^{14}C component. This ratio, between ^{12}C and ^{14}C abundances in material, is described by the δ ^{14}C parameter. This parameter expresses the abundance ratio between the two isotopes relative to a standard value. The value δ ^{14}C depends on various circumstances, such as the ^{14}C production rate as well as chemical fractionation by the physicochemical or even biochemical processes to which the material was subjected, conditions that are well known for determining the normalization standard.

Production of ^{14}C is directly related to incident cosmic-ray intensity, which is due to the Sun's activity. This activity and its fluctuations follow an eleven-year cycle that is directly traceable in measured ^{14}C values over longer periods. These fluctuations, and the intensity of respective ^{14}C activity, can be measured and dated with radiochemical methods, for example in tree rings. Annual wines also allow for long-term measurements (Figure 6.22), as all plant materials ultimately do.

To measure ^{14}C abundance over longer periods, borehole samples are taken from ice in Greenland and the Antarctic. There, annual snowfall cycles have developed annual rings of fresh snow material, which reflect annual variations in air pollution and isotopic ratios. By analyzing such ice rings,

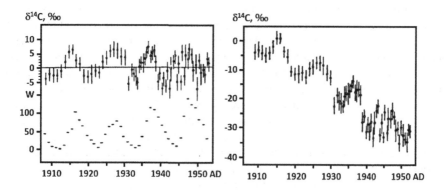

FIGURE 6.22 Left, the correlation between ^{14}C abundances. The left curve indicates ^{14}C in Georgian wines produced annually between 1909 and 1952. The right curve shows annually measured solar activity, which follows the eleven-year cycle. It is clear that increased solar activity translates into increased ^{14}C radioactivity in wine. The uncorrelated increase after 1949 is related to the ^{14}C contribution from the nuclear weapons testing program (see Chapter 14.4, Volume 2). The ^{14}C curve on the left is, however, normalized to a constant value in order to better illustrate the correlation. On the right, the non-normalized ^{14}C curve shows a continuous drop at the ^{14}C equilibrium value. This drop is caused by increasing CO_2 production in the Earth's atmosphere, which is brought about by fossil fuel burning such as coal and oil. These fossil fuels were originally plants subject to normal ^{14}C equilibrium conditions. However, this ^{14}C content has long since decayed and so, coal and oil do not contain ^{14}C components. Industrial combustion processes, and the associated release of $^{12}CO_2$ gases, continuously reduce the relative ^{14}C fraction in the atmosphere as reflected in the declining slope of this ^{14}C curve. Note: [a] A. Burchuladze, S. Pagava, P. Povinec, et al., "Radiocarbon variations with the 11-year solar cycle during the last century," *Nature* 287 (1980): 320–322, https://doi.org/10.1038/287320a0.

it is possible to trace back over several 100,000 years many of the climatic conditions on Earth. This is not, however, useful for ^{14}C analysis due to the short ^{14}C half-life of 5,730 years. Depending on the sensitivity of the detection method, by using for example ^{14}C accelerator mass spectrometry (AMS), one can trace variations back for up to 50,000 years.[38]

Tritium, the neutron-rich hydrogen isotope ^{3}H, has a half-life of 12.3 years. Thus, it decays much faster than the ^{14}C isotope. The β^{-}-decay of tritium releases a total energy of 18.6 keV, with electrons averaging 5.7 keV, corresponding to only a small radiation burden. The decay product is the neutron-deficient helium isotope ^{3}He. Tritium is produced by fast neutron spallation reactions with ^{14}N nitrogen and also with ^{16}O oxygen in the upper troposphere and lower stratosphere. In physical nomenclature, tritium is often simply referred to as symbol "t". Again, the most important reaction is the interaction of neutrons with atmospheric gases: $1n + {}^{14}N \Rightarrow {}^{14}C + {}^{3}H$ or, in nuclear physics terminology, $^{14}N(n,t)^{12}C$. This reaction occurs at neutron energies greater than 4 MeV. A somewhat weaker second production reaction is $^{16}O(p,t)^{14}N$. The average natural production rate is 2,000 ^{3}H nuclei per square meter per second, an order of magnitude lower than the production rate of ^{14}C. The annual production of tritium activity in the atmosphere is thus 72 PBq/a.

Chemically, tritium binds relatively rapidly with oxygen or hydroxyl molecules to form heavy water, $^{3}H_1HO$ or, in lesser amounts, $^{3}H_2O$. The heavy water behaves like regular $^{1}H_2O$ and participates in the normal water exchange between the atmosphere and oceans as well as other bodies of water; this latter environment is called the *hydrosphere*. The average residence time for water vapor in the atmosphere is between 3 and 20 days, before it then falls out as precipitation (rain, snow, hail) and quickly becomes part of the normal groundwater balance. Hydrologists calculate that every 10 days, water vapor stored in troposphere clouds undergoes a full exchange cycle with the hydrosphere.

The natural fraction of tritium in the atmosphere and hydrosphere is estimated to be 26 MCi, or $9.6 \cdot 10^5$ TBq, from the equilibria between cosmic-ray production and the normal hydrologic cycle. About 90% of the tritium fraction is stored in water bodies, with the remaining 10% stored in the atmosphere as water vapor or as part of other gas molecules. There, for every tritium atom, there are $2.5 \cdot 10^{14}$ stable hydrogen nuclei or protons. Due to the large amount of ocean water, the proportion of tritium in the hydrosphere is much lower: there are about 10^{18} stable hydrogen nuclei for every tritium atom. This ratio can, of course, vary greatly by location. Therefore, hydrology uses the tritium unit (TU), a unit that determines the amount of tritium in water. According to this, 1 TU corresponds to one tritium atom in 10^{18} water atoms. Calculating the activity of tritium for one liter of water, 1 TU corresponds to a tritium activity of 0.118 Bq/l. Under normal conditions, the concentration of tritium in rainwater is 5 TU. This corresponds to an activity of about 0.6 Bq/l. However, due to atomic bomb testing programs and tritium emissions from nuclear reactors, the current tritium abundance is considerably higher (see Chapter 13.4, Volume 2).

The radioisotope ^{10}Be beryllium-10 ($T_{1/2} = 1.39 \cdot 10^6$ a) is produced by the spallation reactions of high-energy secondary protons and neutrons on heavier elements in the atmosphere. Laboratory measurements of spallation cross sections have revealed the dominant reactions to be $^{14}N + n \Rightarrow$ ^{10}Be, $^{16}O + n \Rightarrow {}^{10}Be$ at neutron energies below 100 MeV and $^{14}N + p \Rightarrow {}^{10}Be$, $^{16}O + p \Rightarrow {}^{10}Be$ at proton energies greater than 100 MeV. The total production rate is 200 ^{10}Be isotopes per second per square meter. At a comparable rate, ^{14}C is also produced by spallation reactions but this is negligible compared to production by thermalized neutrons. The ^{10}Be decays via β^--decay to ^{10}B boron with a decay energy of 556.2 keV. Rain washes out radioactive ^{10}Be, which precipitates in groundwater.

Similar to tritium, ^{10}Be activity is measured in water bodies as a method for analyzing the hydrologic cycle over long periods of time as well as to determine the age of water in deep groundwater reservoirs. The ^{10}Be precipitates in less alkaline waters and accumulates in the surrounding soil, leading to slight accumulations of activity in the ground. In ocean waters, ^{10}Be precipitates in sediments and therefore can be used to determine the growth rate of sediment precipitation. These and other long-lived radioactive products of spallation reactions, such as ^{26}Al and ^{36}Cl, are, however, largely negligible for general radiation exposure due to their low abundance and low decay energy.

6.1.9 LIGHTNING AS SOURCE FOR RADIOISOTOPES

Lightning during a thunderstorm is a powerful and frequent event. Space travelers circulating Earth would observe that lightning occurs on average more than 40 times per second over the entire globe. This amounts to a total of about 1.4 billion ($1.4 \cdot 10^9$) lightning flashes per year. The distribution of these lightning events does, however, very sensitively depend on location, climatic conditions, and time of year.

Figure 6.23 shows a global map for the distribution of lightning incidents, underlining the close relation to climatic conditions. The figure suggests that lightning preferably occurs with high frequency in humid climatic zones, which include equatorial regions, jungle areas, and steep mountain ranges but also large portions of the United States eastern regions where humid air traveling north along the Mississippi valley interacts with more northern climate regions.

Lightning is caused by electrical discharge between positively and negatively charged volumes. This can occur between different regions in a group of storm clouds and also between storm clouds and the ground, or within the clouds themselves. Different charge states are generated by the separation of positive and negative charges, with positive charges remaining at the bottom of the thundercloud while the negative charge zone distributes over higher-altitude layers of the cloud. These electrical fields can generate potential differences of up to 30,000 volts per meter. The actual charge-production and field-development processes are still not completely understood, but

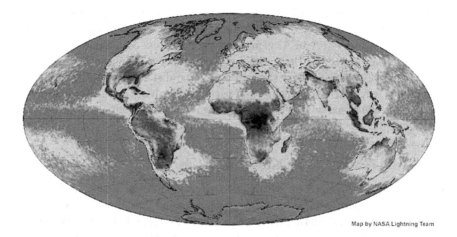

Map by NASA Lightning Team

FIGURE 6.23 Map showing the average of lightning flashes per year and per square kilometer based on data collected by NASA's Lightning Imaging Sensor on the Tropical Rainfall Measuring Mission satellite between 1995 and 2002. Ocean regions, marked in gray or light purple, display on average less than one flash each year. Places with the largest number of lightning strikes are zones of high humidity marked in deep red to black. Source: https://alltecglobal.com/es/nasa-lightning-imaging-sensor-lis/ (July 7, 2023).

potential differences of up to 70 MeV between the different cloud layers can be reached.[39] At high-humidity conditions this can lead to the enormous discharges that we observe as lightning.

Lightning flashes reduce the negative charge above Earth's surface, thus decreasing the electric field; this field does, however, quickly regenerate in the dynamic cloud environment. Several types of discharges are possible including intra-cloud discharges that occur within the confines of a single thundercloud and cloud to ground discharges, which account for 90% of lightning events. Less frequent are inter-cloud discharges, which occur between different thunderclouds, and air discharges that occur between a thundercloud and the surrounding air. Discharges produce electrons that are accelerated to relativistic speed by the electric fields. Inelastic scattering and collision produce high energy X- and γ-rays that can in turn induce the production of radioactive isotopes through photon-disintegration processes of atmospheric nuclei. Direct observation of the reaction products confirmed the production of radioactive nitrogen-13 and oxygen-15 by $^{14}N(\gamma,n)^{13}N$ and $^{16}O(\gamma,n)^{15}O$ photon disintegration, respectively.[40]

This and other experiments record γ-flux from the lightning discharge, seen as a peak in observed neutrons and an afterglow from the 511 keV annihilation radiation resulting from decay of the two isotopes, which resulted when positrons were destroyed by interaction with free electrons in the thundercloud. The initial total-number density of ^{13}N and ^{15}O isotopes is estimated to be about $n_0 = 2.6 \cdot 10^{-2}$ cm^{-3}. These isotopes are relatively short-lived and will quickly decay to stable isotopes ^{13}C and ^{15}O, respectively.

Of more interest is the secondary neutron flux produced by the photodissociation process. It has been argued that during this process, 96% of the neutrons are absorbed by the $^{14}N(n,p)^{14}C$ reaction, producing long-lived ^{14}C in the atmosphere and thus complementing the cosmogenic production of ^{14}C in Earth's equatorial regions. The remaining 4% of neutrons undergo radiative capture $^{14}N(n,\gamma)^{15}N$, enriching the stable ^{15}N component in the atmosphere. Estimates of the total number of neutrons produced in a terrestrial γ-ray flush range from 10^{11} to 10^{15} depending on the volume of the lightning event. These are very crude estimates, since only a fraction of lightning events are powerful enough to generate a γ-ray burst but do demonstrate that atmospheric lightning does generate radioactive elements in our atmosphere.

While this may not have much consequence for the average cosmogenic dose of radiation exposure, recent observations[41] indicate the possibility of a direct correlation between geological activity

– such as volcanic eruptions and earthquakes – and the intensity of lightning-induced gamma-bursts. An unusually high number of intense bursts were observed on February 7, 2023 at the Aragats Cosmic Ray Observatory in Armenia, in very close time correlation to a powerful earthquake in eastern Turkey and Syria. The correlation is explained as an ionospheric response to earthquakes, which enlarges electrical-field strength over large distances from the epicenter.[42] These electrostatic effects may have been enhanced in periods of high seismic and volcanic activity and may have caused an additional high-intensity lightning induced γ-ray flux in the troposphere and at the Earth surface.

6.1.10 Cosmic Radiation and the Geomagnetic Field

The geomagnetic field is induced by the dynamo effect of Earth's liquid outer core. This field deflects the Sun's high-energy charged cosmic-ray particles and thus largely protects Earth's atmosphere and surface. Instead of entering Earth's atmosphere in direct flight, the high-energy charged particles follow a spiral path around the magnetic field lines. Many of the incoming particles are thus reflected back into interplanetary space. The remaining particles focus along the field lines to the polar regions. The average dose of total remaining radiation at sea level is about 0.05 μSv/h, but this doubles with about every 1,500 meters of altitude. However, due to the influence of the geomagnetic field, radiation exposure varies with latitude as well as altitude. Thus, cosmic-ray intensity is considerably higher in the polar regions than in the equatorial region. Figure 6.24 shows cosmic-ray intensity over Earth's different latitudinal lines. Inhabitants of polar regions, as well as passengers on flights over polar regions, will therefore receive a radiation dose more than a hundredfold to a thousandfold higher than people in tropical zones or even inhabitants of the middle latitudes.

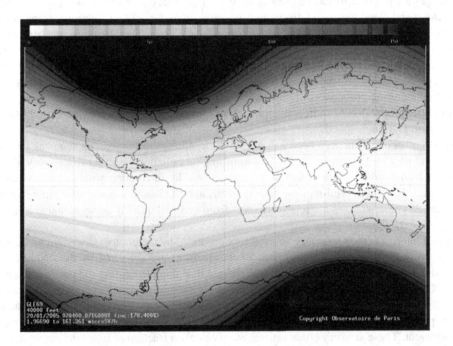

FIGURE 6.24 The hourly radiation dose at 13,000 m altitude, recorded on January 20, 2005 by the Observatoire de Paris during extremely high solar activity. The radiation dose at this altitude varies between 2 and 160 μSv/h with an average radiation dose of 0.04 μSv/h at the Earth's surface. Source: https://en.m.wikipedia.org/wiki/File:Atmosphere_gas_proportions.svg (March 13, 2024).

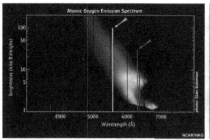 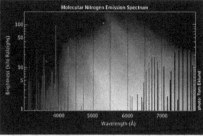

FIGURE 6.25 On the left, the spectrum of excited oxygen (O) atoms with characteristic light emissions, shown with lines in the red and green regions. This explains the red-green color play of the aurora. On the right, the spectrum of molecular nitrogen (N_2), a more complex line structure in the blue-violet and red-infrared regions. This explains the origin of the aura's magenta coloration, since blue hues are difficult for the human eye to detect at night. Source: MetEd/Comet, http://www.meted.ucar.edu/hao/aurora/aurora.1.htm (June 20, 2023).

Cosmic rays entering over the poles cause a strong ionization of the atmosphere's upper regions, from 200 to 600 km above Earth in the so-called ionosphere. This radiation flux is often referred to as the solar wind. In this high-altitude region, particularly the high-energy photon component of this flux – such as X-rays and gamma rays – ionizes the nitrogen and oxygen atoms and molecules. At high radiation intensities, this can produce a large flux of millions of free high-energy electrons per cubic centimeter, which in turn can cause other secondary radiation effects.

This is particularly evident in the well-known aurora borealis or northern lights phenomenon. In the polar regions of the sky, these red-green light shows particularly occur during periods of high cosmic-ray flux from the Sun. During this process, Earth's magnetic field accelerates the charged particles and excites the shells of atoms and molecules in the atmospheric gases at altitudes from 80 to 250 km. During excitation, the characteristic radiation of the particular atoms and molecules is emitted. For oxygen, the wavelength of the emitted radiation is 557.7 nm in the green range. and 630.0 nm and 636.3 nm in the red range (Figure 6.25). This explains the fascinating red-green color play. The emitted radiation of excited nitrogen molecules is in the ultraviolet and infrared wavelength range, which the naked eye perceives as violet.

The protective effect of the geomagnetic field is clearly demonstrated when comparing Earth with conditions on our neighboring planet Mars, where surface and atmospheric conditions have been studied in recent decades by a series of Mars probes. Ground investigations by the Mars rover showed that our sister planet's originally strong magnetic field collapsed about 3.5 billion years ago, more than 100 million years after the great bombardment during the Hadean period. Over the previous 500 million years, the geological development of these planets had been parallel. Surface conditions were comparable, with wide water areas and a dense gas atmosphere. Today, Mars' magnetic field has largely disappeared except for some small selectively distributed residual fields (Figure 6.26). The terrestrial magnetic field has, however, largely retained its original strength.

It is now believed that the Martian dynamo effect collapsed over a relatively short period of time. It is widely accepted that either the metallic (Fe-Ni) core[43] cooled and solidified after only a billion years or cooling was initiated by collision with a larger object.[44] There are, however, numerous speculations about the reason for the sudden disappearance of the magnetic field.[45] Yet, the question remains whether rapid cooling also occurred because the long-lived radioactive material was at a lower level, thereby generating less heat in the Martian interior. For, with a drop in decay heat, the liquid state of the outer metal core could not have been maintained.

Since Mars has lost its magnetic field, the surface is exposed to a much higher radiation level than Earth. This is similar to the cosmic-ray interaction that occurs with Earth's atmosphere at the poles, as described in section 6.1.8. However, Mars also exists without a protective magnetic field at

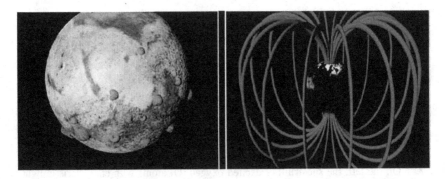

FIGURE 6.26 Comparing today's magnetic-field conditions on Mars (left) and on Earth (right). Earth's magnetic field is generated by a dynamo such that Earth's outer core, of liquid metal, is convecting between the inner solid-iron core and the cooler mantle region. The resulting magnetic field surrounds the Earth and appears as a global phenomenon. Conversely, on Mars the various magnetic fields do not encompass the entire planet and are purely local phenomena. The Martian dynamo no longer exists, and so, its magnetic fields are merely fossilized remnants of an ancient, global magnetic field. Image credit: NASA/GSFC.

FIGURE 6.27 Earth's magnetic field (right) largely deflects high-energy solar radiation. Only small fractions of the radiation are focused on the magnetic poles via the field lines. Mars (left), without its own magnetic field, is exposed to intense solar radiation without protection and so has largely lost its atmosphere due to ionization effects and the associated evaporation. Image credit NASA/GSFC: https://arstechnica.com/science/2015/11/how-mars-lost-its-atmosphere-and-became-a-cold-dry-world/ (June 21, 2023).

a smaller distance from the Sun, so that the intensity of the solar wind in the Martian atmosphere is many times higher (Figure 6.27). New data collected by means of the Mars probe MAVEN, after many orbits through the inner and upper Martian atmosphere, indicates the continuing existence of a strong solar wind due to extreme cosmic irradiation.

As a result, the Martian atmosphere has been slowly eroded by solar wind over 500 million years. This has resulted in a large number of ionized particles in the upper Martian atmosphere, creating a strong electrical field in which particles are accelerated to an escape velocity and then "evaporate" into space. This evaporation of the Martian atmosphere has gradually reduced atmospheric shielding, causing a much higher cosmic-ray field at the Martian surface than is found on Earth. In addition to this regular exposure to cosmic rays and solar wind, there are also frequent and abrupt increases in radiation that are generated by strong solar flares.

NASA's 2001 Mars Odyssey spacecraft was equipped with a special instrument called the Martian Radiation Experiment (or MARIE), which was designed to measure the radiation environment

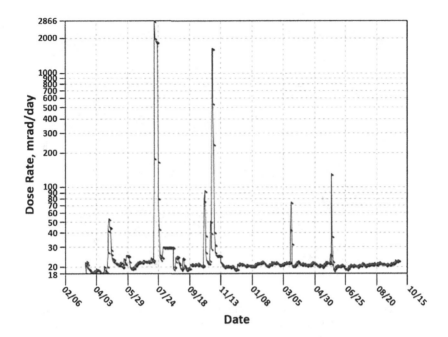

FIGURE 6.28 For the Martian surface, the diagram shows an average radiation exposure per day of 20 mrad/d, or 0.2 mGy/d, as measured by MARIE. This corresponds to an annual rate of 80 mGy/a. Data was recorded from February 2002 to October 2003. Radiation levels measured by MARIE are two to three times higher than those measured on the International Space Station (100–200 mGy/a). Occasional solar prominences produced doses that were hundreds of times higher, with peak values around 0.02–0.03 Gy/d. Source: https://en.m.wikipedia.org/wiki/File:MARIEdoserates.nasa.png (November 4, 2023).

during its orbit around the planet. Because Mars has such a thin atmosphere, the radiation detected by MARIE is comparable to radiation intensity at the surface. Figure 6.28 shows that over the course of about eighteen months, MARIE detected continuous radiation levels that were 2.5 times higher than the observed levels on the International Space Station, where levels were measured at 0.2 mGray per day, which is equivalent to 0.08 grays per year. MARIE also detected two solar-proton events, where radiation peaked at about 0.02 grays per day, and several other events that peaked to about 0.001 grays per day.

At the Martian surface, resulting annual cosmic-ray exposure (Figure 6.29) varies from 10 to 20 rem/a in international units, corresponding to an annual dose of at least 0.1–0.2 Sv. This is two to three orders of magnitude higher than the annual dose received from cosmic rays by the human population, which is 0.3–0.4 mSv depending on altitude; at extremely high alpine elevations, the dose increases to more than 1 mSv.[46] While studies have shown that the human body can withstand a dose of up to 0.2 Sv without permanent damage, prolonged exposure to the radiation levels found on Mars could lead to all sorts of health problems. These include for example acute radiation sickness, increased risk of cancer, genetic damage, and even death. At such levels, even microelectronic elements are at risk; MARIE had to shut down after only a year and a half of operation because the intense radiation damaged its electronic components.

Surface radioactivity was measured using the gamma ray spectrometer (GRS) on the Mars Odyssey, detecting the spatial distribution of characteristic-radiation emission points with high resolution. Measuring the intensity of these γ-rays allows us to calculate how abundant the planet's various elements are and how they are distributed on the surface. Gamma rays emitted by radioactive nuclei show as sharp emission lines in the instrument's spectrum. From the energy of these γ-lines, one can determine which radioactive isotopes are present, while line intensity gives local

Dose Equivalent Values (rem/yr)

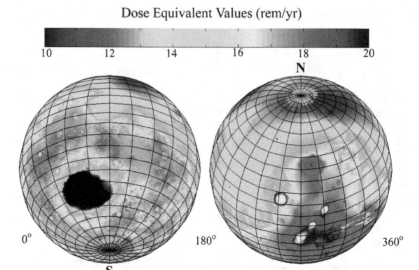

FIGURE 6.29 Intensity and distribution of the cosmic radiation to which Mars' surface is exposed. Units for dose exposure are given in rem/a, which corresponds to international units of 0.01Sv/a. Credit: NASA, https://www.universetoday.com/14979/mars-radiation1/ (June 21, 2023).

isotope concentrations at the surface. Thus, the abundances and distribution of uranium, thorium, and ^{40}K in the Martian material were determined (Figure 6.30).

The existence of radioisotopes emphasizes that Mars as well as Earth were originally formed from the same protoplanetary material containing long-lived radioisotopes. As on Earth, this suggests that these isotopes probably also originally heated the Mars core, where the resulting decay heat produced a dynamo effect that caused the formation of a magnetic field. However, from the radioisotope distributions at the surface, one can only draw conditional conclusions regarding the Mars interior environment.

For the most part, the distribution gives information about the processes that formed and changed rocks on the surface. Lower potassium values, in the polar regions of the map, may be due to dilution of the soil material caused by large amounts of water as reflected by the remnants of ice. Potassium and thorium show a similar pattern of low activity, which correlates with recent, local volcanic activity; distribution suggests that early crustal formation was enriched with these radioactive elements, while magma from later volcanic activity shows lower ^{232}Th and ^{40}K content suggesting an enriched and cooling mantle region.[47]

After this excursion to conditions on Mars, we return to Earth for another look at the effects of cosmic rays on atmospheric processes.

6.1.10.1   Cosmic Radiation and Climate

Over the last two decades, there has been intense discussion about the influence of cosmic rays on our climate. This discussion is based on another effect of cosmic-ray interaction with the Earth's atmosphere. It has been noted that high-energy cosmic rays do not only cause secondary nuclear reactions but also break up existing molecular structures in the atmosphere, producing ionized atoms or even molecular fragments along their path. These fragments can react chemically with each other, producing carbon-nitrogen combinations that provide the basis for aerosol formation in the atmosphere. These aerosols are considered the basic substances for the condensation of water molecules – a process that leads to cloud formation. Clouds increase the albedo, the reflection of solar light

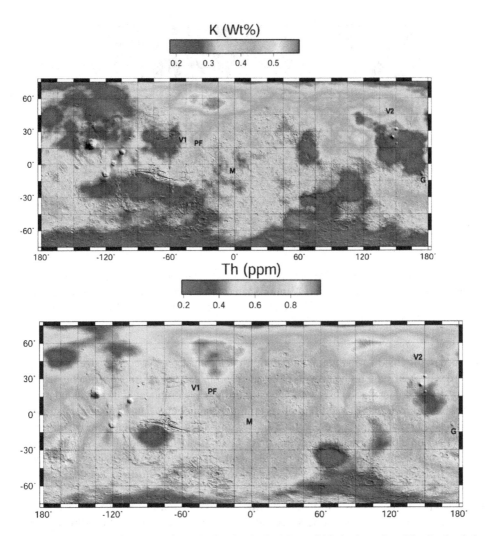

FIGURE 6.30 Top map shows potassium distribution in the Mars mid-latitude region. Distribution is based on 1.46 MeV gamma radiation of the long-lived radioactive isotope ^{40}K, which occurs in small amounts in rock and soil. Regions with high potassium levels appear red and yellow, while regions with low potassium levels appear purple and dark blue. The bottom map shows distribution of the radioactive isotope ^{232}Th. Locations of the five successful lander missions are also marked on the maps: Viking 1 (VL1), Viking 2 (VL2), Pathfinder (PF), Spirit at Gusev (G), and Opportunity at Meridiani (M). Source: http://grs.lpl.arizona.edu/specials/ (June 21, 2023). https://www.lpi.usra.edu/meetings/lpsc2011/pdf/1097.pdf.

back into space. It is being argued that this albedo effect reduces the transfer of radiation heat from the Sun to the Earth and might therefore have a positive effect in our present age of climate change.[48]

This condensation process is well known. For example, aerosols produce contrails from aircraft engines and these contrails form larger clouds that can lead to so-called aviation smog (Figure 6.31). However, new questions arise: Can cosmic rays directly and efficiently contribute to condensation nucleation and thus influence weather and climate? And if yes, to what extent can this effect be of use? These questions are still unresolved. Currently, the efficiency of aerosol formation by high-energy radiation is being studied at the Technical University of Denmark and at CERN, in the so-called cloud experiment. In such experiments, cloud chambers (cf. Chapter 4, Section 4.1.3) containing various atmospheric-like gases are used to study ionization and the resulting condensation nuclei.

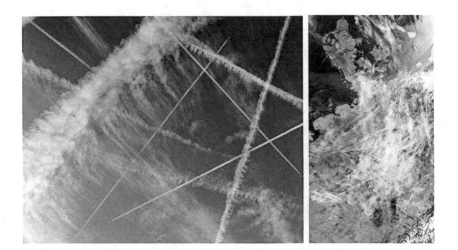

FIGURE 6.31 Condensation trail (contrail) formation and development in air traffic, with its effect on cloud formation (aviation smog) over Frankfurt Airport. Source: https://www.smithsonianmag.com/science-nature /airplane-contrails-may-be-creating-accidental-geoengineering-180957561/ and https://www.grida.no/climate /ipcc/aviation/038.htm (November 27, 2023).

For successful cloud formation, condensation nuclei must have a certain size, about 50 to 100 nm, in order to contribute to the growth of cloud-forming aerosols. Atmospheric chemistry and the chemistry of aerosol formation are, however, extraordinarily complicated. Initial cloud experiments at CERN have shown that the addition of certain gas components causes necessary growth of the aerosol nuclei, allowing for these conditions.[49] Cosmic rays could therefore at certain conditions have an effect on cloud formation and thus on climate, if high radiation flux were to generate the free ions and molecular fragments that provide seed material for aerosols. This is especially true in the deeper layers of the atmosphere where flux would increase through chemical processes and accumulate to form condensation nuclei for clouds.

Extended bright cloud cover, produced by this process, increases the albedo. That is, the back-scattering probability – of incoming solar radiation from space – leads to a direct reduction in surface temperatures. Conversely, when cloud cover is minimal, the Earth's surface – whether land or sea – appears dark, which increases the absorption of incoming radiation and thus also increases surface temperature. Whether or not this occurs with noticeable consequences is still a matter of debate.

The first observations of this effect are related to the Forbush effect in which high solar activity ejects plasma material from the Sun's corona, forming magnetic plasma clouds near the Earth. These plasma clouds provide a temporary shield against cosmic rays, since the cloud's electromagnetic field deflects cosmic rays and thereby reduces their intensity in the Earth's atmosphere. This effect has been well observed and often confirmed.

In cosmoclimatology, it has been suggested that the Forbush effect may also have a direct effect on cloud formation or cloud brightening and thus on Earth's climate.[50] Cosmic-ray fluctuations are postulated as the dominant element in the development of Earth's climate – in contrast to the greenhouse effect, which is caused by an abundance of carbon dioxide in the atmosphere. While other groups have not been able to confirm this correlation,[51] new work again shows evidence of such an effect.[52] There is considerable debate about the veracity of such a claim but such discussions do little to establish the truth since one scientifically validated phenomenon does not exclude another.

Natural developments are often determined by various parameters and influences. If the thesis of cosmoclimatologists is confirmed, this would moderate a rise in temperatures due to the greenhouse

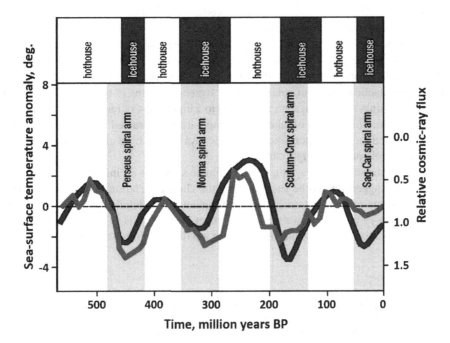

FIGURE 6.32 Four transitions from warm to cold climates during the Phanerozoic era, explained by the solar system's passage through various galactic spiral arms. According to this theory, a passage would be associated with a period of increased cosmic rays, which would lead to increased cloud formation and thus cooler ocean surfaces. The red curve shows the temperature of the sea water and the blue curve shows the cosmic-ray flux. https://paleontology.fandom.com/wiki/Paleoclimate?file=Phanerozoic_Climate_Change.png. Comparisons and simple correlations can, however, often lead to erroneous conclusions. Similar correlations can be observed between the temperature at Earth's surface and CO_2 levels in the atmosphere (Figure 6.33). Here, high CO_2 levels correspond to high temperatures, as a contributor to the greenhouse effect. Precise correlations in the Earth's climate history, as studied by paleoclimatology, are complex and not yet understood in all of their aspects.

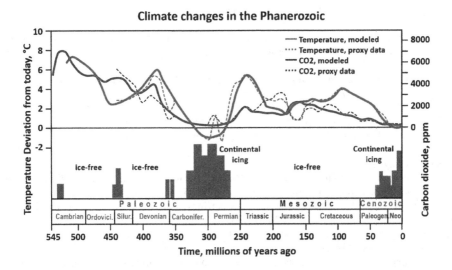

FIGURE 6.33 The evolution of climate in relation to carbon dioxide levels in the Phanerozoic era. Source: http://wiki.bildungsserver.de/klimawandel/index.php/Phanerozoikum (June 21, 2023).

effect and modify statements regarding climate predictions. Although such a cosmic-ray climate dependence may exist, final confirmation is still needed.

At present, there is discussion about an extension of this thesis, correlating climate to the radiation field in our atmosphere. It is being suggested that besides the cosmic radiation from the sun, the high-energy radiation contribution from distant supernovae may also play an important role. In its orbit around the galactic center, our solar system traverses the arms of the Milky Way. During the period of each crossing, the solar system is exposed to a higher flux of radiation. This observation is based on a systematic comparison of lake-water temperatures, extracted from the analysis of fossilized sea animals, and the time dependence of cosmic rays as they have affected the last 545 million years, the Phanerozoic period[53] (Figure 6.32).

The impact of intense cosmic rays on Earth's climate system is an interesting proposition; however, whether this suggestion can be sustained, by the close correlation between the two observational parameters, can only be determined by subsequent, more detailed analysis of observational data and the laboratory measurements that complement them. This is an interesting example of the possible impact of radiation on human life. Whether it will hold up, however, remains to be proven.

NOTES

1. The world's deepest well to date was drilled from 1970 to 1994 on the Russian Kola peninsula near Murmansk, reaching a depth of 12,262 meters.
2. The outer, giant planets are mostly composed of lighter gas material.
3. There are essentially two seismic components, primary or P-waves and secondary or S-waves. P-waves compress both liquid and solid material as shock waves, while S-waves have a deforming effect on solid material and are absorbed into liquid material, thus providing no information.
4. This has also led to speculation about a thermonuclear reactor at Earth's core, which would supply the Earth's heat through slow-fission reactions of concentrated uranium deposits. Philip Ball, "Are there nuclear reactors at Earth's core?" *Nature News* (2008), doi:10.1038/news.2008.822.
5. A. Gando, D. A. Dwyer, R. D. McKeown, and C. Zhang, "Partial radiogenic heat model for Earth revealed by geoneutrino measurements," *Nature Geoscience* 4 (9), (2011): 647–651.
6. It should be remembered that most stars are much more massive than our Sun and therefore undergo their evolution much faster due to the much higher temperature in their interior, which causes an exponential increase in reaction rates.
7. Of course, separation was not geologically absolute as the significant amount of metal in some Earth surfaces indicates.
8. Imre Bartos and Szabolcs Marka, "A nearby neutron-star merger explains the actinide abundances in the early Solar System," *Nature* 569 (2019): 85–88.
9. V. Dwarkadas, N. Dauphas, B. Meyer, P. Boyajian, and M. Bojazi, "Triggered star formation inside the shell of a Wolf-Rayet bubble as the origin of the Solar System," *Astrophysical Journal* 851 (2017): 147, 14pp.
10. M. Lugaro, U. Ott, and A. Kereszturi, "Radioactive nuclei from cosmochronology to habitability," *Progress in Particle and Nuclear Physics* 102 (2018): 1–47.
11. R. C. Greenwood, I. A. Franchi, A. Jambon, and P. C. Buchanan, "Widespread magma oceans on asteroidal bodies in the early Solar System," *Nature* 435 (2005): 916–918.
12. J. F. Kerridge, "Formation and processing of organics in the early solar system," *Space Science Reviews* 90 (1–2), (1999): 275–288.
13. Approximately 2 yotta-becquerel (YBq).
14. R. C. Greenwood, I. A. Franchi, A. Jambon, and P. C. Buchanan, "Widespread magma oceans on asteroidal bodies in the early Solar System," *Nature* 435 (2005): 916–918.
15. G. Bellini, K. Inoue, F. Mantovani, et al., "Geoneutrinos and geoscience: An intriguing joint-venture," *La Rivista del Nuovo Cimento* 45 (2022): 1–105.
16. R. Arevalo, W. F. McDonough, and M. Luong, "The K/U ratio of the silicate earth," *Earth and Planetary Science Letters* 278 (2009): 361.
17. The total mass of the mantle is generally assumed to be $4 \cdot 10^{24}$ kg, although the outer mantle layer or asthenosphere has a lower actinide abundance.
18. https://pris.iaea.org/pris/ (March 27, 2023).

19. J. Korenaga, "Urey ratio and the structure and evolution of Earth's Mantle," *Reviews of Geophysics* 46 (2008): 2007RG000241RG2007.

20. For example, glass is considered to be a very slow-moving liquid.

21. Divergent plate boundaries are mainly found in mid-ocean ridges (Mid-Atlantic Ridge, Iceland), but also in continental rifts. The East African Rift is an example of continental divergence.

22. F. G. Houtermans, A. Eberhardt, G. Ferrara, in H. Craig, S. L. Miller, and G. J. Wasserburg, eds., *Lead of Volcanic Origin* (Amsterdam: Isotopic and Cosmic Chemistry, 1964), 233–243.

23. G. Lambert, A. Buisson, J. Sanak, and B. Ardouin, "Modification of the atmospheric polonium 210 to lead 210 ratio by volcanic emissions," *Journal of Geophysical Research* 84 (1979): 6980–6986.

24. J. T. Bennett, S. Krishnaswami, K. K. Turekian, W. G. Melson, and C. A. Hopson, "The uranium and thorium decay series nuclides in Mt. St. Helens effusive," *Earth and Planetary Science Letters* 60 (1982): 61–69.

25. UNSCEAR 1993 Report, "Sources, effects, and risks of ionizing radiation – United Nations Scientific Committee on the Effects of Atomic Radiation," Report to the General Assembly, New York (1993), 549.

26. G. Cinellia, T. Tollefsen, P. Bossew, V. Gruber, K. Bogucharskaia, L. De Felice, and M. De Cor, "Digital version of the European Atlas of natural radiation," *Journal of Environmental Radioactivity* 196 (2019): 240–252.

27. M. I. Walker and K. S. B. Rose, "The radioactivity of the sea," *Nuclear Energy* 29 (1990): 267–278.

28. See for example: https://www.dermasel.de/ (March 27, 2023).

29. https://www.dermasel.de/ (March 27, 2023)..

30. L. A. Coogan and J. T. Cullen, "Did natural reactors form as a consequence of the emergence of oxygenic photosynthesis during the Archean?" *GSA Today* 19 (10), (2009): 4–10.

31. The Paralana Plateau is one of the most productive uranium-mining areas in the world. It is mined by the Beverly Mine, which continues to be developed as the Beverly Four Mile Uranium Project. The Arkaroola Wilderness Sanctuary is also affected, where the Paralana Springs are located. Source: https://www.energymining.sa.gov.au/industry/minerals-and-mining/mining/major-projects-and-mining-activities/major-operating-and-approved-mines/four-mile-uranium-mine (March 27, 2023).

32. The best-known example of this occurred much later, 65 million years ago. At the end of the Cretaceous period, the Chicxulub meteorite struck near what is now the Yucatan Peninsula in Mexico. Due to the dust it raised, this collision caused a winter that lasted millions of years and is considered the cause of dinosaur extinction. The meteorite was rich in iridium, an element that is clearly evident today. Iridium is present in the so-called K-line, which is considered the geological dividing line between the Cretaceous and Tertiary periods and consists of dust from whirled-up earth masses and meteorite rock.

33. Greenhouse effect is described as warming of the atmosphere by a high proportion of gas that absorbs Earth's thermal radiation. This absorption process occurs mostly through molecular vibrational states, which have been excited by thermal radiation. The particularly critical greenhouse gases are carbon dioxide (CO_2) and methane (CH_4), whose molecular vibrations lie exactly in earth's radiation range.

34. Today, the mean air pressure of the atmosphere (atmospheric pressure) at sea level corresponds to ≈ 1 bar.

35. Today, this fraction of helium is brought to earth's surface using fracking methods, where it is distilled out for use in industry's growing helium consumption.

36. The atmosphere is divided into different layers according to altitude. Beginning at Earth's surface these are: Troposphere, Stratosphere, Mesosphere, and Ionosphere (the latter sometimes called Thermosphere or Exosphere). The layers differ by chemical composition and temperature due to the influence of solar and other cosmic radiation.

37. UNSCEAR 2008 Report, "Sources and effects of ionizing radiation – United Nations Scientific Committee on the effects of atomic radiation," Report to the General Assembly, New York, 2011.

38. https://journals.uair.arizona.edu/index.php/radiocarbon/article/view/609 (June 21, 2023).

39. A. Chilingarian, G. Hovsepyan, E. Svechnikova, and M. Zazyan, "Electrical structure of the thundercloud and operation of the electron accelerator inside it," *Astroparticle Physics* 132 (2021): 102615.

40. T. Enoto, Y. Wada, Y. Furuta, et al., "Photonuclear reactions triggered by lightning discharge," *Nature* 551 (2017): 481–484.

41. A. Chilingarian, G. Hovsepyan, D. Aslanyan, T. Karapetyan, and B. Sargsyan, "Very unusual operation of the electron accelerator above Aragats mountain in Armenia a day after the earthquake in Turkey and Syria," https://arxiv.org/ftp/arxiv/papers/2302/2302 (June 21, 2023).

42. S. Riabova, Y. Romanovsky, and A. Spivak, "Remote response of electrical characteristics of the atmosphere to strong earthquakes," Proc. SPIE 11916, 27th International Symposium on Atmospheric and Ocean Optics, *Atmospheric Physics* 1191655 (2021).

43. On Mars, the iron core has a radius of about 1,700 km, which is comparable to Earth's solid, inner iron core with a radius of 1,300 km. Mars no longer has a liquid outer core, as a basis for the generation of a magnetic field by the dynamo effect. Earth's liquid, outer iron core has a radius of 3,500 km. The dynamo effect is driven by convection between the inner core and the mantle.

44. https://www.sciencemag.org/news/2009/04/did-marss-magnetic-field-die-whimper-or-bang (June 21, 2023).

45. https://www.seis-insight.eu/en/public-2/martian-science/internal-models-of-mars (June 21, 2023).

46. Giorgia Cinelli, Valeria Gruber, Luca De Felice, et al., "European annual cosmic-ray dose: estimation of population exposure," *Journal of Maps* 13, 2 (2017): 812–821.

47. W. V. Boynton, G. J. Taylor, L. G. Evans, et al., "Concentration of H, Si, Cl, K, Fe, and Th in the low- and mid-latitude regions of Mars," *Journal of Geophysical Research* 112 (2007), ID E12S99.

48. M. Tesche, P. Achtert, P. Glantz, et al., "Aviation effects on already-existing cirrus clouds," *Nature Communications* 7 (2016): 12016.

49. J. Kirkby, et al., "Role of sulphuric acid, ammonia and galactic cosmic rays in atmospheric aerosol nucleation," *Nature Letter* 476 (2011): 429–435.

50. H. Svensmark, T. Bondo, and J. Svensmark, "Cosmic ray decreases affect atmospheric aerosols and clouds," *Geophysical Research Letters* 36 (2009): L15101.

51. M. Kulmala, et al., "Atmospheric data over a solar cycle: No connection between galactic cosmic rays and new particle formation," *Atmospheric Chemistry and Physics* 10 (2010): 1885–1898.

52. J. Swensmark, M. B. Enghoff, N. J. Shaviv, and H. Svensmark, "The response of clouds and aerosols to cosmic ray decreases," *Journal of Geophysical Research: Space Physics* 121 (2016): 8152–8181.

53. N. Shaviv and J. Veizer, "Celestial driver of Phanerozoic climate?" *GSA Today* 13 (2003): 4–10.

7 Radioactive Man

7.1 RADIOACTIVE MAN

All living beings are an integral part of nature – this is true from single cell to human being. As such, we live in chemical exchange with nature as we breathe, eat, and drink. In turn, nutrients influence the chemical structure of our bodies. Since the earliest times, biological systems have absorbed not only the stable chemical elements of food intake, which are important for individual bodily functions, but also the long-lived radioactive isotopes that are present. In addition, our systems also absorb non-nutritive radioactive substances through respiration. Once in the body, each of these isotopes becomes part of the body's biochemistry, behaving chemically like all the stable elemental and molecular components that are integrated in biological systems. This process begins in the womb, with the nourishment and growth of the embryo, and continues throughout life in continuous exchange with the environment. Radioactivity is therefore inherent in all biological systems – from the first amoebae to humans.

As we have been discussing, we live in a sea of radiation. Each minute the human body absorbs between 1 and 10 million particles with enough energy to ionize atoms and break up molecules, causing cell damage. Therefore, given this abundance, radiation or radioactive isotopes probably played a central role in the origin of life as well as in the individual phases of evolution up to and including the present day. After all, life evolved in an environment where the natural level of radiation was up to five times higher than it is now.

Some radioactive substances irradiate our body from within. This radiation in biological cells causes mutations and may thus change the genetic code that determines the development of life. Some people argue that this enhances and contributes to cell repair mechanisms, thus instituting hormesis, while others suspect that this same radiation may limit the lifespan of humans through "inevitable" radiation death. In either case, natural radioactivity is part of our biological development and biological systems have developed their own protective measures, which are able to cope with potential overload and danger from radioactive exposure as will be highlighted in this chapter.

7.1.1 THE RADIOACTIVE BODY

The human body is a complex biological and chemical system consisting of a number of basic chemical substances that maintain various bodily functions. The body consists of approximately 60 to 70% water and 16% proteins. In addition, 10% is lipids, of which there are 1% each carbohydrates and nucleic acids. Finally, minerals make up 5%. Proteins are biological macromolecules made up of amino acid chains. As key carriers in every living organism, they are responsible for the body's entire metabolism, controlling cell function, regulating DNA division, and ensuring exchange of information between the different cells. Lipids or fatty acids are also macromolecules, which for example form the cell's membrane structure, storing energy and carrying signals between cell components. Carbohydrates or sugars and starch substances are important for supplying energy to the brain and muscles. Nucleic acids form the aforementioned individual building blocks for the structure and function of DNA and RNA macromolecules. These molecules in turn provide our organs' basic control information and contain the genetic information for reproduction.

Of these molecules, proteins and amino acids contain carbon, hydrogen, nitrogen, oxygen, and sulfur. Lipids and nucleic acids consist of carbon, hydrogen, nitrogen, oxygen, and phosphorus. Carbohydrates, on the other hand, are mainly composed of carbon, hydrogen, and oxygen (Figure 7.1).

Minerals perform various physiological functions in the body. Sodium and potassium play a major role in the nervous system's signal transmission as well as regulating water balance. Other elements such as iodine are hormonal components that ensure the thyroid gland's function. Calcium, magnesium, and phosphorus contribute to the formation and growth of our bones and teeth, while phosphorus is also needed by the nucleic acids. Finally, iron is a central element for blood formation. Trace elements with proportions of less than 1% are very important for enzyme function, including chromium, cobalt, copper, manganese, molybdenum, and zinc.

In total, the human organism consists of about twenty-one elements in different molecular structures. When considering the number of atoms, hydrogen is the most common component with 63%, followed by oxygen with 25.5%; these elements are mostly bound as water molecules. Carbon makes up 9.5% and, together with hydrogen and 1.4% nitrogen, these form the basis for the organic macromolecules that control body function. Calcium is present at 0.3%, with phosphorus at 0.2%. Potassium has a proportion of 0.06%.

Since the body absorbs most substances through food, our bodies contain long-lived radioactive components in addition to the stable atoms of the various elements. These are present in our bodies at similar ratios to the plant and animal nutrients we consume (Table 7.1). Plants take these substances from the soil as nutrients. Soil contains, for example, minerals from weathered rock material as well as dust deposits that the rain washes from the atmosphere. Also included are organic substances from decomposed biological materials such as leaves, plant remains, and other vegetation, including crop residues and fertilizers.

A particular plant's radioactive-material uptake from the soil varies considerably. This depends mainly on the chemical properties of the local soil material and the plant's specific nutritional requirements. So-called transfer factors reflect the uptake probability for specific elements including the radioactive components in different plants and fruits, which vary between per-mille fractions and considerable percentages.[1] For some fruits, especially nuts and bananas, this may cause an enhanced accumulation of radioactive isotopes such as ^{40}K.

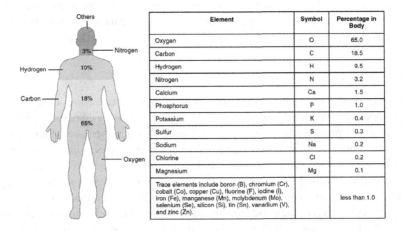

Element	Symbol	Percentage in Body
Oxygen	O	65.0
Carbon	C	18.5
Hydrogen	H	9.5
Nitrogen	N	3.2
Calcium	Ca	1.5
Phosphorus	P	1.0
Potassium	K	0.4
Sulfur	S	0.3
Sodium	Na	0.2
Chlorine	Cl	0.2
Magnesium	Mg	0.1
Trace elements include boron (B), chromium (Cr), cobalt (Co), copper (Cu), fluorine (F), iodine (I), iron (Fe), manganese (Mn), molybdenum (Mo), selenium (Se), silicon (Si), tin (Sn), vanadium (V), and zinc (Zn).		less than 1.0

FIGURE 7.1 Percentage distribution of the most important elements in the human body by weight. Of these elements, carbon (C), hydrogen (H), potassium (K), and chlorine (Cl) possess long-lived radioactive isotopes that can contribute to the human body's internal radioactivity. These elements are components of complex, organic molecular compounds that control bodily functions and processes. In addition to these elements, trace amounts of long-lived actinides are also present but do not perform any independent bodily function. Source: https://commons.wikimedia.org/wiki/File:201_Elements_of_the_Human_Body-01.jpg (July 21, 2023).

TABLE 7.1

Typical Range for Natural ⁴⁰Ka and ²²⁶Ra Radioactivity in Common Foods

Natural Radioactivity in Food					
Food	⁴⁰Ra Bq/kg	²²⁶Ra Bq/kg	Food	⁴⁰K Bq/kg	²²⁶Ra Bq/kg
Bananas	130	0.037	Beer	15	---
Nuts	207	37–260	Beef	110	0.02
Carrots	126	0.02–0.08	Beans	172	0.07–0.19
Potatoes	126	0.04–0.1	Drinking water	---	0–0.63

Source: Johannes Friedrich Diehl, *Radioaktivität in Lebensmitteln* (Weinheim: Wiley-VHC, 2000).

This emitted radiation certainly interacts with body material, as is sometimes dramatized a bit:

> Each decay product attacks the body as though it were possessed by a special desire for a particular living tissue, Radium-226 invaded the teeth and bones, as well as mother's milk, radon-22 assails the lungs, cesium-137 plunders the muscles. Strontium-90 actually bonds to the structure of bone and gathers in the vascular tissues of plants.[2]

While factually correct and nicely formulated, this is nevertheless misleading since it is not the case that specific radioactivity has a special desire to assail various body parts. Rather, the underlying cause of these effects is the natural chemical behavior of the various radioactive isotopes. Therefore, possible specific damage is correlated with the level of activity, as outlined in Chapter 4, as well as the amount of time the particular radionuclide remains within the body.

We absorb radioactive isotopes in the same ratio that they are present in our environment and food. Our body cannot fractionate them, which is to say, it cannot separate stable isotopes from radioactive isotopes in the digestive system. Therefore, respiration and digestive processes as well as blood circulation distribute radioactive substances in the body. However, it should be noted that some radioactive elements that enter the digestive tract are absorbed completely, while others are only partially absorbed. This depends on the chemical characteristics of the particular element and the body's need for that element in order to maintain its functions.

Absorption of radioactive substances in the digestive tract can reach up to 100% for some substances. This is particularly the case with cesium, iodine, potassium, phosphorus, and selenium – elements that the body requires for various functions. However, if there is no need for a particular element, that element is excreted with its radioactive components. With radioactive elements such as uranium, the probability of being permanently absorbed in the digestive process is therefore less than 10%. Requirement also determines the average length of time unabsorbed radionuclides remain in the body. On average, it is assumed that radionuclides will remain in the stomach for one hour and in the intestinal tract for forty hours. Material that is not absorbed by the body within this period is excreted with urine and stool.

The human body's internal radioactivity can be calculated by assuming a balance between absorption and excretion processes, thus determining the frequency of long-lived radioactive isotopes in the body. If we take oxygen – the most common element with 56% by weight – there is no danger. This is because radioactive oxygen isotopes ¹⁵O, on the neutron-poor side, and ¹⁹O, on the neutron-rich side, are extremely short-lived and therefore do not occur in large quantities in either nature or the body.

Carbon is different. As the second most common element with 18% by weight, carbon is a core element found in all organic molecules and thus in all of the human body's functions. In addition to the two stable carbon isotopes ¹²C and ¹³C, there is also the ¹⁴C isotope with a half-life of 5,730 years, which can therefore be regarded as an element that remains integrated in the human body. Carbon is produced in large quantities by cosmic rays in the atmosphere and is absorbed as $^{14}CO_2$ through plant photosynthesis, whereby it enters the food cycle.

The ^{14}C abundance in the atmosphere is subject to temporal fluctuations, which are determined by solar activity and also by anthropogenic processes such as atomic bomb testing and the intensive use of fossil fuels (see Chapters 11 and 12, Volume 2). At present, it is estimated that there are approximately one trillion ^{12}C atoms for every ^{14}C atom.[3] In a human body of 80 kg weight, there are approximately $N(^{12}C) \approx 7.23 \cdot 10^{26}$ stable ^{12}C atoms. So, there are also $N(^{14}C) \approx 7.43 \cdot 10^{14}$ atoms of the radioactive ^{14}C isotope. Activity can be easily calculated by multiplying the number of atoms by the decay constant (see Chapter 3). This gives an average activity of about 3,000 Bq or 3,000 β^--decays per second. Due to the nuclear weapon test program the ^{14}C level has increased significantly and a value of 4,000 Bq is typical for the human body (see Chapter 12, Volume 2).

When converting this to the DNA molecule with 85.7 million carbon atoms, the result is that we have 0.00007 ^{14}C atoms per DNA molecule. This seems like a small fraction but the human body has between 50 and 100 trillion cells, and each cell has two sets of twenty-three chromosomes which are made up of tightly wound DNA molecules. This means that about 320 billion radioactive ^{14}C isotopes are firmly embedded in our DNA material!

Using the same method, the activities of the other radioactive isotopes present in the body can be determined. For example, we can take hydrogen with 9.3% by weight, so that there are $N(^1H) = 4.5 \cdot 10^{27}$ atoms at 80 kg body weight. The radioactive hydrogen isotope ^3H tritium is also constantly produced by cosmic radiation and couples with OH^- molecules to form heavy water. A small natural part of ^3H_2O heavy water molecules enters the body with food, drink, and in water vapor via breathing. In the hydrosphere, there are about 10^{18} stable ^1H hydrogen atoms for every ^3H tritium atom. However, this ratio can be subject to strong local variations. Thus, there are on average $4.5 \cdot 10^9$ or 4.5 billion tritium atoms with a half-life of 12.3 years in the human body. This corresponds to an additional internal activity of 8Bq.

However, as emitters of low-energy radiation, these two radioisotopes have little influence on the internal dose load. In the case of ^{14}C, with 152 keV β-energy, the dose load is about 40 Sv per year while in the case of tritium, with much lower β-energy of 5.7 keV and lower activity of 8 Bq, the dose load is only 3 nSv per year.

Other radioactive components in the body can cause a much higher dose rate due to their higher radiation energy. This is especially true of the radioactive long-lived potassium isotope ^{40}K, which at 0.012% is nevertheless a significant part of the element's natural occurrence. Potassium is one of the most important minerals in the human body. As such, it is involved in most of the physiological processes carried out by each cell. Potassium is a positively charged ion (cation) that electrically transmits information to nerve and muscle cells. It thus controls the muscle system, regulates cell growth, releases hormones, and is indispensable for the production of proteins and the breakdown of carbohydrates.

Potassium is found at up to 98% inside cells and only 2% outside of cell material. Potassium is also an integral part of DNA, as each DNA molecule contains 177,000 potassium atoms and thus 21 radioactive ^{40}K isotopes. This means that the body's total DNA contains up to 10^{17} radioactive ^{40}K isotopes. However, this is far from covering the total amount of ^{40}K in the body.

Women have a total of about 100 grams of potassium in their bodies, while men have about 150 grams. Potassium thus has on average only a small weight percentage of 0.25% of the body's material. This corresponds to a total abundance of $N(K) \approx 2 \cdot 10^{24}$ potassium atoms, of which $N(^{40}K) \approx 2.4 \cdot 10^{20}$ atoms are radioactive. Based on the half-life of $1.25 \cdot 10^9$ years, the total activity amounts to 4,150 Bq.

The decay of ^{40}K proceeds to 90% via β^--decay to ^{40}Ca, whereby the released β^--particles and anti-neutrinos have a total energy of 1.33 MeV. Another 10% of the decay proceeds, via electron capture, to the first excited state in ^{40}Ar and then via subsequent γ-decay to the ground state. During this process, γ-radiation of 1.46 MeV is emitted. The total dose rate is 30 mSv per year, which is a thousand times higher than the dose load from radioactive β-decay of ^{14}C. About 90% of the 10% γ-radiation, resulting from the ^{40}K electron capture, is absorbed while the remaining β-radiation leaves the body. Therefore, each person represents a weak external radiation source from ^{40}K of approximately 40 Bq. However, this is decidedly weaker than the corresponding 1.46 MeV of radiation to which we are exposed following the ^{40}K electron capture that occurs in surrounding building and rock materials.

Over the past few decades, the potassium content has risen sharply in our plant diet. This is the result of an increase in the agricultural industry's use of artificial fertilizers. In addition to nitrogen compounds, artificial fertilizers also contain phosphates and potash salts. In small proportions, actinides such as uranium appear as trace elements in fertilizer through the decomposition of these phosphate salts. Artificial fertilizers may also expose people to irradiation when walking through forests and fields (see Chapter 9), although this is of less importance for the internal radiation exposure.

Although the potassium content and thus the ^{40}K content in plant materials has increased considerably, individual human potassium levels remain carefully balanced by the nutrition cycle itself because potassium values at a level that is too high or even too low can lead to heart and circulation problems. This balance is maintained by physiological-chemical processes, which do not make a distinction between individual isotopes. Therefore, the proportion of ^{40}K will remain more or less constant as well.

Unlike potassium, rubidium with its only stable isotope ^{85}Rb has no biological significance for human body functions. Besides ^{85}Rb, with an abundance of 72.2% of the observed abundance of 72.2%, the second component is ^{87}Rb with an abundance of 27.8%. This isotope has a half-life of 49 billion years – a time that is considerably longer than the age of the universe. It can therefore be considered an almost stable isotope.

A product of the s-process in AGB stars, ^{87}Rb is a component of the interstellar dust from which our solar system was formed. Therefore, ^{87}Rb is very abundant in the earth's crust. As an alkali metal, rubidium chemically behaves in a way similar to potassium and replaces the latter in many known minerals and rocks. The ^{87}Rb decays via β^{-}-decay to the strontium daughter isotope ^{87}Sr. From the abundance ratio of the two isotopes ^{87}Rb/^{87}Sr, geologists can determine the age of various rock materials from different locations and layers of the earth's crust and thereby evaluate the geochemical history of the specific samples.

Due to its chemical affinity with potassium, rubidium is also part of plant material and therefore human nutrients. The human body contains a total of 0.7 g of rubidium and almost a third of this, 0.19 g, is ^{87}Rb. The average activity in the body is 600 Bq. Unlike potassium, its abundance in the body is not self-regulating, which may lead to considerable rubidium accumulation in the body. In anthropology and archaeology, this is useful in forensic analysis of bone material, where the ^{97}Rb-^{87}Sr abundance ratio can be used to determine the origin and age of body components (see Chapter 16.1.4, Volume 2).

In addition to ^{40}K and ^{87}Rb, radiogenic decay products of the uranium and thorium series also make a significant contribution to human radioactivity. These are present in earth's material in large quantities (see Chapter 6) and so enter both the animal and human food cycles as plant nutrients. Radionuclide concentration levels in plants depend on the radioactive content of the local soil and groundwater as well as on chemical composition, availability of nutrients, and other environmental conditions of the particular plant or animal.

Depending on the plant species and the respective developmental and nutritional status of the plants at harvest, minerals are distributed differently. These distributions influence the specific activities in plant foods. For example, radioactivity from the radium isotopes ^{226}Ra and ^{228}Ra is higher in cereal grains than in vegetables or fruit. Apart from plants, these heavy radionuclides also occur mainly as dissolved minerals in drinking water. Thus, they are natural additions from the water cycle in the hydrosphere. This includes commercially available mineral water, which may increase internal radioactivity depending on the amount of dissolved minerals.

TABLE 7.2

Average Activity of the Most Common Long-Lived Radioactive Isotopes in the Human Body

Radionuclide	Activity (Bq)
Tritium	20
Carbon-14	3,500
Potassium-40	4,000
Rubidium-87	600
Uranium-238	0.5
Radium-226	1.2
Lead-210	18
Polonium-210	15
Thorium-232	0.2
Thorium-238	0.4
Radium-228	0.4

Source: Data is taken from the 1982 UNSCEAR report, *Ionizing Radiation: Sources and Biological Effects* provided by the United Nations Scientific Committee on the Effects of Atomic Radiation.

Heavy radioisotopes ^{238}U, ^{226}Ra, ^{210}Pb, and ^{210}Rn as part of the natural uranium decay series contribute to the internal activity of the human body, while ^{232}Th, ^{228}Th, and ^{228}Ra from the thorium series only make small contributions (Table 7.2). This is because food items contain more uranium than thorium in the trace elements.

The typical uranium abundance in the human body is on average only 25 µg. Ingested through diet and respiration, uranium is initially deposited in the digestive system and lungs. Through blood exchange, the material is then quickly distributed throughout the body system but is mainly collected in the kidneys and bones. The 25 µg of ^{238}U corresponds to an activity of about 0.45 Bq in the body because of the long half-life of ^{238}U.

The daughter element ^{226}Ra has a half-life of 1,600 years. It behaves chemically like calcium and also enters the human body from soil via the food chain. The ^{226}Ra accumulates in the bone system, just like radioactive ^{228}Ra from the thorium chain. The activity of the absorbed ^{226}Ra varies greatly depending on geographical location and soil conditions, ranging from 0.3 to 3.7 Bq in the skeleton of an adult human. Assuming an average value of 0.85 Bq, this means an annual dose of 160 µSv, with cells, smaller parts of the bone marrow, and the muscle material surrounding them absorbing most of the radiation.

The radioactive lead and polonium isotopes ^{210}Pb and ^{210}Po belong to the uranium decay chain. With a half-life of 22 years and 128 days respectively, these are the longest living isotopes in the chain after the decay of radium ^{226}Ra. The ^{210}Pb is a β^--emitter and decays via α-emission to the stable lead isotope ^{206}Pb. The main part or 80% percent of ^{210}Pb is deposited in the skeleton,[4] with the resulting ^{210}Pb activity in the skeleton of a person weighing over 80 kg being about 14 Bq. The remaining 20% is evenly distributed over muscles and body organs so that total activity is about 18 Bq.

As ^{210}Pb decays in the body, the decay product ^{210}Po accumulates in the organs. In the decay equilibrium, ^{210}Pb and ^{210}Po have the same activity. However, due to the much shorter half-life of ^{210}Po, the abundance of this daughter element is only 2% of the abundance of the long-lived mother ^{210}Pb. Just as with ^{210}Pb, about 80% percent of ^{210}Po activity, about 12 Bq, is located in the skeleton. The remaining 3 Bq of activity is distributed over the other organs.

The total activity of the human body is therefore approximately 8,000 Bq, which is 8,000 radioactive decays per second. Most of these are low-energy β^--decays like ^{14}C and tritium, of which little energy enters the body and the dose is relatively low. With the decay of ^{40}K, considerably more radiation energy is deposited in the body, about 1.46 MeV per decay. The decay process of ^{40}K therefore represents the body's largest internal dose load.

7.1.2 RADIOACTIVITY IN THE HUMAN FOOD CYCLE

The radioactivity in our body builds up through the nutrition and respiratory cycles. With regular nutrition, a balance is established between the radioactivity absorbed through foods, liquids and respiration and the radioactivity excreted with bowel feces, urine and sweat. Depending on their chemical properties, radioactive elements that aren't eliminated are deposited in various organs, muscles or the bone structure. Of course, this too is related to the biochemical capacity of the individual organ and the physiological importance of the corresponding element for the functioning of the organism.

All of the radioactive substances present in the human body are found in food. This includes the carbon isotope ^{14}C, which is constantly produced by cosmic radiation and has an equilibrium value of approximately 10^{-12} of the stable ^{12}C isotope. The ^{14}C isotope is found in all organic food, which is mainly composed of organic carbon macromolecules. Meat, for example, contains between 25 and 50% carbon by weight, depending on the fat content, with a water content of 30–60%. These values are comparable to protein values in general. Potatoes, bread and other carbohydrate-containing foods such as pasta contain around 50% carbon, with slight variations depending on grain type. Assuming that the carbon content of food is 40%, it can be roughly estimated that humans consume up to 300 g of carbon with 1 kg of food, which corresponds to about 0.3 ng of pure ^{14}C. That is $4.3 \cdot 10^{19}$ ^{14}C atoms, showing a total activity of about 50 Bq or about 1% of an average adult's ^{14}C activity.

Similar estimates can be made for tritium, which occurs at trace levels in natural water such that for every tritium isotope, there are 10^{18} light hydrogen atoms. In most foods, the water content is 50–70% so that with a daily consumption of 1 kg of food, this corresponds to about 600 g of water. With an additional requirement of 2 liters, a person with a normal diet takes in $5 \cdot 10^{25}$ water molecules. This corresponds to approximately 100 million tritium atoms per day with an activity intake of 0.2 Bq, a relatively low value compared to ^{14}C intake (Table 7.3).

Long-lived radiogenic potassium and actinide isotopes again carry the most significant contribution. Potassium is primarily absorbed by plants and accumulates in meat products. Amounts depend on the uptake from soil, by the plants and animals. The efficiency of the exchange is measured by comparing the activities. From the surface to the root depth, activity in soil material is predominantly constant; the ratio of the activity values between soil and plant is the so-called transfer factor, with each factor normalized to one kilogram of material. With regard to the different radioisotopes, the transfer value varies considerably from plant to plant and depends on the respective plant's nutritional requirements. For example, grass has a transfer factor of 30%. The amount of ^{40}K in the grass is then taken up by grazing livestock and distributed throughout the animal's body by the digestive process. The transfer factor for potassium in vegetables is as high as 100%.

Figure 7.2 shows the transfer factors for elements present in different vegetables, as derived from the weight ratios of the respective elements as they are found in dried plant material and dried soil material. The transfer factors of long-lived radioisotopes correspond to weight ratios because their weight correlates directly with the activity. These values correspond well with measurements based on long-lived isotope activity in plant and soil material. Of course, transfer factors only apply when no chemical fractionation[5] between isotopes occurs during the processing of the nutrients, since such fractionation would change the ratio of radioactive and stable isotopes.

In order to maintain the various physiological processes in the body, a person must digest between 2 and 4 g of potassium daily with food.[6] Potassium deficiency or excess can lead to various diseases such as high blood pressure and strokes. In the United States, the average daily potassium consumption for women is 2.3 g, whereas for men it is 3.0 g. On average, the potassium content in

TABLE 7.3

Natural Amounts of Radioactivity from the Most Abundant Radiogenic Isotopes in Various Foods, as well as the Typical Range for the Activity Level as Recorded in the Literature. The range depends on the actual radioactivity level of the ground and the up-take factor of the plants

Food	Potassium-40	Uranium-238	Radium-226	Lead-210	Polonium-210
			Specific activity [Bq/kg]		
Grain	160	0.1	0.3	1.4	0.3
	87–246	0.02–1.54	0.04–1.53	0.04–10.2	0.2–1.94
Flour			0.1	0.4	0.4
			0.05–0.13	0.22–0.87	0.20–0.48
Potato	160	0.8	0.2	0.1	0.1
	122–294	0.02–3.09	0.02–1.30	0.02–0.63	0.20–0.33
Cabbage	130	0.3	0.2	0.3	0.1
	59–196	0.02–0.75	0.01–0.68	0.004–1.28	0.004–1.19
Vegetable		0.4	0.1	0.1	0.1
		0.1–1.28	0.006–0.71	0.007–0.34	0.004–1.19
Carrots	100	0.7	0.2	0.6	0.6
	72–134	0.07–2.31	0.06–0.49	0.02–4.9	0.02–5.2
Fruit	50	0.6	0.2	0.2	0.1
	23–164	0.02–2.89	0.005–2.12	0.02–2.29	0.02–1.1
Berries	140	0.4	2.2	8.4	1.6
	107–190	0.67–1.8	0.03–5.38	1.2–14.8	0.52–2.24
Mushrooms	120	1.3	1.2	1.2	1.3
	8–233	0.18–5.1	0.01–16	0.09–4.1	0.1–5.2
Meat	90	0.01	0.1	0.5	2
	60–120	0.001–0.02	0.03–0.18	0.1–1.0	0.2–4.0
Fish	100	4.1	1.5	0.8	1.1
	80–120	0.5–7.4	0.05–7.8	0.02–4.42	0.05–5.2
Milk	50		0.025	0.04	0.024
	36–65		0.001–0.13	0.004–0.26	0.003–0.07

Source: Data obtained from the UNSCEAR 2000 report, *Ionizing Radiation: Sources and Biological Effects* provided by the United Nations Scientific Committee on the Effects of Atomic Radiation and Johannes Friedrich Diehl, *Radioaktivität in Lebensmitteln* (Weinheim: Wiley-VHC, 2000).

food is 0.5–2% by weight, with a natural proportion of the radiogenic ^{40}K isotope at 0.012%. This results in the specific activity of different vegetable species per kilogram as shown in Table 7.3. Assuming a potassium consumption of 2.5 g, humans consume 0.3 mg of radioactive ^{40}K daily, with an activity of about 80 Bq.

Apart from solid food, smaller amounts of ^{40}K enter the body through drinking water and other beverages, mainly as minerals and salts dissolved in water. The amount legally permitted in the United States is 0.5 Bq ^{40}K-activity per liter. Normal groundwater and drinking water typically contain considerably lower amounts of ^{40}K-activity. Normal water has between 0.5 and 18 mg potassium, which is less than 2 µg of radioactive ^{40}K per liter and considerably less than we consume every day in solid food. Mineral water can contain up to 40 mg potassium per liter while 100 mg/l is the prescribed upper limit. The latter corresponds to 12 µg ^{40}K per liter, corresponding to about 3 Bq.

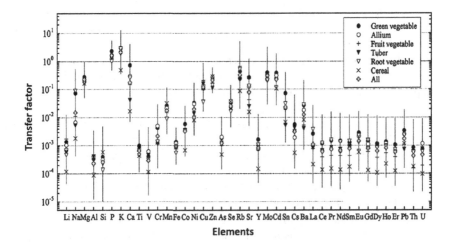

FIGURE 7.2 Transfer factors for different elements, measured by the weight of the different elements in the dried plant material divided by the weight of these elements in the soil material, each normalized to one kilogram of material. Since the isotope ratios are constant and no fractionation occurs during plant uptake, these transfer factors also apply to the corresponding radioisotopes. Prepared after: IAEA: Handbook of Parameter Values for the Prediction of Radionuclide Transfer in temperate Environments; Technical Report Series No. 364, Wien 1994.

The situation is similar with actinides, where the largest contribution comes from [238]U. Just like potassium, [238]U enters the body through the natural diet. However, for both uranium and thorium, the transfer factor of 0.01–0.1% is significantly lower than for potassium. This is because these elements are physically much larger than potassium and the plant does not regard them as food, thus limiting possible biochemical reaction channels for absorption. Nevertheless, an average of 1–5 μg enters the body daily by this route, corresponding to 20–100 mBq.

We can also inhale aerosols or dust particles containing uranium. This applies mainly to uranium mining where workers are in direct contact with uranium-containing material. In areas with uranium mining, amounts can be three to four times higher at 13–18 μg per day. In such circumstances, the daily activity intake would be between 250 and 350 mBq. Assuming an average value of 100 mBq per day in solid food, this results in an exposure of 2.5 μSv. Since uranium decays via α-emission, the effective dose is higher by a factor of 20 and is 50 μSv per year.

Besides through solid food and respiration, uranium can also enter the body in dissolved form with liquids. Legal regulations limit the uranium content in drinking water to 30 μg/l. In general, however, the uranium content is less than 5 μg/l with a few exceptions. In the western and southwestern regions of the United States, values are somewhat higher due to the higher uranium content in rock and soil (Figure 6.11). According to this, the uranium content in water usually correlates with uranium-rich granite deposits in the soil, which release uranium through dissolved minerals in the water. By comparison, the uranium content of commercially available mineral waters varies between 0.1 g/l and 20 g/l.[7] At 30 μg/l there are still $N(^{238}U) = 7.5 \cdot 10^{16}$ uranium particles, making up one millionth of the $N(H_2O) = 3.4 \cdot 10^{25}$ water molecules in one liter. This quantity corresponds to an α-activity of 360 mBq with a potential dose rate of 9 μGr per year. The corresponding effective dose is 180 μSv per year.[8]

In many cases, a balance in the body's chemical elements is maintained by the digestive system through the absorption and excretion of food. Therefore, excretions also contain a radioactive component. In humans, a distinction must be made between solid stool and urine excretion because the elemental abundance ratios are somewhat different. The average daily amount of solid stool is between 150 g and 270 g wet weight, while the water content is between 70 and 80%. Organic and protein substances as well as fats and carbohydrates comprise 21.6%. The solid content of defecation consists of 50% carbon, 5% calcium, 4% phosphorus, and 2% potassium. Urine has a much higher water content at about 95%.

Approximately 60–90 g of solid matter is excreted per day. As a factor of these percentages, on average 15% is carbon, 18% is nitrogen, 4% is phosphorus, 5.5% is calcium, and 4% is potassium. Carbon and potassium also excrete the long-lived radioisotopes ^{14}C and ^{40}K, which are present in a ratio of 10^{-12} and $1.2 \cdot 10^{-4}$, respectively.

Depending on the element, the body absorbs only a small proportion of the radioisotopes in its digestive tract. This means that with the exception of substances that are important to the body, a large part of consumed radioactivity is excreted. Uranium, for example, has no chemical significance for body function. Every day, humans take in on average 2 µg of uranium – through food, drink, dust, or contact – and excrete 98–99% of this through bowel movements. Only 1–2% is deposited in the body for a longer period of time, mostly in hair so that it slowly grows out in the same amount. Thus, the uranium finally accumulates in barber shop dumpsters.

Since the intestinal tract cannot distinguish stable and radioactive isotopes, the average isotope ratio in elemental excretions corresponds to the average natural ratio in the food consumed.

A healthy person excretes about 1–1.5 liters of urine daily. Besides water, this contains 30–40 g of urea and organic acids, 10–15 g of mineral salts, 1 g of ammonia, and 10–18 g of oxides. Of these, 3–4 g of potassium oxide are particularly important in radioactivity considerations. This value does, however, vary since the body's potassium balance is controlled and kept constant by means of potassium excretion and absorption. Considering the natural proportion of 0.012% for the radioactive ^{40}K component, 3 g of potassium oxide contains about $3.9 \cdot 10^{18}$ radioactive ^{40}K-atoms, which corresponds to an activity of 66 Bq. This is a considerably higher value than normal mineral water, for example (Figure 7.4).[9] For comparison, typical uranium amounts in urine are less than 0.1 µg per liter, which is only a comparatively small contribution of 1.2 mBq; this value is lower than the average value of 0.5 µg/l in beverages.

One can estimate that a single person's daily radioactivity deposition into the sewer system is about 222 Bq! The radioactivity accumulated in human feces doesn't seem to be much, since it is part of the natural exchange for biological systems and won't cause the net level of radioactivity to be out of balance. However, due to the tendency to collect and channel our fecal depositions, these levels can cause uncomfortably large radioactivity numbers in our wastewater systems and sewage plants. This is particularly the case when larger crowds and dense population centers are considered, a situation that requires the collection, cleaning, transport, and dissipation of wastewater and the radioactive material in it.

Urine and fecal waste together with normal household wastewater and industrial wastewater form a mixture of many substances, which flows as wastewater into municipal wastewater treatment plants for purification. This wastewater contains a considerable amount of various organic and inorganic compounds, as well as elements from the origin and use of the water itself. In municipalities, the proportion of domestic water present in wastewater is around 70–80%, although this is reduced at industrial sites. Chemical analyses show a typical potassium content of 20 mg/l, such that it contains $3.6 \cdot 10^{16}$ radioactive ^{40}K-atoms per liter with radioactivity of 0.64 Bq/l, a lower level of radioactivity than in human excrement due to the dilution process. Nonetheless, the amount of water used in clarifiers (settling tanks for wastewater) still has a considerable overall activity since in a clarifier of 90,000 m^3 volume, the total activity would be 58 MBq.

Various mechanical and chemical processes clean inorganic and organic pollutants from industrial water with, for example, sieves, settling tanks, precipitation tanks, and trickling filters. Chemical additives gradually precipitate pollutants as flocculants while bacteria are also used, specifically to break down organic substances. How much pollutant remains in the sewage sludge, and how much purified water reaches rivers and groundwater, depends on the effectiveness of mechanical and chemical-biological purification procedures. These procedures may be able to filter out chemical toxins and heavy metals such as uranium and its components, but they cannot separate the radioactive isotope fraction from vital elements such as potassium.

The radioactivity generated and channeled by the human digestive system cannot be cleaned out with traditional chemical or biological methods since radioactive isotopes behave, chemically and biologically, exactly like their stable counterparts. One might be able to develop microscopic fractionation methods but that would represent an enormous technical and logistical challenge, making such an approach financially prohibitive. Radioactive isotopes then remain in the processed sewage sludge that is often used by agriculture, thus closing the cycle of radioactivity in the food chain as will be discussed in a later chapter.

This can be demonstrated by a couple of examples: Considering cities and large population centers, the city of Chicago has a population of 2.7 million people depositing nearly 600 million becquerel daily into the sewer system and water treatment plants. This level of radioactivity cannot be removed, so it gradually drifts down the Chicago Sanitary and Ship Canal before joining the Des Plaines River. Once there, the wastewater travels still farther to the Illinois River and eventually into the Mississippi River. From Chicago alone, this amounts to 214 billion Bq or 0.214 TBq or 6 Ci annually!

Traffic hubs provide another source for human radioactivity release. O'Hare airport in Chicago is the fourth busiest airport in the world with about 77 million passengers annually, providing numerous bathroom opportunities for the needy traveler. Assuming that at least 25% of the travelers use one of these restrooms, this translates into a radioactivity release of 4.4 GBq annually or more than 12 million becquerel daily on average, which is also eventually deposited into the Mississippi River. This adds up to a daily radioactivity deposition of about 612 million becquerel, all of which has an unknown and hardly controllable destiny.

However, to put things in proportion, these numbers have to be compared with the natural radioactivity level in Mississippi River water. The Mississippi's water contains on average 3mg/l of potassium and so 0.36μg/l of ^{40}K. With an average water-flow rate of 320,000 l/s, this translates into a transport rate of 0.12g/s of ^{40}K or 3,060 Bq/s. The annual transport of "natural" ^{40}K-activity is $9.6 \cdot 10^{11}$ Bq or nearly 1 TBq annually, which equals 3 billion becquerel daily. Human excrement therefore adds an appreciable amount of about 20% to the activity load, with about 4% from O'Hare airport facility users.

These observations are intended to show that nuclei and the associated radiation are an integral part of our existence. Radiation is within us, around us, and through us. It is not a foreign hostile entity. It is part of us and our biochemical system. As we have considered in earlier chapters, radiation has been with us since the formation of the Earth, since the beginning of biological life on our planet. Radiation may have even played an important role as an energy source, facilitating the formation of complex organic molecules as the building blocks of biological systems. There are still wide-open questions that are not easy to answer, due to the complexity of the underlying processes. Yet, the discussion continues to present fascinating possibilities, which will be discussed in the following sections.

7.1.3 ORIGIN OF LIFE

The origin of life is one of the great scientific questions of our time. How were the first organic molecules formed in the chaos of the early planetary system? What was the nature of the energy source necessary for this to occur? According to the second law of thermodynamics, a complex closed system without an energy source decays into a state of disorder. The closed system is a physical ideal that does not actually occur because all systems maintain some kind of interaction with the environment and are driven by that exchange. All systems, whether animate or inanimate, consume, deteriorate, rot, and decay to dust. Goethe already lets his Mephistopheles express this in *Faust*: "I am *the* spirit that negates. *And rightly so, for all* that comes to be, *deserves* to *perish* wretchedly."[10] This sentence applies in particular to biological systems, which exist on the basis of an ordered molecular structure.

Without an energy source, biological systems cannot come into being and cannot continue to exist. Plants derive their energy from sunlight through photosynthesis; animals and humans live from the chemical energy conversion of food. These processes are the result of a long biological evolution. This all began in the early phase of our planet with the emergence of complex organic molecules based on chemical compounds of carbon, hydrogen, nitrogen, and oxygen, as well as a few but important heavier elements such as sulfur and phosphorus. These elements are considered the basic building blocks of life, labeled by the acronym CHNOPS – or in its reversed form SPONCH.

Carbon is the backbone of the complex molecules generated by organic and bio-organic chemistry. This is possible due to carbon's ability to form strong chemical bonds with hydrogen, nitrogen, and oxygen; hydrogen forms string bonds to nitrogen and sulfur; nitrogen and oxygen are essential for the building of more complex structures such as amino acids and sugars, which enhance the complexity of organic molecules. Sulfur and phosphorus are important in proteins, and the latter is the basis of a key molecule adenosine triphosphate (ATP), which is labeled as the universal energy currency for life. The energy itself is stored in the bonding of the phosphorus atoms to the adenosine diphosphate (ADP) rest molecule, allowing it to act as a biochemical battery. Thus, CHNOPS are the basic elements that make life possible in its most simple forms.

The increasing complexity of biological systems requires additional elements to handle additional tasks and functions, such as operating the electrical information transfer. These elements include iron, for example, which is necessary for formation of the iron-sulfur clusters that are needed for electron transfer in the cell – with ^{60}Fe as a relatively long-lived radioactive component. Another is potassium for regulating body functions through the development of a nervous system, with the long-lived ^{40}K radioactive isotope as a main source of the body's internal radiation as discussed earlier.

How and where did the first molecules form? Where did the energy come from that caused chemical reactions in elemental material, leading to the basic building blocks of life – amino acids and nucleic acids? All life on our planet, from the smallest micro-organisms to complex creatures such as animals and humans, must have a common origin because all biological life is based on this same structure and the same chemical interactions of these two complex molecular substances. The evidence points to a last universal common ancestor (LUCA) as the common ancestor of all existing life on Earth, including all cellular organisms. Although LUCA may not represent first life on Earth, it is seen as the earliest ancestor of all existing life.

There are approximately twenty amino acids[11] that make up all biological enzymes and proteins. The amino acids also determine the mode of action for these enzymes and proteins. In addition, there are a sugar molecule, a phosphate molecule, and the four base molecules – adenine, cytosine, guanine, and thymine – that form the nucleic acids, which are the basic building blocks of DNA and RNA chain molecules. All of biology and life itself are based on these twenty-six basic chemical molecules and their myriad of possible chemical combinations.

Did the origin of such a complex system take place on Earth as part of or as a consequence of its early formation process? Or, did it develop during the later geochemical evolution of the earth system? Conversely, did this complex system all begin somewhere else, beyond Earth? Almost 150 years ago, in 1874, the famous German physicist Hermann von Helmholtz (1821–1894) asked in his preface to the *Handbuch der Theoretischen Physik* (Handbook of Theoretical Physics), "whether life had an origin, or whether it was as old as matter, and whether its germs, transported from one world celestial body to another, would not have developed wherever they would have found favorable conditions?"[12] This question encompasses the universality and the eternity of life.

This idea about the extraterrestrial origin of life can be traced back to Aristotle, who suggested that life is an all-pervading cosmic force. This idea was supported not only by Hermann von Helmholtz but also by many other eminent natural scientists of the late 19th and early 20th centuries. These include the English physicist William Thomson, 1st Lord Kelvin (1824–1907), and the Swedish chemist and Nobel Prize winner Svante Arrhenius (1859–1927), each with their own more modern formulations. This is also a thesis pursued by the eminent astrophysicist Fred Hoyle (1915–2001).

However, the British naturalist Charles Darwin followed other ideas. In 1871, five years after the publication of his book *The Origin of Species*, he wrote to botanist Joseph Dalton Hooker (1817–1911):[13]

> We could conceive in some warm little pond with some sorts of ammonia and phosphoric salts, light, heat, electricity, etc., present, that a protein compound was chemically formed, ready to undergo more complex changes, at the present day such matter would be instantly devoured, or absorbed, which would not have been the case before living creatures were formed.

This is the idea of the hot primordial soup, which was further developed in the early 20th century by Russian researcher Alexander I. Oparin (1894–1980) and British biologist John B. S. Haldane (1892–1964). According to their ideas, the hot extremely low-oxygen conditions at Earth's surface, after the Hadean period, favored chemical reactions between the reducing gas components of the early atmosphere. In this environment, a lack of oxygen was an important prerequisite because its presence would have immediately broken up the newly formed complex chemical compounds through oxidation processes. In contrast, organic compounds were thought to be synthesized from the reducing chemical components, with electrical discharges (lightning) from Earth's dense cloud layer providing the necessary energy. Simulations have indeed verified that earlier periods of geological evolution were characterized by high seismic and volcanic activity during which lightning, in enhanced electrical fields, would have generated a high flux on gamma radiation as discussed in the previous chapter.

In 1953, Harold Urey and his student Stanley Miller (1930–2007) tested these theses with an experiment that became famous: They exposed a gas mixture of ammonium, methane, and hydrogen to conditions that corresponded to the Oparin-Haldane hypothesis regarding the early planetary atmosphere. After the experiment, traces of more than 20 amino acids were found in this gas mixture, many of which are components of biological enzymes and proteins. Thus, the primordial soup theory was considered confirmed. Follow-up experiments with different gas combinations and cyanide additions increased the reactivity of the mixture and demonstrated that the basic components of DNA, guanine and adenine, and other important biological molecules could also have been produced.

In 1963, Carl Sagan (1934–1996) and his co-workers showed that the addition of phosphorus to the gas mixture even produced adenosine triphosphate (ATP) – the molecule considered the chemical engine for protein, RNA, and DNA synthesis.[14] Further theories posit an additional energy source to lightning in the early Earth's electrical highly charged atmosphere. Besides lightning, intense solar UV radiation at Earth's surface could also have played a role. This energy source would have been possible during the early development of Earth's atmosphere because the lack of O_2 and O_3 oxygen molecules in the atmosphere would have significantly reduced shielding of the intense UV radiation. Besides the possibility of enhanced lightning-induced flux in the γ-radiation, radiogenic radiation has also been considered as an energy source since levels must have been considerably higher in the Earth's early evolutionary phase than are present today (see Chapter 6).

Doubts about the primordial soup theory arose with more recent results regarding the gas composition of Earth's early atmosphere. According to these findings, the atmosphere consisted of fewer reducing gases than those used in Miller's and subsequent experiments. The early atmosphere was composed mainly of carbon dioxide CO_2 and nitrogen N_2 with only minor admixtures of methane and ammonium (see Figure 6.19). Notably absent was hydrogen gas, which would have made important contributions to chemical reactivity in the experiments. In contrast, the presence of only CO_2 and N_2 would have contributed much less to the formation of complex molecules.

Another objection to the primordial soup theory is that the same energy source that led to the formation of amino acids would have prevented the formation of more complex macromolecules. Since high-energy discharges and radiation are known to have been prevalent, these same forces would have broken up and thus blocked the progress of biochemical processes toward the formation of more complex molecules. Nevertheless, it is conceivable that this early Hadean period laid the foundation for the formation of organic molecules, which only developed further in the later, less turbulent Earth ages.

This thesis was seemingly supported by Stanley Miller's later experiments. With these, he demonstrated that gas mixtures stored in a frozen state for twenty years contained increased yields of complex biochemical molecules. How informative this experiment is, we do not know. Based on geological data, the Snowball Earth hypothesis, where periods of absolute planetary glaciation occurred millions and billions of years ago, seems to correlate with the biological evolutionary thrusts suggested by Miller's experiment.

The influences associated with this so-called Icehouse-Greenhouse hypothesis, suggesting alternating frozen and hot environments on earth, remain a subject of intense scientific debate regarding the evolution of life. According to this theory, icehouse conditions would have been driven by a geologically stable environment where CO_2 gases were mostly absorbed by oxidization.[15] Conversely, greenhouse periods are associated with extensive volcanic activity and the breaking up of continents, releasing large amounts of CO_2 and CH_4 greenhouse gases into the atmosphere. It remains to be seen if this would correlate with an increase in the deposition of radioactive material ejected from the deeper mantle layers of earth.

With the presumed failure of Oparin and Haldane's original primordial soup thesis, new ideas about the origin of life developed. Today two theories dominate, which can be labeled heaven and hell as depicted in Figure 7.3. Heaven refers to the postulated astrobiological origin of life on ice-covered comets and asteroids, while hell refers to a possible hydrothermal origin in bubbling lava springs along volcanic rifts, thus producing the new lithosphere in ocean depths.

The idea that the origin of life came from space gave Helmholtz and Arrhenius's ideas a new context. This was particularly propagated by Carl Sagan and Fred Hoyle, who pointed to the numerous organic molecules observed in the cosmic ice on planets and comets.[16] In contrast, the thesis of life's origin and multiplication in hydrothermal environments was propagated by oceanographer Jack Corliss (b. 1936). With his colleagues in 1977, Corliss discovered and studied the diversity of marine fauna along the Pacific Rift.[17] Today, this thesis is supported by many microbiologists who see, in the rich chemical possibilities and diversity of minerals, conditions most suitable for formation of the primordial cell LUCA.[18]

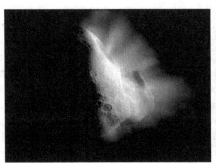 

FIGURE 7.3   Comet Tempel 1, showing in an artist's conception the comet's 3.7-mile-wide icy nucleus. The hazy atmosphere around the comet is made of materials vaporized from its surface by sunlight and expelled from the surface through cracks, leading to fresh ice below the crust. Credit: NASA/ESA – See more at: http://www.pmel.noaa.gov/eoi/

This thesis suggesting the hydrothermal origin of life is based on the idea that Hadean's deep-ocean areas offered a more protected environment than the planet's surface, which was being perpetually bombarded by meteorites. The ocean's chemical conditions also appear to have been more promising than those in the atmosphere, since there are more metallic trace elements in the vicinity of hot springs, which could have catalyzed the chemical processes toward the formation of macromolecules.

Hydrothermal vents are fissures in the earth's crust from which water, heated up to 400 °C by radioactive decay, escapes into the cold seawater environment under high pressure of up to 300 bar. This mineral-rich hot water contains many chemical components – especially sulfides and metals such as iron and manganese – which dissolve as minerals in the water. Observations show that hydrothermal vents form an ideal biotope with an enormous richness of species (Figure 7.3).

Hot water interacting with basalt and mineral rocks initiates chemical processes that lead to the remodeling of the mineral structure and the release of hydrogen gases. For example, iron silicate FeO_4Si and water H_2O are converted into silicon dioxide SiO_2 (quartz), iron oxide Fe_3O_4 (magnetite), and hydrogen H_2. The latter reacts with carbon monoxide CO and carbon dioxide CO_2, which dissolve in ocean water and can produce organic hydrocarbon molecules, up to fatty acids and amino acids, at temperatures above 300 °C. Minerals and metals dissolved in the hot water – such as the Fe ions, FeS, H_2, and H_2S – are highly reductive and serve as rapid catalysts for Fischer-Tropsch reactions,[19] which came to be known by way of industrial fuel production involving coal and biomaterials.

Such theories have led to intensive exploration of the other potential sites where a high radiation level would lead to hydrolysis. With hydrolysis, there is a radiation-induced break up of water molecules, which then serve as radicals toward the building of more complex organic molecules. In particular, deep underground water storage sites are exposed to a much higher radiation level than surface water. First studies at these sites point to an enhanced production of organic molecules in the deep darkness, through complex chemical reaction chains on sulfide-rich minerals as suggested in Figure 7.4.[20]

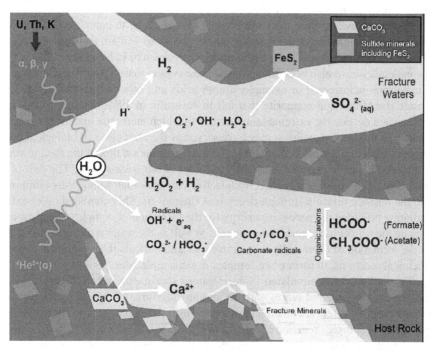

FIGURE 7.4 Radiolytic production of organic formate and acetate molecules in deep water – in rock fissures, on the iron, sulfite, and carbonate minerals found there – via α-, β-, and γ-radiation from the natural decay of ^{232}Th, ^{238}U, and ^{40}K radioisotopes. Source: https://www.sciencedirect.com/science/article/pii/S0016703720307018 (August 4, 2023).

These chemical processes are promising because, during its first two billion years, Earth was much hotter due to meteorite bombardment and the release of large amounts of thermal energy from radioactive decay. Also, there were an enormous number of hydrothermal vents that provided favorable conditions for the first formation stage of more complex organic molecules, combining radiogenic heat, mineral-rich hot water, and cold ocean water. While phosphorus and amino acids favored the formation of amino triphosphate (ATP) – a key energy carrier in the metabolism of biological cells – radioactive decay processes may have been the energy source for these first chemical reactions, reactions which led to complex molecules that enabled chemical processes to maintain metabolism. However, it was still a long way from these first amino acids to the formation of macromolecules, such as RNA and DNA, and even further to the bio-diversified development that surrounds hydrothermal vents today.

It is now being argued that such types of radiolytic chemistry process foster life in deep underground and underwater environments, driven by the energy released in the radioactive decay processes. This would open new vistas and sites for the origin of complex chemical molecules, if not life.[21] These studies also seem promising in expanding the possible origin of life on other planets such as Mars.

A second thesis suggests a cosmic origin of life. This has become an intensive research field in astrobiology in recent years. Studies involve not only the origin of life but also possibilities for biological development as well as the identification of habitable zones on planets that favor this development. This field has taken off, especially with the discovery of numerous exoplanets.

Exoplanets orbit stars other than the Sun in our galaxy. These exoplanets have been discovered through refined indirect and direct observational techniques that consider the brightness variation of their parent stars. Exoplanets are studied spectroscopically to detect chemical structure, in order to determine if the basic chemical elements of life such as water, carbon, and oxygen are present. With regard to the question of the origin of life on our Earth, studies also focus on nearby asteroids, comets, and minor planets in our solar system. These celestial bodies mostly consist of a mineral or metal-rich rocky core that is thickly covered with dust and ice (Figure 7.3). Spectroscopic analysis of comet surfaces reveals more than 150 organic molecules up to complex amino acids. These results have validated unmanned space probe landings such as Philae, from the European Space Agency (ESA), which landed on Comet Churyumov-Gerasimenko in November 2014.

Organic molecules have also been found in our solar system material by direct chemical examination of meteorite inclusions. For example, amino acids and other organic molecules were discovered inside the Murchison meteorite that fell in Australia in 1969. These findings apparently survived unharmed despite the extreme temperatures to which meteorite material is exposed when it falls to earth. In this way, meteorites may have served as the seed cores of biological evolution.

The origin of such organic molecules lies in the intense radiation flux of the Sun, to which interplanetary material has been exposed since the early days of our solar system. The dust of the early solar system contained considerably more radioactive elements than today, thus pumping radiant energy into the dust-ice mixture through decay (see Chapter 6). This combination of external and internal radiation provided the necessary energy for the formation of complex chemical molecules. By breaking up the simple hydrocarbon compounds as well as the water molecules, radiation energy facilitated the formation of chemically very active ions and radicals that combined, through a network of chemical reactions, to form more complex organic molecules.

Such processes have been simulated in accelerator experiments, where high-energy radiation corresponds to cosmic rays and reacts with an ice-dust mixture. Chemical reaction products are then measured either by direct chemical analysis or by infrared spectroscopy. These results seem to confirm the thesis of complex reaction chemistry in cosmic ice. Measurements have been extended, by theoretical analysis, to investigate the network of possible chemical reactions. By means of such a chemical-reaction network, far-reaching predictions can be made about what reactions might have taken place during early planetary development and what molecules might have formed under the influence of radiation. Thus, molecules embedded in meteoric rocks may have provided organic seed material for further biological evolution on earth, especially during meteorite and asteroid bombardment in the Hadean era.

However, there is a long and mostly unexplored step between the formation of organic molecules and the emergence of organic life. It is generally believed that this step must have occurred during the first two billion years of Earth's history. First traces of existing primitive-life forms have been discovered in rocks dating back 2.7 billion years.

The step from chemistry to biology requires the self-organization of molecules, in order to form chain molecules such as RNA and DNA, and also the development of an enzymatic regulatory mechanism for stabilization and self-reproduction. This mechanism is organized by the cell, even in primitive living beings, and can therefore be considered the fundamental unit of all biological life. This entity must be able to generate chemical energy for its existence and for that of the cell's associations. This then is the carrier of metabolism, which enables reproduction by reusing existing building materials, thus laying the foundation for further development of more complex life forms.

We have already discussed the two known cell types that facilitate biological evolution (Chapter 4.1.4). Prokaryotes are considered the oldest known cells. They do not have a nucleus enclosed by a membrane, and so the DNA is freely stored within the cell. A second class of cells, the eukaryotes, developed much later. Their nucleus is enclosed by a membrane, within which the DNA – with the cell's information – is located.

This transition from prebiotic organic molecules to the first biological structures capable of metabolism, such as the cell, is not yet clear. The most important questions are about the energy source for such synthesis and the probability by which it can occur. In the case of hot thermal springs, the energy source for the development of primitive metabolism would be the radiation that heats the water to boiling. Such springs provide not only energy but also, thanks to their high mineral content, possible catalysts such as iron, phosphates, and sulfates, which are necessary for the construction of more complex structures by means of reducing-oxidizing reaction sequences.

Other arguments suggest the evolution of a primitive metabolism that could form predominantly on materials with mineral-containing surfaces, thus providing the prototype for later more complex RNA- and DNA-driven metabolism. The formation of ATP offers an example of these complex possible processes and interrelationships. ATP provides chemical energy for cell function and the assembly of ribosomes, proteins, RNA, and DNA. A combination of observation and laboratory experiments has shown that this molecule could have been formed by chemical reactions, between meteoric phosphorus and hot springs, during the Hadean period.[22] However, although this is a probable explanation, it is not the only possibility.

One concern is the timespan required for such biochemical evolution processes. The probability that a complex self-regulating cell system would form is extremely small. Formation is based on myriads of chemical processes that must occur with some intrinsic probability and under only certain conditions within the limited time scale of earth's early evolution. That this happened on Earth, with its unique habitable conditions, might assign a certain priority to our planet; it implies that life arose only on Earth through a series of random processes, processes that occurred differently or not at all on other planets despite their possibly favorable conditions for life.

Based on such considerations, Fred Hoyle proposed an alternative hypothesis that considerably extends the time period for biological evolution. The panspermia hypothesis[23] assumes the cosmic origin of complex organic molecules but moves its occurrence from our planetary system's primordial time to the primordial time of our universe, when the first star systems formed and produced carbon and oxygen as life-forming elements (Figure 7.5). The spectrum of the oldest discovered star, with the beautiful name SMSS J031300.36-670839.3, is 13.6 billion years old. It reveals a very high carbon and oxygen content but gives no hint of elements heavier than calcium.

According to the panspermia hypothesis, organic molecules were formed from the carbon, oxygen, and hydrogen produced in the first generation of stars. These stars were under the influence of intense cosmic radiation, as the basic material of the biological development of our cosmos. It is

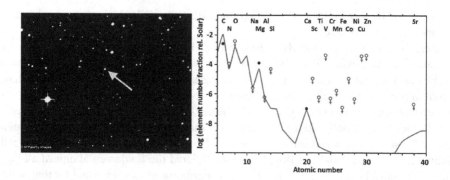

FIGURE 7.5 The oldest known star SMSS J031300.36-670839.3 is 13.6 billion years old and 6,000 light-years from the Sun. The measured abundance of elements present in its stellar atmosphere shows in particular high proportions of carbon and oxygen, the basis of biological life. Elements heavier than calcium ($Z = 20$) could not be observed. Courtesy of Timothy Beers, University of Notre Dame.

further noted that the development period for simple biological systems is ten times longer than the one to two billion years that were available for biological development on Earth.[24]

According to this theory, early organic components were distributed with the interstellar dust – as the ashes of past star and planetary systems – over our Milky Way and other galactic systems. These components represent the first spores for locally developing life on the isolated planets that provided favorable environmental conditions for this purpose.

Simple life forms, such as bacteria, have been found in meteoritic rocks. Beyond this, there is as yet no direct evidence of interstellar microbial traffic. The general attitude of astrobiologists toward the panspermia hypothesis is therefore extremely skeptical and dismissive. This does not, however, mean that the hypothesis is wrong but only that the observational or experimental data necessary for its acceptance has not been produced. On the other hand, bacterial species such as the so-called Conan the Bacterium (*Deinococcus radiodurans*) are known to withstand both extremely low temperatures and extremely high radiation intensities, of up to 10,000 gray, without losing vitality. Thus, they seem to be well equipped for interstellar travel lasting thousands to millions of years.[25]

With respect to the identification of habitable worlds, enormous progress in the collection of observational data has been made in the last years.[26] This is mainly due to NASA's KEPLER observation satellite. Since its deployment in 2009, the satellite, which is equipped with a system of high-resolution CCD cameras, has been continuously observing a field of 156,000 stars in the northern hemisphere, roughly in the range of the constellations Swan (Cygnus) and Lyra.

KEPLER can measure minute variations in the brightness of observed stars. These variations indicate the presence of passing planets. From the frequency of these fluctuations, the orbital period as well as the orbit of the passing planets can be backcalculated. In its first five years, KEPLER had already identified more than 4,000 planets, some of which could possibly support life. This will be investigated in future NASA and ESA missions, where detailed spectroscopic surveys of the planets will look for water as the basis of life.

About 100 exoplanets have been identified that are similar in size to our Earth. However, they are mostly too hot for life, according to our understanding of biology. When comparing the size of the field being observed to the entire Milky Way, astronomers calculate that a total of 40 million near-Earth-sized planets inhabit our galaxy. KEPLER results are, therefore, undoubtedly ground-breaking as we continue to search for exoplanets in distant star systems. These results may also have shown that planet formation, such as our solar system, is not an exception but rather the rule. This means that habitable or even animate worlds may exist and that this likelihood is considerably greater than ever before thought. This possibility presents a challenge for astrobiologists as they work, with better observational data, to find such worlds and detect biological life – life built on the organic chemistry of carbon.

The question about the origin of life as a self-sustaining and self-replicating system has not yet been answered. Radiation energy may well have initiated the first processes by providing an initial energy source in hostile environments, thereby allowing complex molecules to form. There are, however, several pathways by which chemical processes can build up the complex molecules that power our metabolism. The precise physicochemical mechanism for the formation of the first cells, as the precursors of all our biological systems, and the energy sources that underpinned this development are still largely unknown.

During their early evolution, the macromolecules that formed must have been able to assemble into self-replicating cells, which could then develop further through mutation and diversification. This is expected to have been primarily a chemical process that relied on other energy sources such as ATP and efficient energy transport mechanisms in the biological materials. The most important step on this path was the formation of cells that diversified into cells with specific tasks, tasks that were coded within the cell by RNA and DNA information.

The first single-celled organisms, prokaryotes, have shown up as 2.6-billion-year-old fossils in Precambrian rock. Toward the end of the Precambrian,[27] early macro-algae in the prehistoric ocean released oxygen through photosynthesis. This is seen in fossil rocks that contain multicellular organisms or cell colonies. Only 160 million years later, the Cambrian period shows that an explosion of biodiversity occurred as evolution gradually developed new species with more complex metabolisms about which not much is known.

These living structures are thought to have evolved as a consequence of the self-organization of complex biochemical systems, such as cells. Accordingly, structures evolved to maintain the functionality of these cells in their environment. While radiation most likely played an important role as initial energy source, for the development of more complex systems, it would have been replaced by more subtle and less destructive energy sources. In this process, the cell developed features that protected it from the destructive consequences of its internal radiation.

7.1.4   Cells as Self-Organizing Systems

Since its origin, life has been dealing with the existence of radiation and its effects. Some researchers argue that radiation itself contributed to the origin and evolution of life, producing energy through direct irradiation of ice-covered planetary components or indirectly in hot volcanic springs. Radiation also provides energy for the mutation of molecular structures and thus for the possibility of further biological development. This evolution is not directed linearly. Rather, it is subject to random fluctuations that provide a rich spectrum of biologically possible systems. Chances for the survival and growth of these systems depend on the earth's changing environmental conditions.[28] This dynamic evolutionary process is based primarily on the possibility of self-organization within complex systems. The development of the cell into a self-organizing and self-reproducing system is considered the most important step in the evolution of life.

Self-organization is the spontaneous appearance of new structures and behaviors within thermodynamically open systems. In contrast to closed systems, these systems exchange energy, substances, and information with the outside world.[29] According to the second law of thermodynamics, entropy or internal disorder grows steadily toward chaos in closed systems. In contrast, new stable orders can arise in open systems through self-organization. This kind of self-organization is present in numerous processes and systems. Examples of such systems include: thermal equilibrium, the world of elementary particles, physics and chemistry, space with its formation of stars and planets, evolution, and from basic biology to social systems. These systems all arise by themselves through interactions between their components.

In addition to their complex structures, emerging systems also possess new properties and capabilities that the individual components do not have. A self-organized system will therefore change its basic structure, depending on its development process and environment. The interacting components (elements, molecules, agents) act according to simple rules and create order out of chaos without having to have a vision or a goal for the development of the whole. This internal regulatory system also underlies the evolution of life on our planet.

A recent publication pointed to the important role of metabolic cycles that depend on the metabolic activity of single complex catalyst partnering through non-reciprocal interactions in a cyclic reaction "that lead to their recruitment into self-organized functional structures. [...] Different classes of self-organized cycles form through exponentially rapid coarsening processes, depending on the parity of the cycle and the nature of the interaction motifs, which are all generic but have readily tuneable features."[30]

Another classic example is ribosomes, the complex molecules within cells. These were the first biological machines to serve self-development, being formed from a combination of proteins and RNA. Ribosomes are controlled by their RNA coding, generating further proteins from amino acids for the further development and improvement of cell function. Ribosomes need energy to function and obtain it from chemical processes, with the aforementioned amino triphosphates (ATP) playing a critical role as energy carriers.

Without ribosomes, the assembling and reproduction of proteins as well as other cellular materials could not occur. In the early stages of life's evolution, the emergence of ribosomes was perhaps the most important step toward the beginning of self-reproduction. Except for minor differences in detail, all living things have the same ribosome structure. Again, this means that all terrestrial living things are descendants of our single-cell long-gone molecular ancestor LUCA in which ribosome synthesis succeeded.

The cell is the central basic unit of all early unicellular microstructures, from archaea, prokaryotes, and later eukaryotes, up to the complex and diverse life forms that populate the earth today. All of these living things are open systems that seek to maintain their internal structure and functionality through an external energy supply. In humans and animals, this energy supply takes place chemically through food intake, which is further processed in our gastrointestinal system. In plants, sunlight provides the crucial energy supply; through photosynthesis, carbon dioxide and water are converted into sugar nutrients and oxygen.

As systems become more complex and environmental conditions change, chemical possibilities expand. For example, an increase in atmospheric oxygen can be directly correlated with an increase in the physical size of living organisms (Figure 7.6). Oxygen levels in the atmosphere were exceptionally low during the first 2 to 2.5 billion years of Earth's history. The first organisms, such as unicellular prokaryotes,[31] probably arose 3.5 billion years ago and dominated life on Earth for 2 billion years during the Archean Eon.

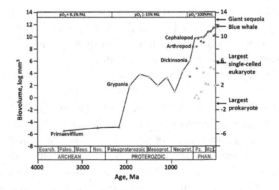

FIGURE 7.6   Correlation between the volume and size of biological systems with regard to changing oxygen content in Earth's atmosphere. From Payne, et al., "The evolutionary consequences of oxygenic photosynthesis: a body size perspective," *PNAS* 107 (2011): 37–57.

Another important step was plant photosynthesis, which increasingly converted CO_2 from the atmosphere into oxygen. This opened another chemical possibility for energy production in cells. Through photosynthesis, oxygen levels grew to 1% of today's levels and defined the Proterozoic Era. As a result of increased oxygen content, eukaryotes evolved; their volume was about 100 to 10,000 times that of prokaryotes. Toward the end of the Proterozoic, 1 billion years ago, oxygen levels continued to rise, reaching 30% of today's levels. With the last 500 million years, biodiversity has developed exponentially in the Cambrian explosion. Based on genomic sampling, it is being proposed[32] that all of the more complex biological systems are based on the formation of eukaryotes, at a time some two billion years ago, from a common progenitor Asgard-Archaeota that is known as the closest archaeal relative of eukaryotes.

An increased oxygen supply provided cell structures with the opportunity for chemical and organizational expansion. The number of species increased as organisms enlarged and life evolved in its tremendous diversity.[33] The development of oxygen-rich cells did, however, present new dangers, since the oxygen effect[34] increased the radiation sensitivity of the cells by two- to threefold.

Similar relationships can be observed in the radiation intensity at Earth's surface. In Earth's history, the average radiation dose to which organisms were exposed was time dependent[35] (see Chapter 6 and Figure 7.7). Here, radiation conditions were reduced to two components. The first is internal radiation, which primarily comes from the radioactive potassium component ${}^{40}K$ in cellular material. As the total amount of ${}^{40}K$ decreases due to decay, this radioactive component decreases exponentially with the earth's age, now contributing only 9% of its original value. Thus, during the early period of Earth's evolution, an organism's internal radiation load would have been many times higher than today.

In addition to internal radiation stress, the second factor to which organisms were exposed was the external stress from actinide and potassium decay in the gradually solidifying crust. Solidification of the crust occurred after the end of the Hadean period and initially progressed slowly. Then, 2.7

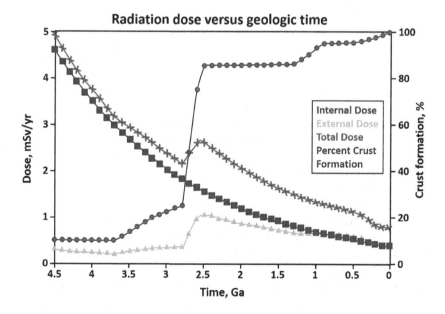

FIGURE 7.7 Internal dose of ${}^{40}K$ in living systems and the external radiation dose provided by actinides in the earth's crust in units mSv/a. The total exposure of living organisms has dropped to one third of the original dose in the last 2.5 billion years. The external dose increased after the Hadean period with crust solidification and growth. The percentage of solid crust on the earth's surface is given. After P. A. Karam, S. A. Leslie, "The evolution of the Earth's background radiation level over geological time," 1996.

billion years ago, the earth's crust rapidly increased, extending from about 20% of surface area to 80%. This led to an increase in the external radiation dose to which early life was exposed. Conversely, the amount of cosmic radiation, which plays such an important role today, was negligible in the Precambrian period. For, despite a lack of oxygen, the atmosphere had a much higher density than today and so much of the cosmic radiation was already absorbed in its upper layers. In addition, the Sun was much weaker in its own early evolutionary phase, which meant a lower intensity of high-energy radiation than is present today.[36] This might have been compensated by a higher intensity of paleolightning-induced γ-radiation flux in the early troposphere, which would decrease with the formation and solidification of the earth crust.

Changes in radiation dose and oxygen concentration directly affected the radiation-induced damage rate to DNA and the resulting mutation rate of early life.[37] These processes correlate with each other and so must be considered together, in terms of their impact. The radiation-dose decrease was compensated by an increase in oxygen concentration. Thus, the damage rate is estimated to have decreased by a factor of 2 on average over the last 2 billion years.

How does the damage rate correlate with the mutation rate? This depends primarily on the efficiency of biological shielding and repair mechanisms in early organisms. During the early evolution of cell structure, very complex but highly efficient molecular defense and repair mechanisms arose in parallel as part of cellular self-organization (cf. Chapter 4, section 4.1.4).

7.1.5 Radiation Defense and Repair Mechanisms

The higher radiation flux and the associated radiation dose to which early biological systems were exposed led to the development of biological defense and repair mechanisms in early cell structures and organism. Since that time, all organisms have developed a variety of repair mechanisms, which each have proteins and enzymes as their base repair carriers. Biological defense can be observed primarily in early organisms since with the increasing complexity of biological structures defense becomes more challenging. This is because modifying the biochemistry of a complex multicellular organism with different cell types and tasks becomes prohibitively difficult.

In this regard, single-cell microbes and bacteria have an advantage because it is much easier for these single-celled organisms to adapt to extreme environmental conditions such as extremely low or high temperatures and high radiation levels. Extremophile organisms have developed particular survival capabilities, which makes their specific environment a requirement for survival, while extremotolerant organisms are able to adapt to extreme conditions. We will focus in the following on organisms that are radiation tolerant, but it should be pointed out that hyperthermophile or high-temperature tolerant organisms are specifically found in hydrothermal vent and hot spring environments, while psychrophile or low-temperature tolerant species are found in permafrost and ice environments including cosmic ice sheets. Both are therefore possible environments for the formation of life (Figure 7.8).

The best-known radiation-hardened organism is the aforementioned Conan the Bacterium (*Deinococcus radiodurans*) that is able to withstand a dose of 10,000 gray, which is a thousand times higher than the lethal dose for mammals. However, there are a number of additional radiation-hardened creatures. Chroococcidiopsis, a bacterium that enjoys living in the cold or in hot deserts, can tolerate γ-radiation up to 15,000 gray. Both organisms are radiotolerant and also can exist in a broad variety of extreme environments.

While these bacterial organisms are roughly 10 µm or 0.01 mm in size, an example of a considerably larger radiotolerant organism is the tardigrade, which has a fossil record that goes back 550 million years to the Cambrian period. Tardigrades are more complex, containing about 40,000 cells and reaching a size of 0.5 mm. They can withstand vacuum and low-temperature conditions and have a radiation resistance of about 5,000 gray. Tardigrades were part of a biological payload on the FOTON-M3 mission[38] in 2007 and for ten days were exposed to high-vacuum conditions in outer space. The reports claim: "radiation had no effect on survival or DNA integrity of active

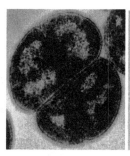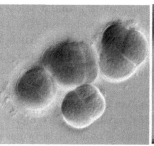

FIGURE 7.8   Three examples of radiation-resistant microscopic organisms, from left to right, the bacteria *Deinococcus radiodurans*, *Chroococcidiopsis thermalis*, and the tardigrade *Ramazzottius varieornatus*. Source: https://en.wikipedia.org/wiki/Deinococcus_radiodurans#/media/File:Deinococcus_radiodurans.jpg https://microbewiki.kenyon.edu/index.php/Chroococcidiopsis, and https://cen.acs.org/biological-chemistry/biochemistry/tardigrades-survive-space/97/i41 (March 14, 2024).

tardigrades. [...] Spaceflight induced an increase of glutathione content and its related enzymatic activities."[39]

Adaptation to high-level radiation environment can be maintained by preserving multiple copies of the genome so that an effective repair based on one of these copies can be ensured.

In the case of *Deinococcus radiodurans*, radiation resistance is thought to be due to an extremely efficient repair mechanism resulting from enhanced enzymatic activity.[40] Other researchers propose that the organism's high manganese concentration may quench the oxygen free-radicals that are released by radiation exposure, which would reduce radiation damage to the DNA itself[41] and help to maintain a more efficient repair mechanism. It is likely that several parallel mechanisms have evolved toward this goal.

In the case of tardigrades, recent studies point to the effects of a damage suppression protein DSUP, which functions as a cell's internal radiation shield. DSUP binds to chromatin, the protein that wraps up DNA inside cells,[42] and as a consequence shields the DNA molecules from hydroxyl radicals produced by ionizing radiation.

While research and discussion about the most effective repair and shielding mechanisms are ongoing, it seems that ancient microbes, bacteria, and even multi-cell systems have been successful in developing defense and repair strategies that have transferred to the enzymatic repair mechanisms of more complex multi-cell organisms.

Recent genetic studies have been successful in applying the tardigrade's DSUP technique of protein shielding to human cells.[43] These studies have shown that the human cells demonstrate approximately 40% more tolerance after exposure to X-ray radiation. These findings have opened discussion about possible medical applications of this technique, namely in cancer treatment.[44] Recent reports in the *South China Morning Post* (from March 23, 2023) suggest experiments toward military applications by claiming, "Chinese team behind extreme animal gene experiment says it may lead to super soldiers who survive nuclear fallout."[45] The experiment was performed using CRISPR/Cas9, a gene-editing tool now available in most bio-laboratories. This report was followed by a storm on Twitter and other social media warning and condemning Chinese militarism. However, considering the complexity of the human genome, and the possible effects on DNA division and replication for protein-wrapped molecules, there is a long way for the proposed step, "From water bear to super soldier"!

After this excursion into future hopes for application, let's return to the discussion about repair mechanisms for radiation damage in complex multi-cell organisms. All cell-repair mechanisms are based on enzymes, as important cell components. Enzymes are proteins whose special structure or shape enables them to act as intermediate carriers for special chemical reactions; this is to say that enzymes act as transports within the cellular domain. These mechanisms must have evolved early because their mode of action is found to be extremely similar across different organisms within the tree of life. Due to the complexity of these repair mechanisms, their mode of action remains poorly understood in many cases despite intensive research in the radiation-biology field. It is, however, known that these mechanisms control random damage or damage caused by radiation and other mutagenic processes that would otherwise potentially result in the miscopying of DNA during cell division.

It is estimated that *each* cell in the human body has on average 10,000 copying mistakes per day, but most of these can be successfully repaired by enzymes.[46] Permanent mutations are few and often only occur as a result of false repair. For repair, enzymes use templates from the second undamaged strand of the damaged DNA molecule or from the accompanying RNA. However, mis-repairs occur when both DNA strands are damaged. These mis-repairs remain in the cellular material and can have long-term effects. If the original damage is too extensive, the cell initiates its death (apoptosis).

In recent years, through a series of studies on various biological systems of varying complexity, molecular genetic research has illuminated how poorly we understand the cell's responses to low-dose radiation. However, it is becoming apparent that susceptibility to radiation is related to the efficiency of enzymatic-repair mechanisms.

Another poorly understood effect is the bystander mechanism, which refers to the biological response of a cell to an event that has occurred in an adjacent or nearby cell. Here, the bystander effect suggests intercellular communication and exchange.[47] This is of particular interest in assessing the possible impact of radiation exposure. The bystander effect is not well understood but seems to occur in two ways, which presumably reflect two different mechanisms. In the case of higher radiation levels, radiation damage in one cell leads to mutagenic or even lethal consequences in neighboring cells that were not directly hit; this might be caused by secondary radiochemical effects, which amplify radiation damage. The other type of bystander effect seems to work as a warning mechanism, causing an adaptive response in cells adjacent to the cell that has been directly damaged by radiation. This response initiates resistive and protective measures by the adjacent cells, as reflected by an increase of preemptive enzymatic activity. At doses below 100 mGy, the beneficial bystander effects outweigh the detrimental effects of radiation.

There is uncertainty regarding what elicits the bystander effect. It may be prompted by chemical agents that are emitted by the cells impacted by radiation. This may cause an upregulation of stress-inducible proteins in the cluster of the adjacent cell. Such an increase can be accompanied by an induction of the DNA response through Serine-15-induced phosphorylation. This refers to the addition of a phosphoryl (PO_3) group, which is released to a particular molecule that is vital for the storage and transfer of free energy within the cell. This then causes the release of tumor-suppressant biological agents.

These are complex biochemical response processes, which cannot be described in detail here. However, the growing body of data suggests that the underlying mechanisms of the bystander effect are multifaceted and depend on the characteristics of the exposed cells, the environment surrounding the cell, and the physical aspects of the radiation exposure, such as dose, rate, and type of radiation. This is, therefore, another clear signature demonstrating the biological and physical nature of radiation response. Although the exact identity of the bystander signal(s) is yet to be identified, published data indicate changes in gene expression – for multiple types of RNA – as being one of the major responses of both cells and tissues within the context of the bystander effect.[48]

It thus seems clear that the evolution of biological systems has included the development of a number of increasingly complex biochemical warning, defense, and repair mechanisms. While many of these are not fully understood, present research in the radiation response of biological systems indicates that this response has evolved to allow for adaptation to particular radioactive environments. In this sense the bystander effect can be understood as an evolutionary mechanism that provides a cellular warning system, which emerged during early cell development toward a self-organizing organism; this development would then energize an increased enzymatic-repair readiness in neighboring cells.[49] This is, however, still a matter of scientific debate and different interpretations of the bystander effect can be found in the literature.

While the bystander effect and repair mechanism maintain control over the extent of radiation-induced changes to the genetic information pool, such changes are important for maintaining the balance necessary for a mutation rate in biological systems. This raises the old question around what drives mutation and biological development. Regarding the contribution that radiation-induced mutations make to the total mutation rate, many statements have been made. Most researchers consider mutations to be random, with only up to 6% being influenced by radioactivity. Accordingly, this proportion would have been considerably larger as the evolution of earth's biological life began.

7.1.6 From Mutation to Evolution

Biological systems are based on cells as the basic building blocks for organisms and their function. Reproducing by cell division, information for cell function is passed on unchanged from the original cell to the daughter cells; this is accomplished by dividing and duplicating DNA and RNA molecules.

Reliable reproduction of the DNA's double-helix structure is indispensable for passing on original genetic information. However, errors do occur that alter the information structure and these are the so-called mutations (see Chapter 4). The mutation rate is the number of mutations per cell division within a genetic unit.

Mutations in germ cells cause changes in the basic genetic DNA structure. These changes are regarded as the microbiological explanation for Charles Darwin's theory of evolution known as the natural selection and development of species. Mutations that positively influence the adaptive abilities and developmental possibilities of living beings give their species a greater probability of advancement and survival. Mutations that weaken the cell or its information structure reduce the chances of further development.

Where do these mutations come from? Mutations can occur as simple random errors in any copying process or can be induced by extraneous effects. These effects include exposure to chemicals or radiation that externally damages or alters DNA structure (see Chapter 4). Experiments on unicellular bacteria, viruses, self-fertilizing worms, fruit flies (Drosophilae), and larger experimental animals such as mice and rats clearly show that higher doses of radiation increase the number of mutations (Figure 7.9).[50]

However, these experiments also show that the mutation rate does not drop to zero in the absence of an external radiation source. This residual mutation is the so-called spontaneous rate and varies considerably between different biological species. Residual mutation does, however, also depend on the number and structure of the enzymes involved in the replication and repair of DNA molecules. Therefore, cells with higher mutation rates are more likely to adapt to new or rapidly changing

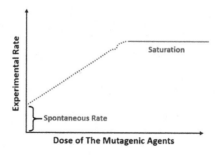

FIGURE 7.9 Increase of the mutation rate with radiation dose. Despite an external zero dose, a fixed mutation rate remains: the spontaneous rate. The reason for spontaneous mutation is still a subject of scientific debate.

environmental conditions. This observation underscores the connection between mutation and evolution.

Cell division or the replication process is seen as a generational change, occurring at different rates depending on the type of replication. Therefore, the mutation rate is expressed as the number of mutations per genome per generation or as the number of mutations per base pair per generation. The number of bases or base pairs varies significantly depending on the genome and this then affects the mutation rate per genome. The mutation rate of viruses with an RNA structure, based on 10^4 base pairs per genome, varies between 1.0 and 0.1 mutations per genome per generation. In microbes with a DNA structure, the rate is 0.003 per genome per generation but the number of base pairs per genome can vary from $6 \cdot 10^3$ to $4 \cdot 10^7$.

In worms, the mutation rate is $2.1 \cdot 10^{-8}$ mutations per base pair per generation. This corresponds to a mutation rate of 2.1 mutations per genome per generation, since the genome has approximately 10^8 base pairs. The spontaneous mutation rate in humans is approximately $1.1 \cdot 10^{-8}$ per base pair per generation. With approximately $7 \cdot 10^9$ base pairs in the human genome, this accounts for seventy-seven mutations per genome per generation, assuming a time span of 20 years per generation.[51] In other mammals, the average spontaneous mutation rate is of the same order of magnitude, about $2.2 \cdot 10^{-9}$ per base pair per year. Thus, the spontaneous mutation rate increases with gene-structure complexity and periodicity or the time frame for generation change, which is to say the rate of DNA replication.

Genetic studies show that a higher mutation rate is likely to cause more cells to die while also making cells more flexible, thereby evolving new capabilities or adapting to new environmental conditions. The number of enzymes in each cell also determines the mutation rate, such that having fewer enzymes limits the repair mechanism's ability and efficiency. Conversely, when there are many enzymes, the mutation rate is reduced because the efficiency for post-copy error repairs is improved. An optimal mutation rate reflects a balance between adaption and repair.

There are, however, many unanswered questions: Is the spontaneous mutation rate based on random (stochastic) copying errors as they occur in daily life (Figure 7.10)? Is the spontaneous mutation rate induced by body-specific conditions? Or, is it a mixture of both components? And, how then does the spontaneous mutation rate relate to the relative frequency of both mutation components?

These questions were already posed in the 1920s by US biologist, geneticist, and Nobel laureate Hermann Muller (cf. Chapter 1.1.5) and chemists Axel Olson (1889–1954) and Gilbert Lewis (1875–1946). Muller induced mutations, in the fruit fly Drosophila, using an increased dose of X-rays. This discovery – and parallel studies involving the effect of gamma radiation on tobacco plants – led Olson and Lewis, in 1928, to suggest that cosmic radiation, discovered in 1912, might be the driving force for radiation-induced evolution.[52]

FIGURE 7.10 Increase in copy errors after 20, 50, and 100 copies of the original title page of Charles Darwin's *The Origin of Species*, shown on the left. From: Scott Freemon, Jon C. Herron, *Evolutionary Analysis*, 4th edition, Pearson Education, Inc. (2007), 151.

This deterministic interpretation was contradicted by Hermann Muller as early as 1930. Muller based his contradiction on measurements taken of the radiation-induced Drosophila mutation rate.[53] These and other measurements were interpreted to mean that an increased radiation dose, of up to 2,000 times higher than that observed naturally, seems to be necessary to produce a higher mutation rate.[54] Therefore, in the following decades, the assumption of stochastic mutations as the driving origin of evolution was favored.

Pure random mutation rates in microscopic systems are again based on the probabilities for quantum-mechanical processes. The probability for the occurrence of different tautomeric molecular states[55] involved in the copying process, and uncertainties regarding the thermal motion of enzymes during the replication or repair process, play a role in miscopy occurrences. The frequency of miscopies is determined by the quantum-mechanical probability for the occurrence of tautomeric processes during the replication process.

Tautomers particularly occur frequently in the base molecule guanine and only infrequently in cytosine, increasing the probability of certain copying errors while reducing others. Typical energy differences between tautomeric states are in the range of 30 kcal/mol, which is in the electron-volt range for a single molecule. According to Heisenberg's uncertainty principle, this results in a lifetime for the tautomeric state of approximately $5 \cdot 10^{-16}$ seconds or half a femtosecond.

While the time scale for a DNA replication depends on the size of the genome, only milliseconds are needed to copy a single nucleotide, forming new base pairs.[56] This means that many transformations may have already taken place in the time of replication – deterministically stimulated by thermal or other energy processes. Therefore, the estimated quantum-mechanical contribution to the spontaneous mutation rate is indeed comparable to the observed rate.

Another spontaneous contribution to mutation stems from internal radioactivity in biological systems. Radiation from these radioactive sources can lead directly and indirectly to chemical changes in DNA (see Chapter 4). These include defectively repaired breaks in DNA strands, but also chemical reactions of the base molecules with water-molecule radicals that have been generated by the irradiation. This has been adequately demonstrated when experimental animals are exposed to high doses of irradiation. In such cases, enzymatic-repair processes are not sufficient to control the many radiation damages. However, at low doses, it is not clear whether existing mechanisms can make repairs. It is also not clear if there is a threshold at which all damage is being repaired, or whether such a threshold could be located.[57]

According to Table 7–2, the internal activity of humans is 8,000 Bq. This means that about every 100 microseconds a radioactive-decay event may cause primary damage to the DNA molecule in its germ cell or the DNA can be damaged in the secondary ionization processes, whose products – free radicals or electrons –- can cause genetic damage. About half of natural human radioactivity has ^{40}K as its source, with 10% of the decay producing γ-radiation of 1.46 MeV energy. This gives an average γ-rate of $8.6 \cdot 10^{-12}$ per second for a single cell.

Postulating that the cross-sectional area of effect, for damage to a base molecule, is the extent of the molecule – $\sim 10^{-14}$ cm^2 – a rough estimate for radiation-induced primary damage can be made (cf. Chapters 3 and 4) by assuming a molecular density of $9 \cdot 10^{19}$ base molecules per cm^3 in a human cell. This gives a primary-damage rate of about 10^{-9} per basic molecule per generation, which is, by order of magnitude, 10% of the mutation rate. This is consistent with the estimated values for internal radioactivity, especially considering that only 5% of total γ-activity resulting from ^{40}K was used in this estimate. In smaller organisms, internal radiation is much lower and therefore causes a much smaller damage rate.

Of course, these are only rather rough estimates, since they do not take into account the secondary chemical effects of the ionized water molecules nor the enzymatic-repair mechanism of the cell. However, damage is often difficult to repair when caused by direct radiation interaction with DNA strands and so, this is likely to be the largest source of error in cell division processes. Secondary chemical reactions occur within a period of about one microsecond, while enzyme repair requires a longer period of more than one second.[58] Therefore, at low irradiation rates, the repair mechanism remains successful. However, the efficiency of enzymatic repair determines the radiation-induced spontaneous rate. Therefore, if radiation increases, at a certain rate the enzymes are no longer able to cope with the amount of repair required. Thus, the mutation rate increases proportionally with the dose.

Enzymes can repair DNA damage, albeit sometimes inadequately, as long as the genetic information is preserved, while mutations are DNA damage resulting in changes to the original DNA sequence. In these situations, enzymes have failed to maintain the information necessary for a reliable replication process and so the mutation is retained – as a change in genetic information – in the germ cells and used for subsequent cell divisions. In this process, the number of mutant cells will either increase if cell function improves after the mutation or decrease if cell survival is reduced. In most cases, the mutation is harmful but in some cases, it improves or expands cell possibilities thus providing an important step in the evolution of the biological species.

This shows that, in addition to random errors in DNA replication, radiation-induced mutations may play a not insignificant role and thus may have contributed to the evolution of life, especially in the early Hadean and Archean phases of our planet.[59]

With this, interesting vistas are opened for considering evolutionary periods when earth, together with her living creatures, were exposed to an enhanced dose of radiation. Such exposure should have increased the general mutation rate, including mutations leading to eventual cell death and extinction of a species but also to positive mutations taking the evolution of life forward a step.

These changes could be proven by the detailed study of mutation rates in the Chernobyl area, which documented variations in many animal and plant species.[60] Through meta-analysis, the study compared the published data sets of several independent analyses, which considered biological materials and examined them for genetic variations. According to the results, mutation rates in general significantly increased for plants, animals, and humans. More than 43% of the observed variations can be attributed to increased radiation exposure. However, the magnitude of the increase is not specified and varies among species. It must also be emphasized that this is not an analysis of the nature and consequence of such mutations. Statements on long-term effects are not yet possible.

7.1.7 RADIOACTIVE CATASTROPHES IN EARTH'S HISTORY

Although it is too early to draw conclusions about the long-term consequences of the increased mutation rate in the wider environment of Chernobyl, results reinforce current understanding of radioactivity's influence on biological evolution, which was already being discussed in the 1950s. This includes historical periods when the earth was exposed to high radiation flux. However, only very limited information is available. By means of geological-paleontological investigation, one can identify several periods of species extinction that today are generally attributed to the impact of larger asteroids.[61]

Mass extinctions of terrestrial life have been described several times with an alternative thesis involving possible supernova explosions in the vicinity of the solar system (cf. Chapter 5.2). These explosions signal the violent end of a massive star whose core collapsed under its own weight, causing the whole system to explode[62] and the outer shell of the star to be hurled into space. Nuclear reactions that take place during such an event produce a considerable amount of long-lived radioactive elements, which are partly ejected in the course of the explosion (cf. Chapter 5); today, these radioactive isotopes can be directly identified by their characteristic radiation. In addition, an enormous amount of radiation energy is released as a consequence of the explosion, ranging from neutrinos to high-energy X-rays and γ-radiation.

According to astronomical observations, there is on average one supernova per century in our Milky Way. However, the actual value is probably higher since large areas of the Milky Way cannot be observed directly because dark clouds of interstellar dust absorb light and other information carriers. This includes especially information from the center of our galaxy, which is considered to be particularly active and rich in stars and stellar explosions. But even if one assumes the observed supernova rate, there have been at least 45 million supernovae in our Milky Way since the formation of the solar system.

Many astronomers even believe that our solar system is the direct result of a supernova explosion. Originating in a region of high stellar density, our solar system may have been hurled into its present orbit by the shock front of an explosion. This thesis is supported by the trace amounts of heavy elements found in the solar system's oldest meteorite material, which must have been formed by explosive nucleosynthesis during the explosion.

Our solar system has been rotating around its galactic center at an average speed of 230 km/s since its formation. In addition, it still seems to move through the local environment with an intrinsic velocity of 15 km/s. This is often interpreted as an effect of neutrino or mass flow caused by an original supernova explosion. French-Canadian astrophysicist Hubert Reeves (b. 1932) has called this the Bing Bang – a designation for the origin of our solar system that refers to the Big Bang hypothesis; Reeves still finds supporters for this theory today. It is quite possible that our solar system passed through clouds of radioactive material, as remnants of stellar explosions, several times on its orbit. This would be reflected by periods of increased radioactivity, which would be difficult to determine today since the residual activity would have mostly subsided or could no longer be filtered out of earth's radioactive subsurface material. [63]

Interestingly, increased amounts of radioactive ^{60}Fe ($T_{1/2} = 2.6 \cdot 10^6$ y) have been detected by test drilling in the manganese and iron-rich deposits at the bottom of the Pacific Ocean.[64] The deposits are the product of hydrothermal vents, which bubble material from the deeper layers of the lithosphere and asthenosphere to the ocean floor. These sedimentation areas are located along the oceans' rift zones and are part of the lithosphere's slow convection process (cf. Chapter 6.1.4). The deposits are of enormous commercial value, which triggered the move to ocean floor harvesting.[65] The drill samples were obtained through efforts to analyze the depth and history of the deposition process.

The observation of radioactive ^{60}Fe deposited in a 2 mm thin sediment layer, 6 mm inside the deposition, has been explained as the consequence of Earth passing through the expanding remnant of a near-by supernova about 2.6 million years ago, thus picking up radioactive dust.[66] Presupposing a constant sedimentation rate, time results have been derived from the depth at which the ^{60}Fe-containing layer was found in the sediment. From the thickness of the layer containing the ^{60}Fe deposition, a time of several 100,000 years was calculated during which Earth passed through the radioactive neighborhood.

Figure 7.11 shows the extent of the ^{60}Fe isotope deposition as a function of depth along the drill core, as observed in two independent studies. The figure also shows the translation of the data into the intensity of deposition as a function of time. Since the event happened about 2.6 million years ago, which corresponds to one half-life of ^{60}Fe, only 50% of the originally deposited amount is left; the rest has decayed to the stable ^{60}Ni isotope.

Radioactive ^{60}Fe is produced in considerable abundances by neutron-capture processes in supernova explosions. While this claim was initially based on the theoretical predictions of nucleosynthesis patterns in supernova events, the spectroscopy of characteristic radiation with ESA's INTEGRAL gamma-ray satellite confirms the prediction by showing a clear increase in ^{60}Fe-induced radiation for supernova-cloud localities along the Galactic plane (Figure 7.12).[67]

If our solar system passed through the radioactive cloud of a supernova explosion, Earth would have been exposed to higher levels of radiation for a period of up to 500,000 years (Figure 7.13). In addition, Earth may have also absorbed larger amounts of radioactive materials through dust deposition. Theoretical work calculating Earth's radiation exposure from such an event indicates that, in order to have had the observed effect, the distance to the supernova itself must have been between 10 and 100 parsecs.[68] At greater distances, the radiation effects would have been negligible; at shorter distances, effects would have been lethal to biological organisms due to the enormous radiation exposure. So, distances in between suggest the observable effects, which might be reflected in biological proxy records.

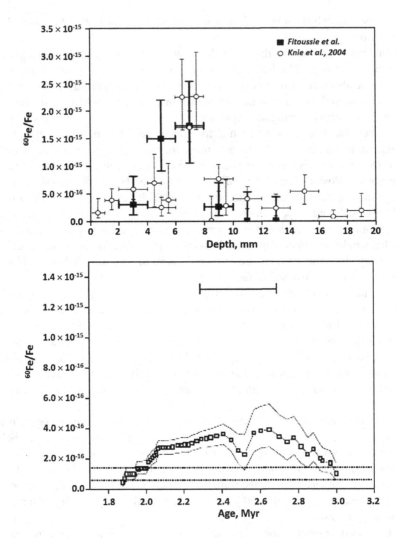

FIGURE 7.11   The upper part of the figure shows the extent of ^{60}Fe deposition as measured by accelerator mass-spectrometry versus total iron as a function of ocean sediment-layer thickness. Source: Knie, et al., *Phys. Rev. Lett.* 93 (2004): 171103. The lower part of the figure shows the same data plotted as a function of time to show the total duration of the irradiation period. The dashed lines characterize the sensitivity limit of the measurement. Source: Fitoussie, et al., *Phys. Rev. Lett.* 101 (2008): 121101.

Radiation flux at Earth's surface does, of course, strongly depend on the absorbing effect of the atmosphere. Generally, a value of 99% is assumed; thus, only 1% of the radiation in the exosphere of the planet would have reached Earth's surface. However, enhanced radiation intensity in the stratosphere can destroy the ozone layer by breaking the O_3 molecule into O_2 and a single oxygen atom. A destroyed ozone layer would significantly reduce absorption of higher-energy UV radiation and subsequently multiply radiation exposure.

Assuming an atmospheric absorption probability of 90%, Figure 7.13 indicates that a distance of less than 35 parsecs would be lethal to life on Earth's surface. At distances between 35 and 100 parsecs, one can expect a radiation dose between 10 and 1 gray, resulting in greatly increased mutation rates in biological systems. This could lead to longer-period extinction in the case of negative mutations and to the improvement of genetic information as well as new evolutionary thrusts in the

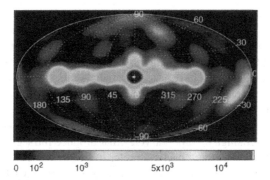

FIGURE 7.12 Distribution of the radioactive ^{60}Fe abundance along the galactic plane. Besides the high γ-flux in the Galactic Center, increased intensity can also be seen in the Cygnus and Vela regions, which are considered to be sites of earlier supernova explosions. Source: W. Wang, et al., "SPI observations of the diffuse ^{60}Fe emission in the Galaxy," *Astron. & Astrophys.* 469 (2007): 1005–1012.

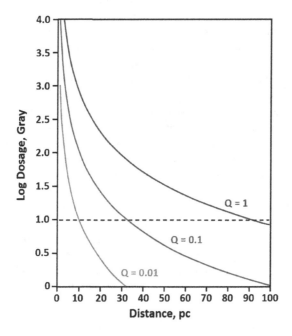

FIGURE 7.13 The radiation dose (in logarithmic representation) to which higher biological life would be exposed during a typical supernova explosion with a luminosity of 1,038 J/s at up to 100 parsecs distance. The dose curve has been calculated for three different shielding factors Q (1, 10, 100 percent) of Earth's atmosphere. At Q = 1, radiation strikes Earth's surface unimpeded; at Q = 0.01, 99% of the radiation is absorbed in the atmosphere. A lethal dose is assumed to be 10 gray. Source: M. Beech, "The past, present, and future supernova threat to Earth's biosphere," *Astrophys. Space Sci.* 336 (2011): 287.

case of positive mutations. At a higher 99% atmospheric absorption, effects would be limited if the supernovae were to explode within 30 parsecs of Earth.

Comparing Earth's evolutionary history with the Sun's path through the Milky Way would reveal whether our planetary system entered the radiation range of earlier stellar explosions and also whether radiation exposure can be correlated with extinction or evolutionary thrusts.

Based on various measurements of the [60]Fe content and its distribution in the deeper layer of the drill sample (Figure 7.14), it appears that Earth entered the supernova cloud about 2.8 million years ago. This corresponds to the end of the Pliocene period and the beginning of the Quaternary, during the Earth age that played a central role in human evolution (homonization). Whether the increased radiation dose, which must have lasted through the first 500,000 years of the Quaternary (the Pleistocene), played a role in the [60]Fe content and distribution considered here may be left open since speculations in this direction presuppose more exact proof of a supernova cloud's existence.

To this end, a key role is played by work on the fossils of magnetotactic bacteria (Figure 7.14). These are a certain class of single-cell prokaryotes that produce intracellular chains of magnetite (Fe_3O_4) nanocrystals. Still today, these bacteria prefer to live below the surface in shallow waters since, due to their special structure, iron and iron-containing dust particles are among their preferred food. They would have, therefore, quickly digested cosmic iron dust. The bacteria's magnetite crystals institute a cellular dipole moment, aligning the cell in the geomagnetic field and helping it to navigate through the water along stratified sediment layers with varying iron content.

The bacteria typically populate layers below the sediment–water interface. This forces the bacteria populations to move upward with the growth of the sediment column. Dead cells are being left behind and remain embedded as magnetofossils within the sediment bulk. Researchers had hoped to use magnetotactic fossil enrichments in order to measure possible [60]Fe isotope enrichments from the early Quaternary period and also to gain more precise understanding of supernova effects.

Indeed, [60]Fe enhancement was found in magnetofossils that were chemically extracted from two Pacific Ocean sediment drill cores.[69] Material samples were taken from the different sediment layers and analyzed for their [60]Fe content using accelerator mass spectroscopy (AMS) methods. AMS is a very powerful tool for identifying spurious amounts of radioactive materials by mass and charge separation in tandem accelerator systems as will be discussed in more detail in Chapter 16 of Volume 2. The [60]Fe signal shows a maximum at a depth that would correlate with a 2.2-million-year-old sediment deposition. This is near the lower Pleistocene boundary, which terminates around 1.7 million years ago and vanishes at about 2.8 million years ago. These time limits are in good agreement with earlier rock analysis. During this time, the solar system entered the supernova remnant cloud.

According to rough estimates, our solar system enters the 30-parsec zone of a stellar explosion every 20 million years, during its long journey through the spiral arms of our galaxy.[70] Every 500 million years, one expects a nearby – only 10 parsec – passing to the remnant of a stellar explosion. In these cases, radiation levels at the earth's surface may increase at a rate that threatens extinction or at least extreme changes in biological life forms. This rate is likely to increase further as the solar system passes through spiral arms or nebulae with high stellar density. As mentioned

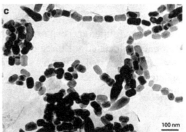

FIGURE 7.14   On the left, a sample of manganese-rich deep-sea sediment from the Pacific Ocean, in which an enrichment of the radioactive [60]Fe isotope was detected and measured at a certain depth. For size comparison, a coin is shown. Source: Knie, et al., *Phys. Rev. Lett.* 93 (2004): 171103; Fitoussie, et al., *Phys. Rev. Lett.* 101 (2008): 121101. On the right, an example of unicellular magnetotactic bacteria. Source: Peter Ludwig, Shawn Bishop, Ramon Egli, et al. *PNAS* 113 (33), (2016) 9232–9237.

earlier, several extinction periods are known but most are currently attributed to asteroid impact. Based on observations, an abrupt decline in biological life appears to occur every 26 to 30 million years (Figure 7.15).[71] Whether this is a random rate, that would for example correlate with expected asteroid impact, or whether this occurs with periodicity is currently the subject of scientific debate.

About 2.6 million years ago, the transition from the Pliocene to the Pleistocene age marked a period of more than 2 million years. This time is characterized by the extinction of marine species and temperatures appear to have dropped as the Arctic polar cap began to freeze over. Until now, these phenomena were thought to be due to altered ocean currents, as a result of the closure of the Panama Isthmus by continental drift. However, during this period, Earth was also within 50 to 100 parsecs from a group of massive hot stars known as the Scorpius-Centaurus association, the best-known member of which is the giant star Antares. These massive stars undergo rapid evolution to supernova and are typically expected to reach a supernova stage within a million years. Such an evolution corresponds to the time period when our solar system was in the vicinity of this stellar association. In fact, several supernova explosions are known to have occurred in this asterism within the last 15 million years, which is considered a region of frequent star formation (Figure 7.16).

One of these supernovae in the Scorpius-Centaurus association may be responsible for triggering the radioactivity shower on Earth that led to the rapid breakup of the protective ozone layer.[72] The accompanying increase in UV radiation destroyed plankton, so that higher marine animals lost their food source, which could have led to extinctions. These symptoms resemble the relatively rapid mass extinction at the end of the Pliocene. Contrary to previous theories involving a change in ocean currents, an increased radiation load due to a supernova could account for these events. Embedded in supernova dust, radioactive ^{60}Fe could have played a secondary role as it was probably deposited on Earth as sediment, when the solar system passed through the cloud. If this was the case, it is quite possible that increased radiation exposure on the earth's surface led to an increased mutation rate, causing some species to become extinct while other life forms moved evolutionarily forward. All of this is speculation. More precise measurements of magnetotactic bacteria could provide more accurate information about the effects on biological life.

Nonetheless, what has already taken place during Earth's journey through space may happen again. Our Sun, with its planets, is currently moving through what is known as the Local Bubble – a relatively dust-free zone with few neighboring stars; this so-called bubble has a diameter of about 150 parsec, which has been carved out by a succession of several supernovae in the Scorpius–Centaurus OB star association.[73]

However, predictions see the solar system heading toward denser gas zones in areas of higher stellar density.[74] This could confront our solar system with the effects of a nearby supernova or other stellar explosions. For such predictions, one must be able to relatively reliably calculate the future route of our solar system as compared with other star systems, which are also moving. Along the expected route, there are indeed some candidates whose violent deaths as supernova are expected within the next million years.

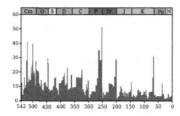

FIGURE 7.15 Periods of mass extinction over the last 550 million years of Earth's history. Plotted is the extinction of biological life as a percentage against time. The largest extinction rates result for five major events, at 66, 200, 250, 375, and 450 million years ago. Source: https://en.wikipedia.org/wiki/File:Extinction_intensity.svg (March 14, 2024).

FIGURE 7.16 This image from the NASA/ESA/CSA James Webb Space Telescope displays the birth of a star. The subject is the Rho Ophiuchi cloud complex in the Scorpius-Centaurus association, the closest star-forming region to Earth. It is a relatively small, quiet stellar nursery, but you'd never know it from Webb's chaotic close-up. Jets bursting from young stars crisscross the image, impacting the surrounding interstellar gas and lighting up molecular hydrogen, shown in red. Some stars display the telltale shadow of a circumstellar disc, the makings of future planetary systems like the one of our Sun. Crossing this region of space may have exposed Earth to enhanced levels of radioactivity. Source: https://esawebb.org/images/weic2316a/ (November 2, 2023).

The most prominent candidate is Betelgeuse in the constellation Orion (Figure 7.17), a red giant that would encompass the asteroid belt if applied to our planetary system.[75] Estimates for the mass of the star range from 10 to 20 solar masses – an uncertainty that translates directly to uncertainty regarding the exact onset of the expected supernova explosion. According to theory, Betelgeuse has already passed the stellar helium-burning phase and, as a result, has a core composed primarily of carbon ^{12}C and oxygen ^{16}O. These two components fuse together, producing the energy that currently stabilizes the star. Observed instabilities are likely characteristic of a helium-burning shell around the stellar core.

The star's current consumption rate of ^{12}C and ^{16}O fuel depends on internal density and temperature conditions, i.e., the mass of the star as well as fusion rates. None of these parameters are known reliably enough to make accurate predictions about the expected supernova occurrence. Current estimates for Betelgeuse predict a remaining lifetime to be between 100,000 and 1 million years. During this burning period, a star runs through the neon, oxygen, and silicon-burning phases in rapid succession before collapsing and igniting a supernova (see Chapter 5). What

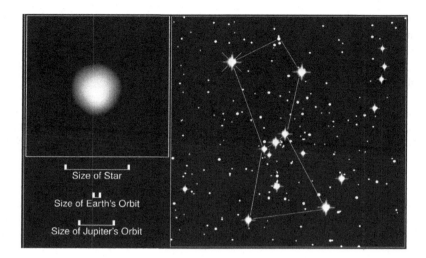

FIGURE 7.17   Image of the red-giant star Betelgeuse in a Hubble telescope image. Betelgeuse is the eighth brightest star in the winter sky of the Northern Hemisphere and the brightest star in Orion. It marks the left shoulder. Source: https://nssdc.gsfc.nasa.gov/image/astro/hst_betelgeuse.jpg (March 14, 2024)

remains is a hot neutron star and a rapidly, radially expanding radioactive dust cloud of ejected matter.

Betelgeuse is currently about 200 parsecs (about 700 lightyears) away from our solar system and is moving away at a speed of 33 km/s. The closest approach, of about 110 parsec between the two systems, occurred about 3.7 million years ago. Thus there is no threat of a radiation catastrophe because, in the period of the expected explosion, the distance of about 300 parsecs is too large to cause problematic radiation exposure. Nevertheless, the explosion of such a nearby star will be a spectacular event in the night sky.

However, Betelgeuse is not the only star whose end as a near-Earth supernova poses a potential threat. Astronomers have identified a number of other candidates[76] that could potentially threaten our biosphere. These include not only supernovae but also, for example, thermonuclear explosions. Such explosions are triggered by accretion processes in binary star systems; these so-called kilo-novae are formed by the collision and merger of neutron stars. Based on calculations, none of these events appear likely to occur along the predicted route of our solar system, at least within the fore-seeable future. The explosions predicted with certainty are all relatively far from the orbit of our planetary system, so that the radiative death of biological organisms by external influences seems rather unlikely for the foreseeable period.

7.1.8   Radioactivity as Part of Our Existence

Radioactivity is a natural part of our human existence. Radioactive elements are part of the first complex molecules. The energy released during their radioactive decay may have fueled the first chemical processes, building amino acids and other biochemical molecules. Radioactive elements are then part of the interstellar material from which our planetary system has formed. Following physical and chemical laws, this material has been distributed inside the earth and in the earth's crust. We can therefore observe remnants of this early radioactivity today both in the environment of bubbling deep-sea springs and on the surface of ice-covered comets, planets, and rock fragments in the asteroid belt.

On a cellular level, in addition to the random errors that occur in DNA copying, radioactive decay provides conditions that facilitate mutation rates in DNA and so contributes significantly to cell evolution. On the other hand, cells have evolved enzymatic processes that largely repair radiation-induced cell damage. There is thereby a balance attained through the mechanism of self-organization. This is also true for radioactivity as part of our natural environment.

Radioactivity has played an important role in the history of our planet and the development of its complex biological organisms. One cannot abolish, deny, or fight the existence of radioactivity, as James Lovelock recognized early on:

> I have never regarded nuclear radiation or nuclear power as anything other than a normal and inevitable part of the environment. Our prokaryotic forebears evolved on a planet-sized lump of fallout from a star-sized nuclear explosion, a supernova that synthesized the elements that go to make our planet and ourselves.[77]

Nevertheless, like all states of equilibrium in nature and society, this equilibrium can also be disturbed. While, in the course of the disruption, consequences are often dramatic and radical, new states of equilibrium do develop. Yet, questions must be addressed: When is the equilibrium between mutation and repair disturbed? At what level of radiation can enzymes no longer make the necessary repairs?

Below a threshold value, radiation has little effect on somatic or other genetic processes, thanks to cellular enzymes and their repair effects. Yet, where this threshold lies is not yet known with sufficient precision; in fact, whether there is a threshold at all is often disputed. The complex radiation-related processes are still too poorly understood insofar as they occur at the microbiological level. Therefore, radiology does not apply a threshold value and the linear no-threshold hypothesis (LNT) is being upheld.

According to the LNT, the probability of radiation damage occurring depends linearly on the radiation dose in sievert units such that, the greater the dose, the greater the probability of a radiation-induced effect. Since there is no threshold for low-dose levels according to this interpretation, radiation damage can occur even at extremely low doses. The LNT hypothesis is a model concept that dictates our current behavior toward radiation. Countless measurements undoubtedly confirm the linear dependence in the range of larger dose values (cf. Chapter 4) but at low-dose values, large uncertainties characterize or even negate the validity of the LNT assumption.[78]

In this regard, important information comes from epidemiological studies in which specific population groups – with firmly defined and, if possible, known natural or occupational radiation exposure – are examined, in order to document possible long-term somatic or genetic consequences. The analysis of such studies requires considerable statistical effort since only large groups with many participants can statistically minimize the influence of individual effects. Radiation conditions, such as intensity and duration, must be well known in order to make statistically relevant statements about radiation consequences and effects. Therefore, Chapter 8 is devoted to discussing a number of epidemiological studies that investigate the effects of low-level irradiation.

NOTES

1. The transfer factors for various radioisotopes in plants are listed by the International Atomic Energy Agency. IAEA, *Handbook of Parameter Values for the Prediction of Radionuclide Transfer in temperate Environments* Technical Report Series No. 364 (IAEA Austria, 1994).
2. David Farrier, *Footprints – In Search of Future Fossils* (New York: Farrar, Straus and Giroux, 2020), 184–185.
3. To be precise, at present $N(^{12}C) = 1.3 \cdot 10^{12} \, N(^{14}C)$, a relation that depends on cosmic-ray intensity as discussed in Chapter 6.
4. UNSCEAR 1982 Report, *Ionizing Radiation: Sources and Biological Effects*, United Nations Scientific Committee on the Effects of Atomic Radiation (1982).
5. Chemical fractionation refers to – often physical – processes in which isotopes of the same element are separated from each other based on their size and mass. For example, when it rains, the somewhat more massive water molecule $H_2^{18}O$ falls to earth faster than the lighter but much more abundant molecule $H_2^{16}O$ – a fact that is used in climate research and also in anthropology. Another example is that ^{235}U atoms

diffuse more easily through fine membranes than the larger ^{238}U atom and so are used for ^{235}U enrichment by the nuclear power industry. During digestion of food material, chemical fractionation might be caused by diffusion through thin membranes in the digestive system. C.f. Rainer Konietzka, "Gastrointestinal absorption of uranium compounds," *Regulatory Toxicology and Pharmacology* 71 (2015): 125–133.

6. https://medlineplus.gov/potassium.html (April 21, 2023).

7. https://www.atsdr.cdc.gov/csem/uranium/docs/uranium.pdf (July 22, 2023).

8. Strict EPA rules guide the amount of radioactive material in drinking water and public water systems in the United States: https://www.epa.gov/dwreginfo/radionuclides-rule (April 21, 2023). These rules are defined by the LNT requirements defined by BEIR: https://www.ncbi.nlm.nih.gov/books/NBK234160/ (April 21, 2023).

9. A. Martín Sánchez, M. P. Rubio Montero, V. Gómez Escobar, and M. Jurado Vargas, "Radioactivity in bottled mineral waters," *Applied Radiation and Isotopes* 50 (6), (1999): 1049–1055.

10. Billionaires, investing fortunes in gen-technology in the vain hope of extending their natural lifespan, should think about this statement.

11. This is remarkable in itself since in principle there exist hundreds of different amino acids. To interpret the fact that all biological systems are based on only 20 amino-acid types, it is understood that the construction of the first cells occurred from just these amino acids and that these cells then formed the foundation of life through evolution, while other amino acids remain excluded.

12. " ... ob überhaupt das Leben je entstanden, ob es nicht ebenso alt wie die Materie sei, und ob nicht seine Keime, von einem Weltkörper zum andern herübergetragen, sich überall entwickelt hätten, wo sie günstigen Boden gefunden?" in the preface to William Thomson, *Handbuch der theoretischen Physik*, German edition, part 2.

13. Letter from Darwin to Hooker on February 1, 1871, https://www.darwinproject.ac.uk/letter/DCP-LETT -7471.xml (August 4, 2023).

14. C. Ponnamperuma, C. Sagan, and R. Mariner, "Synthesis of adenosine triphosphate under possible primitive earth conditions," *Nature* 199 (1963): 222–226.

15. Joseph L. Kirschvink, Eric J. Gaidos, L. Elizabeth Bertani, et al., "Paleoproterozoic snowball Earth: Extreme climatic and geochemical global change and its biological consequences," *PNAS* 97 (4), (2000): 1400–1405.

16. Charles S. Cockell, *Astrobiology: Understanding Life in the Universe*, 2nd edition (Oxford: Wiley-Blackwell, 2020).

17. J. B. Corliss, J. A. Baross, and S. E. Hoffman, "An hypothesis concerning the relationship between submarine hot springs and the origin of life on Earth," Proceedings 26th International Geological Congress, Geology of oceans symposium, Paris, July 7–17, 1980, 59–69.

18. M. Weiss, et al., "The physiology and habitat of the last universal common ancestor," *Nature Microbiology* 1 (2016): 16116 and N. Arndt and E. Nisbet, "Processes on the young Earth and the habitats of early life," *Annual Review of Earth and Planetary Sciences* 40 (2012): 521–549.

19. Robert C. Brady III and R. Pettit, "Mechanism of the Fischer-Tropsch reaction," *Journal of the American Chemical Society* 103, 5 (1981): 1287–1289.

20. B. Sherwood Lollar, V. B. Heuer, J. McDermott, S. Tille, et al., "A window into the abiotic carbon cycle – Acetate and formate in fracture waters in 2.7-billion-year-old host rocks of the Canadian Shield," *Geochimica et Cosmochimica Acta* 294 (2021): 295–314.

21. https://www.quantamagazine.org/radioactivity-may-fuel-life-deep-underground-and-inside-other -worlds-20210524/ (July 24, 2023).

22. David E. Bryant, et al., "Hydrothermal modification of the Sikhote-Alin iron meteorite under low pH geothermal environments. A plausibly prebiotic route to activated phosphorus on the early Earth," *Geochimica et Cosmochimica Acta* 109 (2013): 90–112.

23. A very readable summary of observational data and arguments for the panspermia hypothesis can be found in a book by former Hoyle collaborator, Chandra Wickramasinghe (b. 1939): *The Search for Our Cosmic Ancestry* (Singapore: World Scientific Publishing Inc., 2015).

24. It should be noted that Fred Hoyle was the leading proponent of steady-state theory, the cosmology of a constant universe whose observed expansion he sought to explain by the constant production of new matter at the center. Hoyle was known for his unconventional but often successful thoughts. However, he remained an avowed opponent of the Big Bang theory until his death. Thus, within the framework of the steady-state theory, life could have had a much longer, if not infinite evolutionary time.

25. *Deinococcus radiodurans*, https://en.wikipedia.org/wiki/Deinococcus radiodurans (July 27, 2023), is not the only example for radiation-resistant life forms. We will discuss the specific evolutionary features of radiation-resistant biological systems at a later point.

26. Neil Fine, ed., *The Search for Life in the Universe* (New York: Time Books, 2015).

27. The Precambrian era covers the period from 4.5 billion years ago to about 550 million years ago, from the origin of Earth to the rise of oxygen in the atmosphere.

28. John Tyler Bonner, *Randomness in Evolution* (Princeton: Princeton University Press, 2013).

29. For example, Earth itself is an open system since it receives external energy, through solar radiation, and is in exchange with other systems through accretion of interstellar dust. The British natural scientist James Lovelock (b. 1919) even goes a step further with his so-called Gaia hypothesis by considering Earth to be a self-organizing and self-regulating system.

30. V. Ouazan-Reboul, J. Agudo-Canalejo, and R. Golestanian, "Self-organization of primitive metabolic cycles due to non-reciprocal interactions," *Nature Communications* 14 (2023): 4496.

31. Prokaryotes differ from the later appearing eukaryotes in that they have neither nucleus nor mitochondria. Their DNA is therefore freely located in the cytoplasm. The DNA of prokaryotes consists of a single double-stranded, densely structured, self-contained molecule. Eukaryotes, on the other hand, are much larger and are characterized by a solid nucleus enclosed in a membrane. Most known multicellular organisms are based on eukaryotes, including plants, animals, and humans.

32. Laura Eme, Daniel Tamarit, and Eva F. Caceres, et al., "Inference and reconstruction of the heimdallarchaeial ancestry of eukaryotes," *Nature* 618 (2023): 992–999.

33. For an easy-to-understand description of the evolution, role, and importance of the oxygenic atmosphere in the development of biological life, see Donald E. Canfield, *Oxygen: A Four Billion Year History* (Princeton: Princeton University Press, 2014).

34. Cf. Chapter 4: Radiation-induced primary reactions release aggressive oxygen radicals that cause secondary radiation effects.

35. P. A. Karam and S. A. Leslie, "The evolution of the Earth's background radiation level over geological time," *International Congress on Radiation Protection* 9 (2), (1996): 238–240 and P. A. Karam and S. A. Leslie, "Calculations of background beta and gamma radiation levels over geologic time," *Health Physics* 77 (1999): 662–667.

36. P. A. Karam, "Inconstant Sun: How solar evolution has affected cosmic and ultraviolet radiation exposure over the history of life on Earth," *Health Physics* 84 (2003): 322–333.

37. P. A. Karam, S. A. Leslie, and A. Anbar, "The effects of changing atmospheric oxygen concentrations and background radiation levels on radiogenic DNA damage rates," *Health Physics* 81 (2001): 545–553.

38. https://www.nasa.gov/ames/research/space-biosciences/foton-m3 (April 22, 2023).

39. Lorena Rebecchi, Tiziana Altiero, Roberto Guidetti, Michele Cesari, Roberto Bertolani, Manuela Negroni, and Angela M. Rizzo, "Tardigrade resistance to space effects: First results of experiments on the LIFE-TARSE mission on FOTON-M3 (September 2007)," *Astrobiology* 9 (6), (2009): 581–591.

40. Anita Krisko and Miroslav Radman, "Biology of extreme radiation resistance: The way of Deinococcus radiodurans," *Cold Spring Harbor Perspectives in Biology* 5 (2013): a012765.

41. M. J. Daly, "A new perspective on radiation resistance based on *Deinococcus radiodurans*," *Nature Reviews Microbiology* 7 (2009): 237–245.

42. Carolina Chavez, Grisel Cruz-Becerra, Jia Fei, George A. Kassavetis, and James T. Kadonaga, "The tardigrade damage suppressor protein binds to nucleosomes and protects DNA from hydroxyl radicals," *eLife* 8 (2019): e47682.

43. T. Hashimoto, D. D. Horikawa, Y. Saito, H. Kuwahara, H. Kozuka-Hata, T. Shin-I, et al., "Extremotolerant tardigrade genome and improved radiotolerance of human cultured cells by tardigrade-unique protein," *Nature Communications* 7 (2016): 12808.

44. K. Ingemar Jönsson, "Radiation tolerance in Tardigrades: Current knowledge and potential applications in medicine," *Cancers (Basel)* 11 (9), (2019): 1333.

45. https://www.scmp.com/news/china/science/article/3215286/chinese-team-behind-extreme-animal-gene-experiment-says-it-may-lead-super-soldiers-who-survive (July 29, 2023).

46. For a recent example of enzymatic repair mechanisms as DNA damage response, see: A. Moretton, S. Kourtis, A. Gañez Zapater, et al., "A metabolic map of the DNA damage response identifies PRDX1 in the control of nuclear ROS scavenging and aspartate availability," *Molecular Systems Biology* 19 (2023): e11267.

47. R. E. Mitchel, "The Bystander effect: Recent developments and implications for understanding the dose response," *Nonlinearity in Biology, Toxicology, and Medicine* 2 (3), (2004): 173–183.

48. M. Sokolov and R. Neumann, "Changes in gene expression as one of the key mechanisms involved in radiationinduced bystander effect," *Biomedical Reports* 9 (2018): 99–111.

49. W. F. Morgan, "Non-targeted and delayed effects of exposure to Ionizing radiation: II. Radiation induced genomic instability and bystander effects in vivo, clastogenic factors and transgenerational effects," *Radiation Research* 159 (2003): 581.

50. These observations apply not only to radiation-induced mutations but also to mutagenic chemicals that can lead to DNA damage.

51. Jared C. Roach, et al., "Analysis of genetic inheritance in a family quartet by whole-genome sequencing," *Science* 328 (2010): 636–639. Before this study, a slightly higher mutation rate of 2.5•10^8 per nucleotide and generation was considered standard, cf. Michael W. Nachman and Susan L. Crowell, "Estimate of the mutation rate per nucleotide in humans," *Genetics* 156 (2000): 297–394.

52. A. R. Olson and G. N. Lewis, "Natural reactivity and the origin of species," *Nature* 121 (1928): 673–674.

53. H. J. Muller and L. M. Mott-Smith, "Evidence that natural radioactivity is inadequate to explain the frequency of "Natural' mutations," *Proceedings of the National Academy of Sciences* 16 (4), (1930): 277–285.

54. Since the annual natural-radiation exposure of Drosophila corresponds to about 10 pico-Sv, this interpretation would establish an annual threshold of about 20 nano-Sv, below which no increased mutations could be observed. Scaled to humans, this would be more than 5 Sv per year, a value that contradicts all recent empirical values.

55. Tautomers are molecular states that change into each other so quickly, through the exchange of individual atoms or groups of atoms, that they are in chemical equilibrium with each other. This means that they often cannot be distinguished or isolated from each other.

56. In the prokaryotic unicellular organism *E. coli*, cell replication occurs at a rate of 1,000 nucleotides per second. By comparison, replication occurs in human DNA eukaryotic cells at a rate of 50 nucleotides per second.

57. This again brings up the question of hormesis – whether, below a certain radiation dose, enzymatic-repair processes have sufficient capacity to effectively perform necessary repairs.

58. H. Nikjoo, "Radiation track and DNA damage," *Iranian Journal of Radiation Research* 1 (2003): 3–16.

59. In addition to natural radiation, other chemically induced changes in base structure also contribute to the mutation rate, but these will not be considered here. For a detailed discussion of these effects, the reader is referred to other resource books including, A. J. F. Griffith, J. H. Miller, and D. T. Suzuki, *Introduction to Genetic Analysis* (New York: W. H. Freeman, 2000).

60. A. P. Møller and T. A. Mousseau, "Strong effects of ionizing radiation from Chernobyl on mutation rates," *Nature, Scientific Reports* 5, 8363 (2015).

61. The best-known example is the Chicxulub asteroid, which slammed into the sea 66 million years ago near the Yucatan Peninsula in Central America. Dust thrown into the atmosphere, as well as ash from huge wildfires, triggered a period of darkness and cold that lasted for years and led to the extinction of most dinosaur species.

62. This event is characterized as a Type II supernova. Other types of supernovae have different explosion mechanisms, such as Type Ia (read as "Type one-A") supernovae which are based on an exploding binary star system.

63. However, recent analysis of meteoritic material from the early phase indicates a distinct lack of radioactive ^{60}Fe, which suggests that it is less likely that a supernova was the trigger for the formation of our solar system. Reto Trappitsch, Thomas Stephan, and Andrew M. Davis, "The curious case of ^{60}Fe in the early solar system," 2021 Nuclei in the Cosmos, Conference Proceedings, 2022.

64. K. Knie, G. Korschinek, T. Faestermann, E. A. Dorfi, G. Rugel, and A. Wallner, "^{60}Fe anomaly in a deep-sea Manganese crust and implications for a nearby supernova source," *Physical Review Letters* 93 (2004): 171103; and C. Fitoussi, G. M. Raisbeck, K. Knie, G. Korschinek, T. Faestermann, S. Goriely, D. Lunney, M. Poutivtsev, G. Rugel, C. Waelbroeck, and A. Wallner, "Search for supernova-produced ^{60}Fe in a marine sediment," *Physical Review Letters* 101 (2008): 121101.

65. Kevin W. Mandernack and Bradley M. Tebo, "Manganese scavenging and oxidation at hydrothermal vents and in vent plumes," *Geochimica et Cosmochimica Acta* 57 (1993): 3907–3923.

66. Accurate studies of the annual deposition rate of cosmic dust and meteorite material on the Earth's surface have calculated about 40,000 tons per year. This amount may increase substantially if the Earth moves from its present position, in the so-called local bubble, into gas clouds of greater dust and particle density (S. G. Love and D. E. Brownlee, "A direct measurement of the terrestrial mass accretion rate of cosmic dust," *Science* 262 (1993): 550).

67. W. Wang, M. J. Harris, R. Diehl, H. Halloin, et al., "SPI observations of the diffuse ^{60}Fe emission in the galaxy," *Astronomy & Astrophysics* 469 (2007): 1005–1012.

68. 1 Parsec = 3.26 lightyears.

69. S. Bishop and R. Egli, "Discovery prospects for a supernova signature of biogenic origin," *Icarus* 212 (2011): 960 and Peter Ludwig, Shawn Bishop, Ramon Egli, and Georg Rugel, "Time-resolved 2-million-year-old supernova activity discovered in Earth's microfossil record," *PNAS* 113 (33), (2016): 9232–9237.

70. A. L. Melott and B. C. Thomas, "Astrophysical ionizing radiation and Earth: A brief review and census of intermittent intense sources," *Astrobiology* 11 (2011): 343.

71. It should be mentioned, however, that cosmologist and theoretical physicist Lisa Randall presents an alternative theory to account for the periodic occurrence of extinction in her book, *Dark Matter and the Dinosaurs*. Here, the gravitational influence of a disk of black matter surrounding the solar system dislodges boulders or larger rock masses from the Oort Belt, thereby posing a regular asteroid threat.

72. Narciso Benítez, Jesús Maíz-Apellániz, and Matilde Canelles, "Evidence for nearby supernova explosions," *Physical Review Letters* 8 (2002): 081101.

73. The origin of this empty zone is still debated. The general view is that it is the effect of a massive supernova explosion 10 million years ago in the Scorpius-Centaurus association, which swept space clean. However, other possibilities have not been ruled out. New NASA satellite observatories are planned to investigate: http://science.nasa.gov/science-news/science-at-nasa/2003/06jan_bubble (July 21, 2023).

74. Currently, the density of the dust cloud through which the Earth moves is about 10^{-6} dust grains/m^3. This translates into a daily deposition of about 60 tons of cosmic dust in the Earth's atmosphere, which slowly gravitates to the surface. An increase in the interstellar medium's density would translate to an increased deposition rate.

75. M. M. Dolan, G. J. Mathews, D. D. Lam, et al., "Evolutionary tracks for Betelgeuse," *The Astrophysical Journal* 819 (2016): 7, (15).

76. See Martin Beech, "The past, present and future supernova threat to Earth's biosphere," *Astrophysics and Space Science* 336 (2011): 287 and Brian J. Fry, Brian D. Fields, and John R. Ellis, "Astrophysical Shrapnel: Discriminating among near Earth stellar explosion sources of live radioactive isotopes," *Astrophysical Journal* 800 (2015): 71, as well as publications quoted in these papers.

77. James E. Lovelock, *The Ages of Gaia* (Oxford: Oxford University Press, 1989).

78. Charles L. Sanders, *Radiation Hormesis and the Linear-No-Threshold Assumption* (Berlin: Springer, 2010).

8 Radioactivity in Low Doses

8.1 RADIOACTIVITY IN LOW DOSES

After discussing the origin of radiation in both humans and our environment as well as its influence on biological systems, this chapter will address the question of risk to human health at low-level radiation doses. There is no doubt that high doses of radiation cause somatic and genetic effects that can seriously damage cell material and lead to death. The question is about the low levels of radiation to which we are all exposed. Considering the fact that this exposure level might be pretty significant and often above the legal limits as defined by the LNT (linear no-threshold) theory, it is important to explore the actual impact of low or natural radiation exposure.

The LNT theory assumes a linear correlation between dose and damage and further assumes that there is no threshold below which radiation may not be harmful. However, as mentioned in Chapter 4, other possibilities seem to be much more reasonable, particularly in view of the enzymatic repair mechanisms discussed in Chapter 7. Therefore, the critical questions remain: Is there a threshold below which radiation is not harmful? Or, does radiation at even the smallest dose carry the risk of damage and even death? This is what Helen Caldicott (b. 1938) believes. Caldicott is an Australian pediatrician, author, and internationally known opponent of nuclear power. In an interview regarding the nuclear accident at Fukushima in 2011, she noted, "Some say the doses are too low but you only need one hit on a regulatory gene in a single cell to get cancer – it's a random game of Russian roulette."[1]

Despite her somewhat dramatic presentation, which ignores or underestimates the cell's enzymatic repair capacities, questions about potential dangers at low-level radiation exposure remain.

Attempts to measure effects directly on biological organisms from fruit flies to mice have failed to bring statistically relevant results. Uncertainties in the resulting data are just too large to allow for a reliable extrapolation – from the widely accepted linear relation between dose and damage at larger exposure to effects at the very low dose range. To conclusively study low-dose effects would require an epidemiological approach in which large population groups with professional exposure to radiation are compared with control groups with the average level of exposure to radiation.

For environmental or occupational reasons, some people are exposed to higher radiation levels. These include aircrew, miners – especially uranium miners – radiologists, and workers in nuclear power plants, to name just a few. In each of these professions, elevated exposure levels are considered acceptable. However, to what extent are these exceptional values to be regarded as an acceptable risk?

These questions can only be answered by long and careful data collection with subsequent statistical analysis, which compares exposed and non-exposed groups for possible evidence of radiation damage. This is particularly necessary since radiation is not the only potential cause of cancer and genetic mutation. There are in fact a number of environmental causes, including living conditions and smoking, that are ascribed significantly higher mutagenic properties and so contribute to these conditions.

In addition, larger population groups must be consulted as part of systematic epidemiological investigations so that statistically reliable assessments can be obtained, in order to compensate for possible individual differences. Such a study must necessarily include control groups for whom the effect that is being investigated is not expected. However, this is particularly difficult in studies to determine the influence of low radiation doses because we are all constantly exposed to external and internal natural radiation in addition to mutagenic chemicals. In the final assessment, such factors combine to cause large statistical uncertainties that are often interpreted at will.

Mark Twain's classic quote, "There are lies, damn lies, and statistics,"[2] reflects how statistical data is often applied in this emotionally heated discussion on long-term radiation effects. However, it is also true that the statistical data that has been compiled so far is insufficient and sometimes contradictory, especially regarding the effects of low occupational or environmental doses. Available data does not seem to underline the imminent health impact that the population is often warned against in cases involving the voluntary or involuntary release of lower amounts of low-level radioactive materials. Again, the data is characterized by large uncertainties that allow for a broad spectrum of interpretation.

For this reason, multiple studies have not led to a convincing assessment of the impact of low-level radiation exposure on biological systems, and the question remains subject to ideological belief rather than scientific evaluation. Because the expected or projected impact of such scientific evaluation is so small, radiologists consistently fail to convincingly demonstrate the impact of low-level radiation doses. This is not, however, taken as sufficient evidence by the LNT believers. Staying "on the safe side" and maintaining regulations that are more than half a century old, without further critical review, has meant that other, different and more inclusive assessments are not available. While medical people frequently suggest a re-assessment, professional radiologists seem to oppose it. This situation leads to heated debates in the scientific literature, while the majority of the population never knows that there are questions or debate.

8.1.1   RADIATION DOSE AND EFFECT – HISTORICAL OVERVIEW

While the dangers of high radiation doses are well known and described by the linear correlation between dose and effect, the influence of low-dose radiation is largely unknown and therefore subject to interpretation. Since long-term measurements failed to yield reliable results, the low-level effects can only be understood by obtaining precise knowledge through reliable simulation of the biochemical processes, a rather hopeless task with our currently limited knowledge. To accomplish this, the possible interactions – between radiation and the affected macromolecules – would have to be known over a wide energy range. The various repair mechanisms and their efficiency would also have to be considered, in order to simulate the probability of all possible subsequent repair processes; this requires using the so-called Monte Carlo method.[3] While the mathematical methods for such analyses and calculations have been developed and are applicable in principle, the information gained currently has only very limited validity. Although much has been learned in recent decades about radiation effects on selected cell cultures and other biological carriers, our knowledge is still too incomplete to simulate more complex biological systems, such as human organ cells, with sufficient accuracy.

Alternatively, the impact of low-level radiation has been extracted from the statistical analysis of large-scale epidemiological studies. Most risk assessments of this kind are based on statistical investigations of selected sample groups whose radiation exposure is thought to be known. This phenomenological method tries to draw conclusions from observed effects at higher doses, in order to understand non-observable effects at lower doses. But how reliable is such an extrapolation of the linear correlation between dose and dose effect? For an answer, data must first be independently recorded with the highest possible accuracy.

Reducing errors in statistical statements requires large groups of people who have been exposed to radiation. Their medical background, together with their exact time-fixed radiation dose must be known in order to avoid misinterpretations. This is logistically almost impossible. Therefore, most studies of this nature are based on relatively small control groups, with reliable health status and lifestyle information, or on large groups where uncertainties are statistically leveled due to a lack of background knowledge. Whichever statistical method is used, significant uncertainties remain in the final results and conclusions.

The extrapolation of such data into the unknown and mostly not directly measurable range of the low-level dose should be subject to a model conception that as far as possible takes radiation effects and enzymatic repair processes into account. As long as low-dose data remains uncertain, interpretation will continue to be guided by cultural prejudices and assumptions. Therefore, one must first build the experimental data base and develop the theoretical foundation that form the basis for an unbiased interpretation of radiation effects.

With the discovery of X-rays and other types of ionizing radiation, interest in applying this new phenomenon grew rapidly. Numerous experiments in Europe, the United States, and other countries tested the influence of X-rays, and in some cases also gamma radiation, on cell cultures, sperm, plants, insects, and other living organisms. Much of this has already been mentioned in previous chapters. The effects of these irradiation experiments were not easy to determine. Since they focused on the mortality rate,[4] a significant dose of radiation on living creatures was required to produce visible results within a given time frame.

Measurements were mostly concerned with determining the dependence of a biological effect on a known amount of radiation[5] or, with improvement of dosimetric techniques, on the amount of absorbed radiation energy, the radiation dose. This dependence was called the *damage curve*. The results showed a linear relationship between the length of irradiation at a given radiation intensity and the mortality rate of the irradiated biological samples, but only if they were exposed to larger amounts of radiation. At low dose, the effects disappeared and were considered insignificant. Quantitative analysis, together with development of a theoretical understanding of the mechanism of action, proved necessary in order to make further statements in other areas.

Initially, linearity was only derived from these early observations and based on the assumption of mathematical simplicity. This assumption states that any radiation reaction with a DNA molecule leads to ionization and thus to the molecule's change – to mutation. This approach does not consider the biological effects of enzymatic repair processes, which were unknown at the time. The limitations of this interpretation were, however, known and other possibilities were postulated that corresponded with the threshold-effect principle for chemical toxins, as formulated by Paracelsus.

The biological possibility of dose effect as proposed by German geneticist Richard Glocker (1890–1978) – a student of Wilhelm Conrad Röntgen – corresponds to the so-called threshold theory[6] (Figure 8.1). Since according to observations, no major effect on biological systems can be observed below a certain threshold dose, Glocker argued that there must be a natural threshold below which radiation does not represent a danger to biological systems. However, he also considered a physical

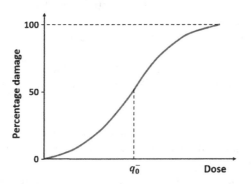

FIGURE 8.1   Percentage radiation effect on biological systems as a function of dose in a presentation according to Richard Glocker, 1929 in "The effect of X-rays on the cell as a physical problem," in *Festschrift der Technischen Hochschule Stuttgart zur Vollendung ihrer ersten Jahrhunderts, 1829–1929*, edited by T. H. Stuttgart, Berlin.

interpretation of the so-called target theory, according to which the probability of hitting sensitive cell components as targets does determine cell damage.

Glocker assumed that both components played a role: "Looking at the current state of the problem reveals that there are undoubtedly two causes that determine the shape of the damage curve, a biological, the fluctuating variability, and a physical, the probability of the radiation-sensitive area of the cell being hit." This sentence demonstrates considerable insight because from today's point of view, the biological component can now be interpreted as resulting from the enzymatic repair processes. The physical aspect, on the other hand, reflects the chemical-physical effect of radiation on the cell's DNA and RNA components.

However, Glocker's was a purely qualitative interpretation. The difficulty lay in translating it into a formalism, which would help to provide predictive power regarding the effect at low doses. This was of particular interest since irradiation techniques were becoming increasingly fashionable for medical applications at that time. The most important breakthrough in this direction was in further development of the target theory by geneticist Nikolaj Vladimirovich Timofeev-Resovskij (1900–1981), radiation physicist Karl Günther Zimmer (1911–1988), and theoretical physicist and later biologist Max Delbrück (1906–1981).

Timofeev's work, "Some facts of mutation research" (Einige Tatsachen der Mutationsforschung), Zimmer's work, "Hit theory and its relation to mutation induction" (Die Treffertheorie und ihre Beziehung zur Mutationauslösung), and Delbrück's work, "Atomic physics model of mutation" (Atomphysikalisches Modell der Mutation) comprised the first parts of a collaborative work discussing this theory, while the final fourth part, "Theory of gene mutation and gene structure" (Theorie der Genmutation und der Genstruktur) was jointly written by all three. This work, published in 1935, dominated the development of early molecular genetics and determined, as the so-called *Green Pamphlet* or *Dreimännerwerk* (three-man work), the scientific discussion on the effects of radiation in following decades. Their essential statement was: "We [imagine] the gene as an atomic association within which the mutation, atom rearrangement or binding dissociation (triggered by the supply of radiation energy) can take place, and which is largely autonomous in its effects and in its relationships to other genes."[7]

In a purely physical interpretation, the gene – or later the DNA molecule – was considered the largest molecule. The assumption had been that mutation was triggered when a single hit was made to the gene by an ionizing radiation quantum or particle.

The theory presented in *Dreimännerwerk* is based on the idea that when biological material of a certain volume V is irradiated with a certain dose D, the probability of critical molecules being hit and induced to mutate can be determined. The theory was refined in the following years[8] because the prediction of a linear relationship between dose and dose effect seemed to describe the observed data well (Figure 8.2). This approach dominated mutation considerations for many years due to the dominant influence of physicists.

Physicists saw the phenomenon of mutation as a purely physical interaction process and long ignored the complex relationships within the cell and the macromolecules.[9] Nevertheless, this work is considered an important step toward the intellectual unification of biology and the mathematical natural sciences, a breakthrough in the quantum-physical explanation of biological phenomena.

Prior to this work, biological considerations had been based on a threshold value, below which radiation could be neglected. However, new theoretical predictions made possible by the hit theory anticipated that even the smallest dose values could do harm with a certain probability. Therefore, radiation can always be dangerous.

Experiments to determine the effects of X-rays were continued during the war, in order to advance measurements toward lower dose levels and to experimentally test the thesis of linearity. In the United States, particularly geneticist Hermann Muller advocated the thesis of linearity. As already mentioned, he was awarded the Nobel Prize in 1946 for his experiments on the irradiation of fruit flies (Drosophilae) and the observation of radiation-induced mutation (cf. Chapter 7).

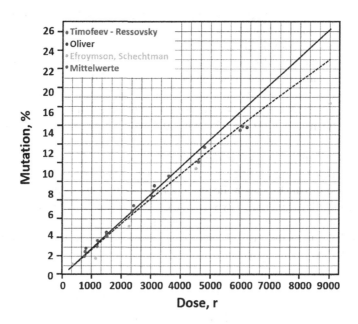

FIGURE 8.2   Taken from the 1935, *Dreimännerwerk*, this diagram of the damage curve demonstrates the agreement between the prediction of linearity, according to the hit theory, and the observed data for a range from 300 to 9,000 R. This corresponds approximately to a biological dose of 3 to 90 Sv. The deviation to lower effect at high dose corresponds to Glocker's presumed saturation effect toward higher doses. According to N. W. Timofeev-Resovskij, K. G. Zimmer, M. Delbrück, "On the nature of gene mutation and gene structure," *News from the Gesellschaft der Wissenschaften zu Göttingen: Mathematical-Physical Class, Biology* 1(13), (1935).

However, these measurements often produced contradictory results that deviated from the thesis of linearity without a threshold value.

Toward the end of the 1940s, highly respected German-American geneticist Curt Stern (1902–1981) and his colleagues Warren P. Spencer (1898–1969), Ernst Caspari (1909–1988), and Delta E. Uphoff (1922–1992) measured the effects of X-ray and gamma-ray quantities at 25 R, which corresponds to about 0.24 Sv. These measurements were particularly important to the Atomic Energy Commission (AEC) in the development of nuclear weapons and the need to work with radioactive materials. In 1949, the last of three publications summed up their results as follows: "Viewing all experiments together, it appears that radiation at low doses, administered at low intensity, induces mutation in Drosophila sperm. There is no threshold below which radiation fails to induce mutations."[10] However, this statement contradicts the results of Ernst Caspari and Curt Stern's earlier work in 1947, which led to a differing statement: "From the practical viewpoint, the results open up the possibility that a tolerance dose for radiation may be found, as far as the production of mutation is concerned."[11] This statement was, however, deleted in a subsequent publication. Yet, further contradictions still maintain doubt about the validity of linearity claims, fueling discussion to this day.[12]

It appears that against Caspari's opposition, Stern's mind was changed regarding the existence of threshold values by the claims and opinion of the influential Hermann Muller. Also, during these critical years between 1950 and 1953, Curt Stern was a member of the AEC's advisory commission, which supported and promoted radiation-physics research.[13]

Hermann Muller developed into a strong critic of the AEC's policy, which was sympathetic to the threshold theory at the time.[14] He was one of the first to warn about the dangers of radioactivity.[15] This view finally gained acceptance with scientific commissions, especially the influential National Academy of Sciences which produced the first official report on the *Biological Effects of Atomic*

Radiation I (BEIR I) in 1956. Despite numerous contradictions and heated internal debates, the committee made the official decision to recommend the linear dependence between dose and dose effect as a basis for radiation levels in medical examinations.[16]

This recommendation was also maintained in principle by the subsequent BEIR committees, although later results deviated from the original slope of the damage curve at lower-dose values and indicated a flattening. To this day, the 1956 decision still determines many views on radioactivity and the discussion is not yet concluded. Increasing criticism is evident in the BEIR VII report published in 2009, which debates the validity and reliability of the experimental statements.[17]

Such criticism is not only expressed in the United States but also in Europe. In particular, physicians point to the growing contradiction between biological and experimental data and the LNT predictions. This means that clear scientific proof of the LNT hypothesis has not yet been provided, while observational evidence increasingly seems to contradict the LNT. The question regarding the effects of radiation at low doses is therefore still open. More than sixty years after the official introduction of the LNT thesis, as the basis for predicting radiation damage, emotions still dominate the discussion, overriding scientific statements.

8.1.2 Statistical Investigations of the Natural Radiation Dose

The LNT model is recommended in BEIR reports as the method for extrapolating low-dose irradiation effects. However, in BEIR VII, it is recommended that extrapolation also use correction factors that can attenuate linearity in the relationship between radiation dose and effect, to better align with observations. Mathematically, the linear relationship is replaced by a more linear-square relationship, which de facto more closely resembles the shape of the threshold model.

In addition to radiological studies on cells and smaller organisms, the LNT model is increasingly based on epidemiological studies of population groups that have been exposed to low radiation doses over a long time, including members of the professional groups mentioned above. New investigations focus on exposure to the radioactive noble gas radon, as this relates to affected parts of the population. However, despite a large amount of data, interpretations and conclusions remain fuzzy.

Increasingly, alternative possibilities for extrapolation are being discussed. If implemented, this would have a significant effect on the prediction of low-dose radiation consequences, particularly the possible relationship between radiation effect and dose with regard to, for example, a cause of cancer. For now, however, the linear relationship between dose and effect remains essentially unchallenged, although initial findings have been modified by the influence of early warning mechanisms through the bystander effect (cf. Chapter 4) and cell repair mechanisms (cf. Chapter 7), which can modify the impact of radiation at low doses.

The question remains: How efficient are the bystander and repair effects? As depicted in Figure 8.3, the quadratic-linear model reduces the negative effect of possible radiation damage and negates any radiation effect at low doses, basically mimicking the existence of a threshold for radiation effects. The hormesis model goes even further, predicting damage-prevention and regeneration measures by cell enzymes at low radiation levels. Current data does not adequately allow for a statistically acceptable description to be provided by any of these very different models, underlining the difficulties of reliable prediction. Therefore, continuous investigation of low-dose effects is necessary in order to reduce large statistical uncertainties.

Before we present some examples of epidemiological studies on the dangers of ionizing radiation, some standard statistical terms and parameters must be introduced. These parameters allow for the quantitative comparison of different danger zones and danger thresholds.

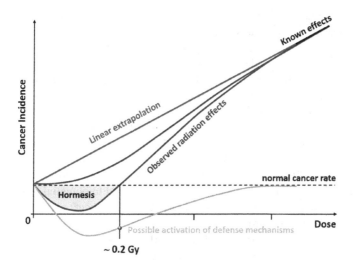

FIGURE 8.3   A series of alternative mathematical models for stochastic effects at low doses. On the left, the linear dose-effect relationship on which the legally prescribed limit values are based. The diagram also shows the quadratic-linear curve that is more consistent with observed data. Also shown is the so-called radiation hormesis, which at low radiation doses <0.2 Gy is even reported to have a healing effect since it can activate additional defensive mechanisms in the cell area. The dotted line shows the average cancer rate without any external radiation dose. Prepared by author.

8.1.2.1   Epidemiological Concepts

Large-scale epidemiological investigations aim at reducing a multitude of data and possible inter-pretations toward a simplified and generalized model description. This is done with mathematical-statistical methods, extracting generally valid laws and predictions from a complex set of data. Since there are no absolutely valid statements in statistics, but only a distribution of possible solutions and consequences, all predictions are afflicted with a range of uncertainties or errors and deviations in the data and statistical uncertainties associated with the amount of data being interpreted, as well as errors emerging from the mathematical methodology itself.

Therefore, the results are reduced to risk factors, which is to say, to questions about the prob-ability that certain events or circumstances can have negative consequences. Well-known examples include the estimation of accident risk when using transportation systems, from car to plane, or the risk of the occurrence of cancer from smoking or inhalation of asbestos dust. For radioactivity, risk factor refers to the probability or risk of increased radiation doses causing cancer or genetic muta-tions. For this purpose, formal quantities and units are defined that allow the risk to be represented mathematically.

The first variable to be defined for the risk assessment of radiation exposure is the standardized mortality ratio, generally abbreviated as SMR. It shows the ratio between the number of deaths in an experimental group observed over long periods of time and the expected number of deaths in a control group based on the aforementioned population statistics. If the ratio is greater than 1: excess deaths are present, which indicates additional possible deaths caused by radiation. If the ratio is significantly greater than 1: the circumstances result in an increased probability of death or an increase in the mortality rate. The actual cause of this increased mortality is thus not determined and must be correlated between the possible circumstances and the specific causes of death.

For example, the general mortality rate in the United States in 2013 was 731.9 deaths per 100,000 people. This should also be the case for the various individual states. In the state of Mississippi, however, mortality is much higher, with 959.6 deaths per 100,000 people in the comparable period. The SMR for Mississippi would thus be 1.31 compared to the total population of the United States, so 30% higher. However, this does not identify the reason for the greatly increased mortality, which can be due to poorer health or inadequate health care in the state of Mississippi.[18] The statistic can have its origin in poorer nutrition, cultural factors such as smoking, the use of firearms, traffic behavior, or simply economically poorer living conditions. To find out, one needs to address well-chosen subgroups from the appropriate target areas.

This perhaps somewhat crude example demonstrates that an attempt to determine higher mortality due to increased radiation is only permissible if many living conditions are considered. Therefore, in better structured epidemiological studies, other parameters are used to determine risk.

Relative risk (RR) is defined as the probability ratio of a negative subsequent development (such as cancer) occurring in a group of people (or animals) exposed to potential hazards (such as radiation) as compared to the probability of a similar development (cancer) occurring in a control group of people without additional hazard exposure.

A classic example is smoking: Male smokers are 50% likely to develop lung cancer as compared to non-smokers at 2%. The relative risk of developing cancer would then be RR = 50/2 = 25. This means that twenty-five times more smokers develop lung cancer than non-smokers. Conversely, if an observation group is exposed to a suspicious pollutant and comparison with the control group results in a relative risk of RR = 1, then there is statistically no difference between the exposed group and the control group with regard to suspected effect resulting from the suspicious pollutant. Therefore, the suspected pollutant has no effect and is declared harmless under the experimental conditions.

Excessive relative risk (ERR) is a measure defined by the World Health Organization that indicates the percentage by which the risk of persons exposed to pollutants exceeds the risk of non-exposed persons.[19] This measure is preferably being used in epidemiological studies and is calculated as a percentage of ERR = 100% for RR of 1. For an RR of 1.5, the ERR is 150%; for an RR of 2.0, the ERR is 200%; and for an RR of 25, the ERR = 2,500%. If exposure to a suspected pollutant has no effect, then ERR = 0%.

These statistical terms are generally applicable. They will be presented here in examples of epidemiological studies on the effects of radioactivity. These representations are not comprehensive. Such accuracy is attempted in official, regularly published reports such as the US National Academy's BEIR report and the UN's UNSCEAR report. Examples given here are only intended to highlight the difficulties and contradictions that arise in such statistical investigations.

8.1.2.2 Mountain and Valley Dwellers

On average, cosmic or solar radiation contributes 8% to the annual radiation exposure of humans (see Chapter 4, Figure 4.7). Exposure increases exponentially with the altitude of a person's residence, doubling every 1,500 m. People living at high altitudes, including flight crews and frequent flyers, absorb considerably more cosmic radiation than people living at low altitudes. These population groups therefore offer favorable conditions for studying the influence of increased exposure. Several studies are available on this subject.

Mountain-region inhabitants are often additionally exposed due to decay of radiogenic actinides in the mountain rock. Radioactive decay of radon ^{220}Rn and ^{222}Rn makes a considerable contribution to human radiation exposure, averaging 22.5%. Increased radon emissions, from the uranium and thorium decay chains, mainly occur in granite mountain ranges.

Radon is inhaled as a gas and immediately exhaled. During this short stay in the lungs, it causes almost no damage since only 0.2% of the inhaled radon particles decay within the lungs. Radon decay products such as polonium and bismuth do, however, adhere to tiny dust particles present in the air we breathe and these settle in the lungs when inhaled. It is claimed that the decay of these particles releases radiation that can permanently damage lung tissue and contribute to carcinogenesis.

This means that not only can radon be harmful, but the dust content in the air can also be harmful as a carrier.

A systematic comparison of the cancer incidence in populations with natural radiation exposure in the high-altitude US states of Idaho, Colorado, and New Mexico (the so-called Mountain states) with populations in the low-altitude states of Louisiana, Mississippi, and Alabama along the Gulf coast (the Gulf states) shows a direct correlation.[20] Due to increased radon emission from rocky material in the Mountain states, the annual radiation exposure of 5,500 Sv is on average three to four times higher than the radon emission of 1,400 Sv in Gulf states at sea level.

In terms of the annual cancer rate, however, the ratio is reversed. Averaged from 1990 to 1994, the cancer rate in the Mountain states with 147 cases per 100,000 inhabitants is clearly below the rate of 185 cases per 100,000 inhabitants in the three Gulf states. If one limits oneself to lung cancer, the cancer rate is forty-seven cases in the three Mountain states and sixty-eight cases per 100,000 inhabitants in the Gulf states. This result is completely different than what is traditionally expected and predicted. Other reasons than radiation exposure must be the cause for the enhanced cancer rate in the Gulf states.

In 2012, the observed anti-correlation between radiation and cancer rate was confirmed in another study[21] that covered the period from 2002 to 2006. This study included all fifty US states and also considered other possible carcinogenic effects such as smoking and lifestyle. The 2012 result was confirmed by a US-wide comparison between the geographical distribution of the annual cancer rate, published by the National Cancer Institute, and the radon emission rate of the Environmental Protection Agency (Figure 8.4). Increased cancer rates occur predominantly in states with low radon exposure.

Some researchers regard this observation as an indication of hormesis. However, results based on these observations might also be influenced by other factors, such as smoking, differences in medical care, poverty, and other differences in living conditions. Nonetheless, observations also show that natural radiation has a much lower effect on the tumor rate than is generally assumed.

It should be pointed out that similar studies have been made for the high altitude and urban areas in the western states of the United States for the period between 1950 and 1969. Despite the higher expectation for radioactive fall-out from the nuclear test program in the Nevada desert at that time, no significant enhancement in cancer rate was observed.[22]

Over the last decades, a multitude of ecological studies of this kind have been performed to identify the correlation between altitude and an increase in cancer rate and mortality. The results of forty-eight epidemiological studies on the inhabitants of mountainous regions worldwide demonstrate that the results remain at best inconclusive. The results do, however, seem more likely to

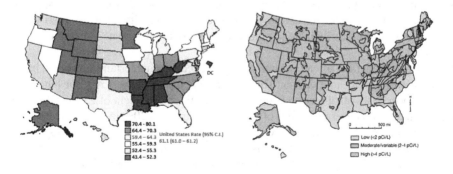

FIGURE 8.4 On the left, geographical distribution of the cancer rate (per 100,000 inhabitants) in the United States according to the National Cancer Institute; source: http://statecancerprofiles.cancer.gov/ (July 29, 2023). On the right, radon emission distribution in various US states. Source: http://www.kgs.ku.edu/Publications/PIC/pic25.html (July 29, 2023). The figure points to a strong anti-correlation between local natural radiation from radon and the cancer rate.

underline the health benefits of living in high altitude regions than to underscore radiation dangers.[23] No obvious correlation with the enhanced radiation level at high altitudes is recorded.

Radon gas exposure is listed by the Environmental Protection Agency (EPA) as one of the major causes of cancer in the nation, "Radon is the number one cause of lung cancer among non-smokers, according to EPA estimates. Overall, radon is the second leading cause of lung cancer. Radon is responsible for about 21,000 lung cancer deaths every year."[24] As previously mentioned, radon is a natural odorless noble gas that is emitted as part of the ^{238}U (^{222}Rn) and ^{232}Th (^{220}Rn) decay chains. It is chemically inactive and because of its weight it remains primarily on the ground. Radon gas can, however, accumulate in basements and closed rooms. Only a small fraction of inhaled radon decays in the lungs. The EPA assessment goes back to two studies, which were made in closed rooms with higher radon accumulation. The results are listed on the EPA website as proof for the assertion made by the Surgeon General in 2005,[25] which became an integral part of the overall claim that radon kills.

While the studies relied on a careful accumulation of data in terms of radon levels and ecological conditions, analysis and derivation are based on the so-called model-dependent approach. This means that the stated effects rely on the theoretical extrapolation of the health impact of the radiation dose. While the actual dose that participants received could be calculated from the respective radon data, the impact was derived by using the aforementioned linear no-threshold LNT extrapolation suggested in the BEIR-VI report. This translates into a higher risk than the alternative low-level extrapolation techniques, as demonstrated in several studies.[26] In turn, this result leads to the mathematical prediction of the anticipated high rates for lung cancer. A direct correlation between lung cancer death and lung cancer cause as an effect of radon has not, however, been demonstrated. Again, the uncertainty is entirely in the extrapolation technique toward low-dose impact, a technique that is called into doubt by numerous researchers who suggest alternative extrapolation methods.[27]

Overall uncertainties due to the model assumptions in these predictions have been presented in a study that compares radon-risk assessment techniques and the resulting prediction of annual lung cancer deaths. These findings are presented with respect to the radon average activity in becquerel per cubic meter, for many countries worldwide.[28] The average radon activity per volume ranged from 20 Bq/m³ (UK) to 232 Bq/m³ (Romania), with the United States at a more medium level of 46 Bq/m³.

A number of different models were used for the risk calculations. Most of the studies used the two LNT models (EAC and EAD)[29] suggested by BEIR-VI, while some of the studies used the different EPA model which assesses the risk somewhere between the BEIR-VI models. Several studies used the European model, which generally predicted less risk per dose. Only two studies (Netherlands and Sweden) used the two-mutation carcinogenesis model, an approach which included biological arguments for cancer developments. The results reflect a broad variety in the prediction of lung cancer attributable to radon exposure and the estimated annual number for lung cancer cases.

The results are rather far spread, which basically reflects the uncertainty involved in attempting to link radon exposure to lung cancer. The Netherlands has the lowest risk for lung cancer, with 4% of lung cancer cases attributed to a low radon density environment of 23 Bq/m³. Sweden, with one of the highest radon average values of 110 Bq/m³, attributes 20% of lung cancer to radon; they claim an annual lung cancer rate of between 150 and 420 cases. The European approach ascribes between 4% and 8% of the lung cancer cases to radon, claiming a number for lung cancer cases caused by radon exposure that ranges between 200 and 3,000 for respective nations. The United States attributes between 10% and 14% of their lung cancer cases to radon depending on the model, with an estimated total number of deaths caused by radon to be between 15,400 to 21,800. Although the population of the United States is only four times larger than Germany's, this range is ten times higher than the number of deaths attributed to radon for Germany, where the range is between 1,806 and 2,000 depending on the model assumptions.

There are clear and significant variations in these numbers, reflecting uncertainties in the eco-logical assessments, with the main uncertainty resulting from the simplified assumptions of the LNT theory. While it is not surprising that a government agency such as the EPA adopts the rec-ommendations of a government-appointed panel such as BEIR – so as to be "on the safe side" – an increasing number of doctors and scientists are demanding a more sophisticated and science-based approach in the extrapolation model.[30]

EPA studies were made in closed room and house environments, where radon may accumulate to higher excess levels. We will come back to the issue of radon exposure in closed dwellings at a later point.

Within the context of radon exposure, there are further studies of particular interest that dem-onstrate the influence of radiation in those areas with extreme radon exposure. This particularly applies to Ramsar in northwestern Iran (see Chapter 6). There, the annual dose is up to 260 mSv/a, significantly higher than the global-average human annual dose of about 25 mSv/a. The local popu-lation has been exposed to this increased radiation dose for generations and therefore offers a unique opportunity to investigate the effects of this dose on cancer rates as well as changes in DNA and mutation rates in genetic material. Both Iranian and American scientists have conducted such stud-ies and presented various aspects of long-term irradiation in numerous scientific publications. In these studies, no significant increase in cancer or cancer mortality rates has been found compared to control groups.[31]

In Ramsar, average SMR values were found in two areas. In a local area with high radiation exposure, the SMR value was 0.9 for men and 1.3 for women. In a local area with lower radiation exposure, the SMR value was 0.8 for men and 1.2 for women. The slightly higher value for women was explained by the fact that they were more frequently indoors. However, due to the relatively low number of cancer-related deaths, the 0.35% statistical errors are so large that, statistically, there is no difference between the SMR results for both men and women. SMR values of 1 mean that despite extremely increased radiation exposure, residents in highly radioactive areas were found to not have significantly higher cancer and mortality rates.[32]

These findings, in conjunction with DNA mutation rates and chromosomal aberrations, provide strong supporting evidence for increased enzymatic repair mechanisms as outlined in Chapter 7. These results for Ramsar are often criticized because the number of investigated residents was relatively small and results show considerable statistical uncertainties. However, other areas with high levels of local radioactivity also provide similar results. Nevertheless, the mere fact that there are no clear results demonstrating increased cancer or mutation rates, despite an extreme increase in the dose load, contradicts LNT predictions of a direct linear correlation in the low-dose range. A re-evaluation of the LNT-based extrapolation seems appropriate.

In summary, no clear correlation could be observed between low radon exposure and somatic or genetic cell changes. This seems to be true even in extreme radon conditions, as for example within closed rooms in Ramsar.

8.1.2.3 Radiation Exposure at Depth

Within the earth, average radiation exposure increases with increasing depth, due to the growing occurrence of actinide-rich rocks. There are, therefore, strong geological differences determined by rock occurrence and rock chemistry. Salt deposits, for example, are radioactivity-free as long as they are not potassium salts. Uranium deposits, on the other hand, show a high radioactivity level. Apart from uranium itself, the radiation is produced by primary decay processes of the various radioactive daughter elements in the decay chain, as well as by secondary processes triggered by nuclear reac-tion processes induced by the decay products. This has been identified as a problem for the mining industry, in particular in uranium mines or mines with high actinide content.

The decay of actinides and their daughter elements in the decay chain bear a risk, despite being safely bound in the mineral structure of the rock material. Besides β-radiation, which is mostly absorbed in the rock material, they release α-particles with an average energy of 5 MeV. Light isotopes trap these particles in the surrounding rock masses and convert them into neutrons through nuclear reactions such as $^9Be(\alpha,n)$, $^{11}B(\alpha,n)$, $^{13}C(\alpha,n)$, $^{17,18}O(\alpha,n)$, $^{25,26}Mg(\alpha,n)$, and $^{29,30}Si(\alpha,n)$, depending on the available abundances of these elements. This results in a significant radiogenic neutron flux in actinide-rich rocks.

The tunnels of a 300-m-deep salt repository, in Llano-Estacado in southern New Mexico, are part of the Waste Isolation Pilot Plant (WIPP) that is the US Army's interim radioactive waste repository (Figure 8.5, left).[33] In addition to this disposal site, there is also an independent salt-tunnel system where scientists from nearby Los Alamos National Laboratory conduct experiments in a nearly radiation-free environment. The salt in these tunnels does not contain actinides and only the potassium chloride (KCl) deposits in the salt dome – which is dominated by sodium chloride (NaCl) – measure a 1.46 MeV-line of ^{40}K.

Figure 8.5 on the right shows, for comparison, a drift more than 1,500 m deep at the Homestake mine in Lead, South Dakota, which served as a gold mine for more than 100 years. Figure 8.6 shows the neutron flux in the gold mine compared to the earth's surface; the red line also represents the neutron flux in the WIPP. The neutron flux in the salt mine is hundreds of times smaller than at the Earth's surface because the neutrons originating from cosmic rays have been absorbed at this depth. On the other hand, the neutron flux in the gold mine is a hundred times higher than the neutron flux in the salt mine, even though it is less deep. This is because the salt mine is free of heavy actinides while the normal gold-bearing rock layers contain a high actinide content.

However, it is not so much the neutrons produced in the mine that are considered a danger – since, due to the low reaction probability, they barely make up a millionth of the number of α-particles being released – but rather, it is the α-particles themselves. Here again, it is not the α-particles produced by the decay of the original uranium and the solid daughter elements but rather the α-particles from the decay of radon, which are emitted into the atmosphere of the mining environment.

Most of the decay products remain chemically bound in the rock, thereby stopping and neutralizing the α-particles, which subsequently diffuse to the surface as helium gas. Radon, on the other hand, is itself a noble gas and so cannot form a chemical bond. Therefore, it diffuses into the air, which is breathed in the mine tunnels. As mentioned before, with the high radon concentration in the mining tunnels, there is a risk that the α-decay of this inhaled radioactive gas will lead to long-term radiation damage, such as lung cancer.

For this reason, radon has been identified as the main problem because it diffuses into mining shafts where it accumulates, reaching a substantial radiation concentration of up to 100,000 Bq/m^3 depending on rock composition and ventilation conditions. These values clearly exceed by far the exposure at above-ground conditions.

FIGURE 8.5   On the left, a 300-m-deep tunnel in the WIPP mine whose walls are made of pure salt. On the right, a 1,500-m-deep drift in the former Homestake gold mine in the Black Hills of South Dakota. Photo: Michael Wiescher.

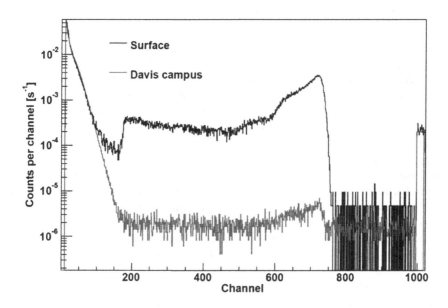

FIGURE 8.6   The neutron spectrum over an energy range measured with a ³He counter via the reaction ³He(n,p). The slope on the left is electronic noise in the detector. The region in the middle of the spectrum, between channels 200 and 800, indicates the neutron flux measured at the surface and at the Homestake gold mine in South Dakota. At the surface, the neutron flux is about 100 times greater than at the 1,500 m depth. According to A. Best, et al., "Low energy neutron background in deep underground laboratories," *Nuclear Instruments and Methods in Physics Research* A 812 (2016): 1–6.

FIGURE 8.7   Uranium miners in an illustration from the 1950s. Courtesy and with permission: David McMacken, 2014.

In the town of St. Joachimsthal in the Bohemian mountains (today: Jáchymov), physician and mineralogist Georg Agricola (1494–1555) observed an unusually high incidence of the severe, mostly fatal lung disease Schneeberger Bergkrankheit (Schneeberge mountain sickness) among miners. Later diagnosed by Paracelsus as a typical miner's disease, it was particularly prevalent in the mines of central Germany and Bohemia where galena, copper, and silver were mined. At the end of the 19th century, Schneeberg mining disease was diagnosed as lung cancer. In these mining areas, the rock is particularly rich in uranium and represented the main source for uranium during the turn of the 19th century. Marie Curie obtained her pitchblende from St. Joachimsthal. In fact, former silver mines such as the one in Jáchymov were converted into lucrative uranium mines in the 20th century, supplying in particular the Soviet weapons program.

The high incidence of lung cancer in these areas was explained by the increased presence of radioactive radon as part of the uranium decay chain.[34] In some tunnels more than 100,000 Bq per cubic meter of air were measured. But an accurate quantitative analysis is not easy because many other factors, such as dust particles in the air that miners breathe, can also lead to lung cancer. The combination of radon and dust content in the breathed air constitutes a combined hazard factor for the development of miner's disease. Therefore, careful differentiation must be made between possible exposure factors.

Due to the higher radiation exposure in mines and other underground environments (Figure 8.8), many studies have been devoted to the long-term epidemiological analysis of miners in different countries and under different working conditions. Results consistently show that the risk of developing lung cancer increases linearly with accumulated dose. There are, however, other environmental conditions as well. These may include for example smoking habits and the excessive silica-dust concentration in a working mine, which is also known to be carcinogenic. The risk increase varies depending on the different conditions and on the radon and dust levels in each mine. For radon, the accumulated dose is measured in units of working level month (WLM). This unit not only considers the radioactive radon gas ^{222}Rn contribution to radiation exposure in the lungs but also considers radioactive daughter isotopes entering the lungs through dust respiration.

The working level (WL) unit is considered the collective unit representing the radiation energy in one liter of air, which corresponds to an energy of 130,000 MeV/l or 20.8 J/m³ of energy that is thereby released. The unit of time is assumed to be the month, which corresponds to a traditional time of 170 hours underground. Thus, a dose value of about 5 mSv can be estimated for a WLM, assuming a working time of about 2,000 hours per year.

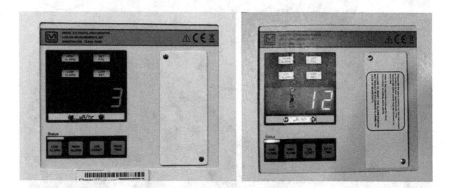

FIGURE 8.8 Measured values for natural radiation dose; on the left, the wall in the accelerator room of the Nuclear Science Laboratory at the University of Notre Dame, without beam operation; on the right, at the CASPAR accelerator 1,500 m below ground in the Homestake mine of South Dakota, without beam operation. Natural radiogenic radiation at a depth of 1,500 m is four times higher than at the Earth's surface. Photo: Michael Wiescher.

Miners for the former SDAG Wismut (Soviet-German Joint Stock Company) have been well documented by a number of studies. SDAG Wismut was a mining company in the GDR that became the world's fourth largest uranium producer after the USSR, the US, and Canada during the Cold War period between 1946 and 1990. The successor company, Wismut GmbH, is a federal company entrusted with the rehabilitation and recultivation of Wismut mining's legacies. This includes systematically investigating the causes of miners' occupational diseases, including silicosis (pneumoconiosis) and lung cancer. As part of this work, all Wismut miners who had worked underground for at least 6 months between 1946 and 1989 were recorded. This involved an annual average of about 15,000 people, with employment figures rising to 20,000 miners in the 1950s during the heyday of the Soviet nuclear-armament program.

The number of miners and the average WLM levels to which they were exposed peaked at 120 WLM in the early 1950s but dropped rapidly in the early 1960s to averages of 20 WLM, decreasing further to constant levels around 2 WLM over the remaining thirty years of mine operation (Figure 8.9). This rapid decline can be attributed in part to tighter legislation that improved miner's respiratory-protection conditions but more importantly, technological innovations such as better mine ventilation can be credited. While improved ventilation significantly reduced the concentration of radon underground, the gradual switch to wet drilling in the 1960s was also helpful.[35] While this considerably reduces silica-dust content in the air, thereby reducing this aspect as possible carcinogen, wet drilling has a lesser effect on the radiation level. This is because the radon concentration will remain unperturbed even when the decay products associated with the dust fall-out have been removed. Nonetheless, a combination of these measures explains the sharp drop in radiation exposure among Wismut miners.

The impact of radon and possible other environmental carcinogens on Wismut workers was studied over a 50-year time period.[36] The occupational group included 58,987 miners. By 2003, 3,016 of these had died of lung cancer, 3,355 of other cancers, 5,141 of heart disease, and 1,742 of stroke. Fifty years after peak exposure in the early 1950s, 20,920 miners had died, nearly 36% of the former workforce.

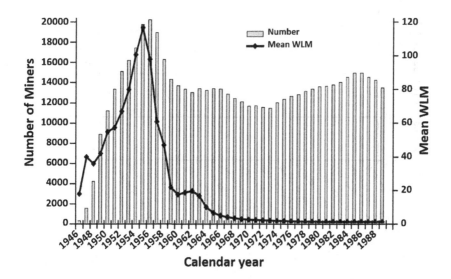

FIGURE 8.9 Annual number of miners working for SDAG Wismut from 1946 to 1989 and the average annual radiation exposure in units WLM (for description see text). Note: M. Kreuzer, M. Schnelzer, A. Tschense, L. Walsh and B. Grosche, "Cohort Profile: The German uranium miners cohort study (WISMUT cohort), 1946–2003," *International Journal of Epidemiology* 2010 (39), (2010): 980–987.

For statistical risk analysis, miners were divided into groups with different levels of WLM. Information on disease development and mode of death yielded the average excess relative risk (ERR) per WLM for different types of cancer. Only for lung cancer, the ERR showed a statistically relevant value of 0.19% per WLM exposure, with an uncertainty range of 0.02%.[37] There was a somewhat increased risk value for respiratory-tract cancer noted at 0.05% with comparable uncertainties. For all other cancers, no correlation between radon exposure and disease risk was found (Figure 8.10). This listing is not conclusive; a similar EER correlation between elevated silica and quartz dust in the air we breathe and the incidence of lung cancer has also been demonstrated.[38]

A peculiar environment with high radon concentrations is given in radon spas that are often located in abandoned mines. In the years before the Second World War, these spas had been a great success in the United States as well as in Europe. However, radon spas have pretty much disappeared in the United States, while having a renaissance in Europe. These spas offer a radiation therapy in which radon is inhaled, mineral water containing radon is consumed, and baths are taken in water with a high radon content. Already in ancient times, radon was attributed with having a therapeutic effect.[39] Epidemiological studies confirm that for rheumatic diseases and rheumatoid arthritis, as well as for degenerative spinal and joint diseases, long-lasting pain relief and improved joint function can be achieved by radon therapy.[40] However, according to critics, patients are at risk of cancer excitation from exposure to doses that are too high.

It is therefore interesting to study conditions in a typical radon health spa such as Bad Gastein in Austria.[41] The breathing air, in the inhalation rooms provided there, contains about 3,300 Bq/m³ of which almost half are radioactive decay products such as polonium and lead. However, since no rock is mined, the dust content is considerably lower than in a mine. Typical inhalation periods last 3 hours, which corresponds to an α-particle lung dose of 19.5 μSv. In a spa week, this accumulates to 0.116 mSv, if the increased radon content in the spa's atmosphere is neglected. This would be approximately 0.02 WLM, when compared to the exposure units of miners. In the bath, the radon concentration in the water is 555,000 Bq/m³ with typical length of patient stay being 20 minutes.

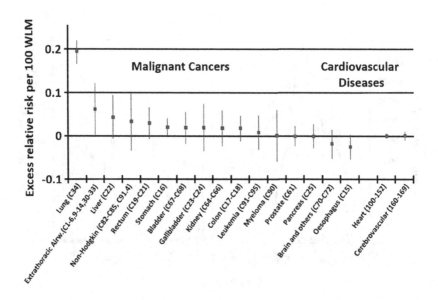

FIGURE 8.10   The excess relative risk ERR per 100 WLM for the occurrence of various types of cancer in miners working for SDAG Wismut. Note: M. Kreuzer, L. Walsh, M. Schnelzer, A. Tschense, and B. Grosche, "Radon and risk of extrapulmonary cancers: Results of the German uranium miners' cohort study, 1960–2003," *British Journal of Cancer* 99 (11), (2008): 1946–1953.

This corresponds to a dose of 0.27 μSv per bath, including the radon content in the air. This dose accumulates to about 1.6 μSv per spa week for daily bathing, considerably less than for inhalation bathing. From these estimates, the percentage by which the risk of persons taking the annual radon cure exceeds the risk of non-exposed persons for getting cancer (the so-called ERR) is calculated to approximately 0.04%. Because of the time-limited exposure, this is well below the risk limits indicated in Figure 8.10.

8.1.2.4   Radiation Exposure with Altitude

For mountain dwellers, a direct correlation between enhanced-altitude radiation and possible somatic and genetic consequences could not be established. However, flight personnel are on average exposed to much higher radiation levels. At regular flight altitudes of 10 km, conditions exist for a more than tenfold increase in the irradiation dose. In 1990, the International Commission on Radiological Protection[42] recommended epidemiological studies to investigate possible consequences for mortality and cancer development. The studies focused mostly on American and European airlines that had collected the necessary data. Despite the limited pool, resulting observations can probably be generalized.

Average radiation exposure for airline personnel is broad, ranging from 1 to 6 mSv/yr, although the 1993 UNSCEAR report gives a somewhat narrower range of 2.5 to 3.5 mSv/yr. Thus, flight personnel are among occupational groups with the highest exposure rate, while the exact dose primarily depends on flight altitude and route. Working 700 hours on the Los Angeles–Frankfurt route, a worker has an annual exposure of 4.1 mSv/y during average solar activity. By comparison, 700 hours on the Frankfurt–Lagos route, with comparable flight duration, corresponds to a much smaller annual dose rate of 1.5 mSv/y because cosmic-radiation intensity is lower in the equatorial region than in northern latitudes.

First summary results are particularly aimed at the mortality rate and incidence rate for various types of cancer in flight employees, compared with reference groups from the non-flying population. Results show no cosmic-radiation effect with regard to either cancer incidence or mortality.[43] Studies include a variety of other forms of employment and demonstrate that aircrew mortality rates are generally lower when compared to the average population – this includes cancer as the cause of death. This is attributed to the exceptionally good health care provided to airline employees. The incidence of cancer, for most types of cancer, is comparable to that in the general population.

One significant difference is the incidence of skin cancer (melanoma). In the studies, this is, however, attributed less to the influence of highly ionizing high-altitude radiation and more to increased UV radiation levels between flights, when crew members are exposed to the sun during long stays in tropical zones. Additional occupational or environmental conditions were also identified as health hazards that could possibly influence results. This included electromagnetic fields from flight instruments, cabin-air pollution from engine gas, and unnatural sleep patterns among aircrew.

A more detailed study focused on the possible effects of cosmic radiation on cockpit crews of German airlines, such as Lufthansa and Air Berlin.[44] This study involved a targeted survey of 6,006 individuals between 1960 and 2004. The study showed that the average annual flight time increased by 37%, from 460 hours in 1960 to 630 hours in 2004. This then led to an increase in the annual dose exposure (Figure 8.11). Nonetheless, observations made in this study are mostly in agreement with statements from previous work; general mortality was lower than in the control group and this was also the case for the cancer-related mortality rate.

Overall, mortality in pilots is significantly lower than that of the general population, with a standardized mortality ratio (SMR) of 0.5; this means that the mortality rate of the aircrew was only half of the mortality rate of the control group at the ground. This includes deaths from cancer. The relative risk (RR) for mortality per 10 mSv is 0.85 within a 95% uncertainty range of 0.79 to 0.93; the RR per 10 mSv for cancer-related mortality is slightly higher: 1.05 in the 95% confidence interval of 0.91 and 1.20 (see Chapter 6.1.2.1). This means that airline personnel have roughly the same chance of dying from cancer as members of the ground crew as a control group. This does

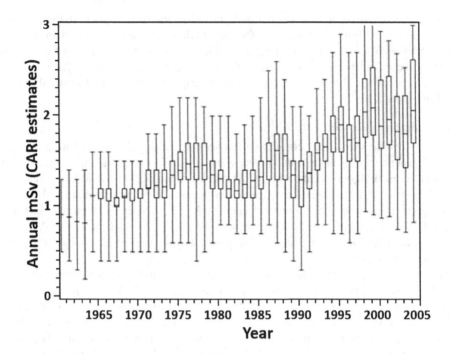

FIGURE 8.11   Distribution of annual effective dose (mSv) for German cockpit crews (male) from 1960 to 2004. Boxes show the mean-range value, bars the 5 to 95% reliability rate. Adapted from: G. P. Hammer, M. Blettner, I. Langner, H. Zeeb, "Cosmic Radiation and Mortality from Cancer among male German Airline Pilots," *Eur. Journal for Epidemiology* 27 (2012): 418–429.

not include radiation-related cancers but implies that no statistically significant radiation-related disease outcomes were identified.

However, a study published in 2018[45] shows an increased cancer rate among airline personnel. For this study, disease data from 5,366 US flight attendants was analyzed and compared to data for the general population. The study shows a 15% higher cancer rate in flight attendants than in the comparison sample from the general population. Specifically, 3.4% of flight attendants developed breast cancer, compared with only 2.3% of women in general. For tumors of the gastrointestinal tract, the rate of disease among cabin crew was 0.47% – almost twice as high as the rate of 0.27% for the general population. Flight attendants were also almost twice as likely to contract black-skin cancer (melanoma) – 1.2% as compared with 0.69%. The study does not, however, commit itself as to the possible cause of these increased levels. In addition to increased radiation exposure, the study emphasizes the influence of other factors such as air pollution in flight cabins, lack of sleep, and shift work, each of which have long been suspected as risk factors that increase cancer risk. As this study revealed, shift work decreases melatonin levels, thus interfering with cell DNA repair. And, if DNA damage is not repaired, the cell can degenerate and cancer can develop.

Long-term epidemiological studies of the possible consequences of natural-radiation exposure in populations are not straightforward: often the dose is small, the statistical relevance is questionable, and separation from other possible causes of disease or cancer is difficult to achieve. Published results often contradict each other and depend on interpretation of the data sets. Radiation-related cancers can undoubtedly occur, but the risk generally appears to be much lower than is often assumed. Studies of other carcinogens such as tobacco and exhaust-related air pollution show a much higher risk rate. In addition, people who are also increasingly subject to man-made or anthropogenic radiation exposure are found to be at high risk.

NOTES

1. http://www.emfacts.com/2014/10/interview-with-helen-caldicott-on-fukushima/ (January 22, 2023).
2. Mark Twain attributes this sentence to British Prime Minister Disraeli, in his autobiography published in 1924. However, in Disraeli's written estate, no corresponding sentence could be found. A more careful analysis about the origin of the phrase, albeit without conclusions was recently posted on the History of Statistics site of the University of York, UK, website: https://www.york.ac.uk/depts/maths/histstat/lies.htm (April 23, 2023).
3. A Monte Carlo simulation is a mathematical technique that predicts the outcome of uncertain events on the basis of statistical probabilities. A reliable simulation must include all possible (known) physical, chemical, and biochemical processes and interactions associated with randomized radiation impact.
4. Mortality is defined as the number of deaths observed per number of observed species. It is usually determined in comparison with a control group.
5. The amount of radiation was mostly measured in the unit X-ray (R), which originally corresponded to the duration of radiation at a certain intensity and energy. It is quantified as the ionization probability of radiation in air. One R corresponds to a dose of 0.00877 gray (0.87 rad) in dry air or 0.0096 gray (0.96 rad) in biological materials. The unit is no longer used today.
6. Richard Glocker, "Das Grundgesetz der physikalischen Wirkung von Röntgenstrahlen verschiedener Wellenlänge und seine Beziehung zum biologischen Effekt," *Strahlentherapie* 26 (1927): 147–155 and Richard Glocker, "Die Wirkung der Röntgenstrahlen auf die Zelle als physikalisches Problem," in *Festschrift der Technischen Hochschule Stuttgart zur Vollendung ihres ersten Jahrhunderts, 1829–1929*, ed. T. H. Stuttgart (Berlin: Springer, 1929).
7. N. W. Timofeev-Resovskij, K. G. Zimmer, and M. Delbrück, "Über die Natur der Genmutation und der Genstruktur," *Nachrichten von der Gesellschaft der Wissenschaften zu Göttingen: Mathematische-Physikalische Klasse, Biologie* 1 (13), (1935): 189–245.
8. K. G. Zimmer, *Studien zur quantitativen Strahlenbiologie, Abhandlungen der Akademie der Wissenschaften* (Wiesbaden: Franz Steiner Verlag, 1960).
9. A. v. Schwerin, *Medical Physicists, Biology, and the Physiology of the Cell (1920–1940)*, MPI für Wissenschaftsgeschichte, Reprint 393 (2010), 231–258.
10. D. Uphoff and C. Stern, "The genetic effects of low intensity irradiation," *Science* 109 (1949): 609–610.
11. E. Caspari and C. Stern, "The influence of chronic irradiation with gamma rays at low dosages on the mutation rate in Drosophila melanogaster," MDDC-1200, US Atomic Energy Commission (1948), 1–8: http:www.hathitrust.org (July 29, 2023).
12. E. J. Calabrese, "Key Studies used to support cancer risk assessment questioned," *Environmental Molecular Mutagenesis* 52 (2011): 595–606.
13. A good overview of developments in radiobiology can be found in the contemporary annual reports of the AEC; cf. "Atomic Energy and the Life Sciences," Report of the United States Atomic Energy Commission, 1949. The Atomic Energy Commission was later replaced by the Department of Energy (DOE). A summary of these events is given by E. J. Calabrese, "Linear Non-Threshold (LNT) historical discovery milestones," *Med Lav* 113 (4), (2022): e2022033.
14. Muller's motivation is not quite clear. In a recent book, it is argued that he might have been motivated by a large grant from the Rockefeller Foundation, which provided him with considerable funding for his new research program at Indiana University. Evidence for this may be circumstantial, since at that time Muller's scientific reputation was mixed due to his work before the war in Nazi Germany and the Soviet Union. His teaching position at Amherst had been terminated when too many undergraduate students complained about his style and lack of interest in their training and education. Indiana offered Muller graduate students and the Rockefeller Foundation provided funding. Was this because the latter discovered his capabilities and intellect (Elof Axel Carlson, *Genes, Radiation, and Society: The Life and Work of H. J. Muller* (Ithaca: Cornell University Press, 1981), 274–288) or was it the foundation's interest in his interpretation of his genetic observations? Cf. Jack Devanney, *Why Nuclear Power Has Been a Flop*, 2021, 61–64; for the latest edition, see: https://gordianknotbook.com (July 29, 2023). This question is well beyond the scope of this book; a more detailed study of both the Rockefeller Foundation and Indiana University archives might elucidate actual historical development.
15. Until Muller's death in 1967, this confrontation was often carried out in the pages of the *Bulletin of the Atomic Scientists*, devoted to the political and economic aspects of nuclear energy; cf. H. J. Muller, "How radiation changes the genetic constitution," *Bulletin of the Atomic Scientists* XI (9), (1955): 329–338 and T. T. Balio, "The public confrontation of Hermann J. Muller," *Bulletin of the Atomic Scientists* XXIII (1967): 8–12.

16. E. J. Calabrese, "The road to linearity: Why linearity at low doses became the basis for carcinogen risk assessment," *Archives of Toxicology* 83 (2009): 203–225 and "Origin of the linearity no threshold (LNT) dose-response concept," *Archives of Toxicology* 87 (2013): 1621–1633.

17. E. J. Calabrese and M. K. O'Connor, "Estimating risk of low radiation doses" and K. D. Crowley, H. M. Cullings, R. D. Landes, R. E. Shore, and R. L. Ullrich, "Comments on estimating risks of low radiation doses – A critical review of the BEIR VII report and its use of the Linear No-Threshold (LNT) hypothesis by Edward J. Calabrese and Michael K. O'Connor," *Radiation Research* 183 (4), (2015): 476–481.

18. Possible correlations can be compiled by readers themselves on the basis of the following website: http://www.worldlifeexpectancy.com/usa/all-races-death-rate (July 29, 2023).

19. World Health Organization, *WHO Handbook on Indoor Radon: A Public Health Perspective* (World Health Organization, 2009).

20. John Jagger, "Natural background radiation and cancer death in Rocky Mountain states and Gulf Coast states," *Health Physics* 75 (1998): 428–434.

21. John Hart and Seunggeun Hyun, "Cancer mortality, state mean elevation, and other selected predictors," *Dose-Response* 10 (2012): 58–65.

22. Thomas J. Mason and Robert W. Miller, "Cosmic radiation at high altitudes and U.S. cancer mortality, 1950–1969," *Radiation Research* 60 (2), (1974): 302–306.

23. Martin Burtscher, "Effects of living at higher altitudes on mortality: A narrative review," *Aging and Disease* 5 (4), (2014): 274–280.

24. https://www.epa.gov/radon/health-risk-radon (April 24, 2023).

25. http://www.adph.org/radon/assets/surgeon_general_radon.pdf (August 3, 2023).

26. Ludwik Dobrzyński, Krzysztof W. Fornalski, and Joanna Reszczyńska, "Meta-analysis of thirty-two case–control and two ecological radon studies of lung cancer," *Journal of Radiation Research* 59 (2), (2018): 149–163.

27. Krzysztof W. Fornalski, Rod Adams, Wade Allison, Leslie E. Corrice, Jerry M. Cuttler, et al., "The assumption of radon-induced cancer risk," *Cancer, Causes & Control* 26 (2015): 1517–1518.

28. R. Ajrouche, G. Ielsch, E. Cléro, C. Roudier, et al., "Quantitative health risk assessment of indoor radon: A systematic review," *Radiation Protection Dosimetry* 177 (1–20), (2017): 1–9.

29. Both models are linear excess-risk models taking into account the time since exposure, the attained age, and either the duration of the exposure (exposure-age-duration, BEIR VI-EAD), or the level of the concentration (exposure-age-concentration, BEIR VI-EAC).

30. A. M. Zarnke, S. Tharmalingam, D. R. Boreham, and A. L. Brooks, "BEIR VI radon: The rest of the story," *Chemico-Biological Interactions* 301 (2019): 81–87.

31. While the various publications of objective and scientific results are not individually listed here, these papers and those on other areas with high natural-radiation exposure are summarized in, A. S. Aliyu and A. T. Ramli, "The world's high background natural radiation areas (HBNRAs) revisited: A broad overview of the dosimetric, epidemiological and radiobiological issues," *Radiation Measurements* 73 (2015): 51–59.

32. Alireza Mosavi-Jarrahi, Mohammadali Mohagheghi, Suminori Akiba, Bahareh Yazdizadeh, Nilofar Motamedi, and Ali Shabestani Monfared, "Mortality and morbidity from cancer in the population exposed to high level of natural radiation area in Ramsar, Iran," *Elsevier International Congress Series* 1276 (2005): 106–109.

33. National Research Council, *The Waste Isolation Pilot Plant: A Potential Solution for the Disposal of Transuranic Waste* (Washington, DC: National Academies Press, 1996).

34. S. Darby, D. Hill, A. Auvinen, J. M. Barros-Dio, H. Baysson, F. Bochicchio, H. Deo, R. Falk, F. Forastiere, M. Hakama, I. Heid, L. Kreienbrock, M. Kreuzer, F. Lagarde, I. Mäkeläinen, C. Muirhead, W. Oberaigner, G. Pershagen, A. Ruano-Ravina, E. Ruosteenoja, A. Schaffrath Rosario, M. Tirmarche, L. Tomášek, E. Whitley, H. E. Wichmann, and R. Doll, "Radon in homes and risk of lung cancer: Collaborative analysis of individual data from 13 European case-control studies," *BMJ* 330 (2005): 223–227.

35. In wet drilling, water instead of compressed air flushes the cuttings out of the borehole, significantly reducing dust formation. This does, however, have only limited effect on radon release.

36. M. Kreutzer, B. Grosche, M. Schnelzer, A. Tschense, F. Dufey, and L. Walsh, "Radon and risk of death from cancer and cardiovascular diseases in the German uranium miners cohort study: follow-up 1946–2003," *Radiation and Environmental Biophysics* 49 (2010): 177–195.

37. As a study of six major cities shows, this figure is comparable to the risk of lung cancer for metropolitan residents exposed to daily traffic exhaust. D. W. Dockery, C. A. Pope III, X Xu, et al., "An association between air pollution and mortality in six U.S. cities," *The New England Journal of Medicine* 329 (1993): 1753–1759.

38. J. Heinrich, V. Grote, A. Peters, and H. E. Wichmann, "Gesundheitliche Wirkungen von Feinstaub: Epidemiologie der Langzeiteffekte," *Umweltmedizin in Forschung und Praxis* 7 (2), (2002): 91–99.

39. Z. Zdrojewicz and J. Strzelczyk, "Radon treatment controversy," *Dose Response* 4 (2), (2006): 106–118.

40. A. Franke, L. Reiner, and K. L. Resch, "Long-term benefit of radon spa therapy in the rehabilitation of rheumatoid arthritis: A randomised, double-blinded trial," *Rheumatology International* 27 (2007): 703–713.

41. E. Pohl and J. Pohl-Rüling, "Die Strahlenbelastung der Bevölkerung von Bad Gastein, Österreich," *Ber. Nat. Med. Ver. Innsbruck* 57, (1969): 95–11.

42. International Commissions on Radiological Protection, *Recommendations of the International Commission on Radiological Protection* (Oxford: Pergamon Press, 1991).

43. J. D. Boice, M. Blettner, and A. Auvinen, "Epidemiological studies of pilots and aircrew," *Health Physics* 79 (2000): 576–584 and G. P. Hammer, M. Blettner, and H. Zeeb, "Epidemiological studies of cancer in aircrew," *Radiation Protection Dosimetry* 136 (2009): 232–239.

44. G. P. Hammer, M. Blettner, I. Langner, and H. Zeeb, "Cosmic radiation and mortality from cancer among male German airline pilots," *European Journal of Epidemiology* 27 (2012): 418–429.

45. Eileen McNeely, Irina Mordukhovich, Steven Staffa, Samuel Tideman, Sara Gale, and Brent Coull, "Cancer prevalence among flight attendants compared to the general population," *Environmental Health* 17 (2018): 49–58.

9 Summary and Outlook

Radioactivity is a natural phenomenon resulting from changes in the quantum structure of atomic nuclei and, in the case of X-rays, the atomic shell. In the microscopic world, radioactive decay corresponds to the transition from an energetically excited configuration to a more stable state of the nucleus with the release of the energy difference as radiation. In our macroscopic world, this corresponds to the fall of a stone from a certain height to the ground. Lifting the stone to that height corresponds to the nuclear or atomic excitation process, while the falling stone corresponds to radioactive or atomic decay, where the released energy manifests itself in the compression and deformation of the impact site. The time scale for radioactive decay is determined by its probability, which in turn depends on the quantum-mechanical configuration of the initial and final states. This can result in short-lived radioactive-nuclei decays, lasting from nanoseconds to minutes, or long-lived decays occurring over hours to billions of years.

All of these radioactive transitions and phenomena are part of our natural environment. In part, they are remnants of past stellar generations, embedded in the materials from which our solar system formed. They are also produced by high-energy cosmic rays originating from our Sun or the distant stellar explosions that incessantly bombard our planet.

The solar system's internal radioactivity has influenced not only the formation of planets but also, as a source of energy, their further evolution and stabilization, as well as possibly the formation of complex biological molecules, leading to the emergence of life as a self-organized system. Earth's internal radioactivity produces the dynamo effect that generates our geomagnetic field and thus protects us from the deadly dangers of cosmic rays, which are largely deflected by this field. The gradual decay of internal radioactivity signals the uncertain future of this diminishing magnetic field, with significant consequences for the Earth habitat – signaling a future that has already become a reality for our neighboring planet Mars.

High-intensity radiation from radioactive sources is dangerous and lethal in very large doses. It destroys the structure of biological molecules, especially information-carrying molecules such as DNA and RNA. Depending on the dose, this causes modifications in somatic and genetic cell development and can even lead to the rapid breakdown of all biologically controlled functions.

The natural dose of radioactivity present in our environment is far below the levels that are considered radiologically dangerous, although the effects of low long-term doses are poorly understood since the nature and impact of this exposure has been extrapolated from existing data for higher radiation levels. Depending on the mathematical function underlying this extrapolation, enormous differences in predicting low-level radiation impact emerge, which in turn manifest in heated medical and political discussions. In short, we don't know the long-term medical and genetic consequences of low-level radiation.

The natural radioactivity level to which humanity is exposed today may cause cell damage but this is mitigated by enzymatic-repair mechanisms, introducing a kind of equilibrium that humans and other biological systems have adapted. There is still controversy among scientists as to whether long-term low-dose irradiation, to which humans are exposed on a daily basis, acts as a motor for evolution – actually serving as a cure for medical ailments – or as a cause of early death.

Although Earth's biological equilibrium has existed for at least the millennia of the Holocene, and possibly through earlier periods of biological evolution, this balance is now being disturbed. New developments in the Anthropocene, the age during which *Homo sapiens* have taken over the shaping of Earth and her environment, have not necessarily been for her or our own good. The Anthropocene has its origins in human society's industrial development, which has emerged hand in hand with technical-scientific development at universities and research institutes. This has also

DOI: 10.1201/9781003435907-9

given rise to the field of nuclear physics, whose experimental successes and findings led to the birth of anthropogenic radioactivity. Today, man-made radioactivity – from bombs to medical devices, reactors, and accelerators – dominates human radiation exposure. This raises questions about the sources and effects of anthropogenic radioactivity, as well as about the extent to which the millennia-long balance of biological systems may have been disturbed. This in turn raises more socially relevant questions about how to deal with this new type of radiation exposure and what technical developments are required for its long-term control.

These questions will be discussed in the second volume on anthropogenic radioactivity, which will identify additional causes of radiation exposure – from economic, military, and medical activities – and will also discuss the increasing use of radiation probes in a wide range of scientific-industrial applications.

Subject Index

Author Index

Printed in the United States
by Baker & Taylor Publisher Services